Creative (Climate) Communications

Conversations about climate change at the science–policy interface and in our lives have been stuck for some time. This handbook integrates lessons from the social sciences and humanities to more effectively make connections through issues, people and things that everyday citizens care about. Readers will come away with an enhanced understanding that there is no "silver bullet" to communications about climate change; instead, a "silver buckshot" approach is needed, where strategies effectively reach different audiences in different contexts. This tactic can then significantly improve efforts that seek meaningful, substantive and sustained responses to contemporary climate challenges. It can also help to effectively recapture a common or middle ground on climate change in the public arena. Readers will be equipped with ideas on how to harness creativity to better understand what kinds of communications work where, when, why and under what conditions in the twenty-first century.

Maxwell Boykoff is the Director of the Center for Science and Technology Policy, which is part of the Cooperative Institute for Research in Environmental Sciences at the University of Colorado Boulder. He is also an Associate Professor in the Environmental Studies program at the University of Colorado. Max has ongoing interests in cultural politics and environmental governance; science and environmental communications; science–policy interactions; political economy and the environment; and climate adaptation. He has authored many peer-reviewed journal articles, book chapters and books on these subjects, including *Who Speaks for the Climate? Making Sense of Media Reporting on Climate Change* (2011, Cambridge University Press).

Creative (Climate) Communications
Productive Pathways for Science, Policy and Society

MAXWELL BOYKOFF

University of Colorado Boulder

CAMBRIDGE
UNIVERSITY PRESS

CAMBRIDGE
UNIVERSITY PRESS

University Printing House, Cambridge CB2 8BS, United Kingdom

One Liberty Plaza, 20th Floor, New York, NY 10006, USA

477 Williamstown Road, Port Melbourne, VIC 3207, Australia

314–321, 3rd Floor, Plot 3, Splendor Forum, Jasola District Centre, New Delhi – 110025, India

79 Anson Road, #06–04/06, Singapore 079906

Cambridge University Press is part of the University of Cambridge.

It furthers the University's mission by disseminating knowledge in the pursuit of education, learning, and research at the highest international levels of excellence.

www.cambridge.org
Information on this title: www.cambridge.org/9781107195387
DOI: 10.1017/9781108164047

First published 2019

Printed in the United Kingdom by TJ International Ltd. Padstow Cornwall

A catalogue record for this publication is available from the British Library.

ISBN 978-1-107-19538-7 Hardback
ISBN 978-1-316-64682-3 Paperback

I dedicate this book to Monica Boykoff, Elijah Boykoff and Calvin Boykoff.

I also dedicate this work to the memory of Max Thabiso Edkins (1983–2019).

Contents

A word cloud showing frequency of words (four letters or more) used in this book.

Preface

Creativity, Collaboration, Confrontation

Climate change has become a defining symbol of humans' collective relationship with the environment. Since the 1990s, climate change has become a high-stakes, high-profile and highly politicized venture involving science, policy, culture, psychology, environment and society. Confronting climate change is essentially a collective action problem. Addressing climate change gets to the heart of how we live, work, play and relax in modern life, shaping our everyday lives, lifestyles, relationships and livelihoods.

The bad news is that this is a daunting venture, bigger than any one particular way to solve it or one particular way to even communicate effectively about it. The good news is that there are many ways that individuals, collectives, businesses, organizations and institutions are stepping into the challenge and working in different ways to creatively find resonant ways to connect with different sectors of society. Through new and enterprising communication approaches, these address a range of objectives. Among them, goals of communication efforts include improving education and literacy, helping mobilize more effective advocacy efforts, prompting individual- to collective-scale awareness raising and behavior change, and promoting cultural change. These are burgeoning spaces of engagement. Today, many people, collectives, businesses and institutions are creating content, giving advice on what content to create and researching the efficacy of this content for different segments of public citizens.

In this book, I address key themes in creative climate communications as I track, appraise and evaluate various creative communications on climate change. I highlight how and why certain approaches find success with selected audiences as I critique approaches that fall short in a variety of critical ways. This work is motivated in part by an argument put forward by Dan Kahan (2015a) to gain "satisfactory insight" into the science of science communication through scientific approaches to evaluation (p. 1). Dan Kahan has taken up the "science of science communication" while this project takes a different tactic through creative climate communications. Moreover, Stephen Schneider (2001) – one of the most effective climate communicators in the past decades – argued for moving beyond platitudes to show, through detailed empirical examples, what works in different circumstances. Similarly, Dan Kahan and Katherine Carpenter (2017) have commented that empirical research is "essential to distinguish the mechanisms that are true from the vast set of those that are merely plausible but untrue, lest researchers and real-world communicators drown in a veritable sea of just-so stories" (p. 310).

Reckless speculation masked as scholarship can be damaging; however, there is a danger of too narrowly defining legitimate scholarship through quantitative (over qualitative) approaches to research in these areas. As such, while this book values and highlights empirical research into these arenas of creative climate communication, it also values storytelling and other ways of examining and knowing about these phenomena. For example, it is a mistake to impose hypothesis testing on creative artists and practitioners as the requisite pathway to knowing what communication strategies are effective in selected audiences. To do so is to alienate a key set of communicators needed to confront these communication challenges. In this book, I therefore take up the position that hypothesis testing and storytelling both can contribute substantively, and at times complementarily, to better understand the efficacy of creative climate communications as both arts and the sciences together.

This project is also motivated in part by comments from Amy Luers (2013), who has also called on researchers to "evaluate what works and share what we learn" (p. 13). In that spirit, I deliberately deploy extensive citations to take advantage of this long-form book artefact and provide opportunities for you, the reader, to follow up on research, ideas, concepts and cases that pique your interest. While this might be a little cumbersome for the casual reader, I hope it ultimately makes this text a more useful resource for all readers. Misplaced name-dropping (and hero worship) can be both distracting and annoying. But I aim for this pathway of crediting many researchers and practitioners by name to help readers more capably dig into how they may then choose to approach their own communications efforts. I also recount numerous research projects that I have been involved in in recent years. I hope that this cataloguing – evidenced by an extensive reference list at the end of this volume – can catalyze further explorations and research endeavors into these spaces. This pathway also aspires to help you to see how diverse and multimodal approaches have found success in creatively communicating about climate change on multiple scales with different people in the public arena.

Many scholars (including myself) have (vigorously) researched and debated the extent to which media representations and portrayals are potentially conduits to attitudinal and behavioral change. However, there remains a dearth of systematic analyses regarding how creative climate communications elicit varying levels of awareness and engagement. Beyond examining the mechanisms of news media communications on climate change (Boykoff, 2011), we need to continue to expand our experimentation with ways that conversations about climate change can become present and meaningful in our everyday lives. Adam Corner from the Climate Outreach in the United Kingdom (UK) has commented, "There's a real kind of absence of inspiring programming or engagement to go with all this amazing science we're producing" (Sobel Fitts, 2014).

Creative and participatory communications and representations can be ignored or dismissed in shaping climate science, governance and everyday cultural politics at our peril. Denigrating views and demeaning utterances about creative climate communications as a sideshow or mere "jawboning" are the views of yesteryear. Through interdisciplinary engagements, this book takes stock of lessons learned from the past few decades of research and practices, in order to inform effective ways to move forward. I have therefore endeavored to write a cogent and central text to anchor us in creative climate communications research and practice going forward.

In the chapters that follow, I wrestle with various dimensions of climate communication, exploring ways to harness creativity to better understand what endeavors work where, with whom (what audiences), when and why. I explore elements and realities that constitute shared twenty-first-century communications ecosystems. In Chapter 1, I lay a foundation for effective communication by understanding intersecting dimensions of intended and perceived audiences. I also consider elements of trust along with who might be creative and effective messengers in the context of a post-truth Anthropocene era. After I initially explore these notions, Chapter 2 then considers how we have come to know what we know about climate change. In this chapter, I explore the value of narratives and stories in meeting people where they are and finding common ground on climate change. I also contend with an argument invoked through the title: effective communications about climate change sometimes may importantly involve *not* invoking the term "climate" or "climate change" explicitly. In Chapter 3, I interrogate how and why the deficit model of communication persists, and how this persistence stands in the way of more effective and creative climate communication. The chapter is animated by considerations of how dissent from climate contrarian (or "denier") voices persist and find traction. I make the case that dominant information-deficit model approaches to climate communication counterproductively provide oxygen to breathe more life into counterproductive claims. An expanded approach can then stifle the efficacy of outlier assertions. In Chapter 4, I focus on ways in which experiential, visceral, emotional and aesthetic learning informs scientific ways of knowing about climate change. To illustrate, I explore how comedy can be an effective vehicle for creative climate communications. Through Chapter 5, I consider the importance of framing in context. I explore how framing for selected audiences has functioned through the "More than Scientists" collaboration with "Inside the Greenhouse." I also look at different scalar approaches where climate communication seeks to create change. This offers opportunities to consider consumption issues and the oft-focused locus of agency at the individual. It also provides a good space to consider climate communication strategies such as "consensus messaging." In Chapter 6, I analyze the flavors of climate advocacy in today's highly politicized communications environment. Here I trace a current engagement gap, drawing on survey research and exploring the influence of certain climate science communicators to better make sense of the promises and pitfalls of advocacy through climate communications. In Chapter 7, I situate the value of experimentation in these areas of creative climate communication. I explore numerous examples of forays into these spaces and then assemble features on a "road map" along with "rules of the road" to help guide ongoing creative climate communications. In Chapter 8, I ponder how younger people today are grappling with these issues and consider how they may face them in the decades to come. I link these inquiries to intersecting routes of communication about decarbonization and sustainability. In taking up this set of considerations, I pull in intergenerational and intragenerational equity questions about who has a voice and how, going forward in creative (climate) communications.

As the book proceeds, case study examples in Chapters 4 and 5 reveal that I am not only an (armchair) analyst and researcher but also a participant in experimentation, mainly through the Inside the Greenhouse (ITG) project at the University of Colorado Boulder. With Professor Beth Osnes from the Theatre Department and Professor Rebecca Safran from the Department of Ecology and Evolutionary Biology, we cofounded ITG in 2012. We designed ITG to facilitate and support creative storytelling about issues surrounding climate change through video, theatre, dance and writing to help connect wider and new audiences to climate change in resonant and meaningful ways. In the process, we have also worked to build competence, confidence and capacity of undergraduate and graduate students as emergent communicators and leaders in the new millennium. As such, this project has sought to create cultures of participation and productive collaboration among students, interfacing with the larger community and world in retelling the stories of climate change and to become meaningful and sustaining content producers. In 2018, Professor Phaedra Pezzullo from the Department of Communication joined the project, adding insights from her experiences and research. The chosen title of the ITG initiative acknowledges that, to varying degrees, we are all implicated in, part of and responsible for greenhouse gas emissions into the atmosphere. Through the development and experimentation with creative modes to communication, we treat this "greenhouse" as a living laboratory, an intentional place for growing new ideas and evaluating possibilities to confront climate change through a range of mitigation and adaptation strategies. Through commitments to meet people where they are on climate change, the ITG project draws on students' strengths and perspectives to consider the complexity of climate change in new ways. In so doing, ITG offers direct links between the natural and social sciences and arts to communicate, imagine and work toward a more resilient and sustainable future.

Overall, by systematically scrutinizing these linkages and fissures in awareness as well as engagement with climate mitigation and adaptation themes, I hope this book will be valuable to you: researchers, students, practitioners and members of the public citizenry who are interested in **creatively** and **collaboratively confronting** persistent (climate) communication challenges and improving climate communication outcomes.

Acknowledgments

I am grateful to networks of collaborations and support that have made the writing of this book possible. Among them, direct collaborators are Beth Osnes, Rebecca Safran, Justin Farrell and David Oonk. Parts of Chapter 3 come from work coauthored with Justin Farrell, forthcoming through wider collaboration with Nuria Almiron at Universitat Pompeu Fabra in Barcelona, Spain. Portions of Chapter 4 come from work coauthored with Beth Osnes. Portions of Chapters 4 and 5 come from collaboration with Rebecca Safran and Beth Osnes through the Inside the Greenhouse project. Parts of Chapter 6 come from work coauthored with David Oonk.

Secondarily, I have many colleagues to thank who have (often selflessly) helped support research therein. They include Bienvenido Leon, for supporting early thinking on this project during my 2015 sabbatical at the University of Navarra in Pamplona, Spain. I also thank Jennifer Fluri and Jessie Clark for prompting writing in Chapter 4 through their organization of American Association of Geography (AAG) sessions on "Humor Amid Adversity" and the resultant 2018 special issue of the journal *Political Geography*. Thanks as well to *Political Geography* associate editor Kevin Grove for his measured handling of the process of review for the manuscript. I thank Mike Goodman, Julie Doyle and Nathan Farrell for catalyzing the writing of Chapter 6 through their planning and coordination of a special issue on "Everyday Climates" in the journal *Climatic Change*. Thanks as well to colleague Amanda Carrico for her help in the early stages of development of the research embedded in Chapter 6. And thanks to Michael Bruggeman, Stefanie Walter, Fenja de Silva-Schmidt, Ines Schaudel and others in CliSAP (the Cluster of Excellence Integrated Climate System Analysis and Prediction) at Universität Hamburg who led a 2017 "Re-defining the Boundaries of Science and Journalism in the Debate on Climate Change" workshop where I presented a working version of Chapter 4. I also appreciate and thank Eric Michelman from the Climate Change Education Project for his partnership through More Than Scientists that influenced work described in Chapters 4 and 5. And I thank Brian Daniell and Vicki Bynum for their support of Inside the Greenhouse endeavors, with consequent perspectives, insights and outputs that show up throughout the book.

I also thank colleagues at the University of Illinois, Susan Koshy and Trevor Birkenholtz in particular, for hosting the "Unnatural Disasters" conference; the University of Chicago, Elizabeth Chatterjee and Greg Lusk in particular, for hosting the "Neubauer Collegium for Culture and Society on Climate Science and Democracy"; the Colorado School of Mines, Adrienne Kroepsch and Shannon Mancus in particular, for hosting me in the "Hennebach Lecture Series in the Humanities, Arts and Social Sciences Division"; the Universitat Pompeu Fabra, Departament de Comunicació, in Barcelona, Spain, Nuria Almiron in particular; and Reading University, Catriona McKinnon and Mike Goodman in particular, for hosting the "Communicating Climate Change in Troubled Times" workshop. At these conferences, guest lectures and work-shops in 2018, I presented drafts of various portions of the manuscript for this book.

Thanks go to the many colleagues and co-conspirators who have given permission to reproduce their figures, schematics and photos in this volume. These people include Sarah Barfield Marks, Una Chaudhuri, William Daniels, Fritz Ertl, Gordana Filipic, Madeleine Finlay, Giovanni Fussetti, Matthew Goldberg, Lisa Goulet, Justin Brice Guariglia, Meaghan Guckian, Abel Gustafson, Ole Christoffer Haga, Iain Keith, Oliver Kelihammer, Anthony Leiserowitz, Heather Libby, Ed Maibach, Michael Mann, Ezra Markowitz, Andrews McMeel, Julia Metag, Sarah K. Miller, Ami Nacu-Schmidt, Beth Osnes, Amanda Overton, Hannah Phang, Bill Posters, Tejopala Rawls, Seth Rosenthal, Mike Schäfer, Tom Toles, Solitaire Townsend, Sander van der Linden and Marina Zurkow.

I also appreciate the collegiality of co-workers including Mike Goodman, Phaedra Pezzullo, Lisa Dilling, Bruce Goldstein, Steve Vanderheiden, Cassandra Brooks, Eve Hinckley, Carol Wessman, Heidi Vangenderen, Pete Newton, Suzanne Tegen, David Ciplet, Matthew Burgess, Ben Hale, Björn Ola-Linner, Victoria Wibeck, Roger Pielke, Jr., Steve Nerem, Waleed Abdalati, Christine Wiedinmyer, Mike Hardesty, Jen Kay, Katy Human, Robin Moser, Rebecca Stossmeister, Dave Zakavec, Nate Campbell, Matthew Price, Abigail Ahlert, Ryan Harp, Aditya Ghosh, Pablo Suarez, Janot Mendler de Suarez, Katie Chambers, Denise Fernandes, Emily Nocito, Jerry Peterson, Robert Ferry, Alice Madden, Matt Druckenmiller, Fernando Briones, Ryan Vachon, David Kang, Rob Schubert, Linda Pendergrass, Emily Coren, Josh Wolfe,

Anne Gold, Erin Leckey, Susan Lynds, Susan Sullivan, Michael Kodas, Chuks Okerekee, Diana Liverman, Paul Chinowsky, Noah Finklestein, Dustin Mulvaney, Paty Romero-Lankau, Brian Gareau, Tara Pisani-Gareau, Jill Harrison, Jill Litt, and Susan Avery. Thanks as well to all the students in the "creative climate communications" classes who bravely performed in the "Stand Up for Climate Change" comedy events in 2016 and 2017. Thanks to the teaching and support teams for those experiences as well as the More Than Scientists work including Dan Zietlow, Scott Gwozdz, William "Max" Owens, Garrett Rue and Barbara MacFerrin.

I also thank Ami Nacu-Schmidt at the University of Colorado Boulder for her great help throughout, from securing permissions for images, figures and schematics to designing the book cover. And I thank James Balog and the Earth Vision Trust for their contribution of the photograph represented on the front cover. In addition, I thank Jennifer Katzung for her great support and encouragement during the process, and for keeping our Center for Science and Technology Policy Research (CSTPR) buzzing productively during some of my necessarily cloistered writing times. Celeste Maldonado and Andrew Benham from CSTPR deserve my gratitude as well for their support and assistance with various aspects of this project, including figure designs and reference formatting.

Thanks as well go out to Matt Lloyd, Lisa Bonvissuto, Theresa Kornak, Gayathri Tamilselvan, Zoë Pruce, those involved in the anonymous peer review and others at Cambridge University Press for their gracious assistance and support throughout the process.

I have many to thank in my personal support system including Monica Boykoff, Susan Schoenbeck, Leah Moore, Thomas Boykoff, Gitta Ryle and my sons Elijah and Calvin. Thanks to Monica, Elijah and Calvin too for the practical (yet somewhat sad) birthday gift of a real office chair to get this book manuscript over the line to completion.

In addition to these key individuals, I also thank many groups and programs (with associated people) that have supported and catalyzed my writing of this book. In particular, I thank the Leadership Education for Advancement and Promotion (LEAP) Growth Grant at the University of Colorado. I also thank the University of Colorado Hazel Barnes flat in London, England for support during some field research and writing for this project. And I thank the ATLAS Black Box Experimental Studio in the Center for Media, Arts and Performance as well as the Program Council and Old Main at the University of Colorado for hosting the comedy events. I also extend thanks to the Spanish Ministry of Science and Technology for travel support associated with presentation of work appearing in Chapter 4 at the 2017 Association of American Geographers meeting. And thanks to the International Collective on Environment, Culture and Politics (ICECaPs) and the Media and Climate Change Observatory (MeCCO) including key folks such as Patrick Chandler, Olivia Pearman, Jeremiah Osbourne-Gowey, Lucy McAllister, Marisa McNatt, Meaghan Daly, Kevin Andrews, Lauren Gifford, Midori Aoyagi-Usui, and Rogelio Fernández-Reyes. I thank a growing group called the Boulder Faculty Science and Education Committee, with Shelly Miller, Kris Karnauskas, Jim Meiss, Seth Hornstein, Steve Nerem and Sam Flaxman as particular inspirations. Thanks as well to the Albert A. Bartlett Science Communication Center, with inspiring people such as Shelly Sommer, Marda Kirn, Wynn Martens and Brett KenCairn.

I also appreciate the influence that several organizations had on my work in this book. These include the Cooperative Institute for Research in Environmental Sciences (CIRES), the Center for Science and Technology Policy Research (CSTPR) and the Environmental Studies Program at the University of Colorado Boulder; Grounds Guys snowplowing; the Environmental Change Institute (ECI); and the School of Geography and the Environment (SoGE) at the University of Oxford, and the International Order of Oddfellows Lodge #9 in Boulder, Colorado.

1

Here and Now

We are looting both the past and the future to feed the excess of the present. It's the dictatorship of the here and now.
– John Schellnhuber, director of the Potsdam Institute for
Climate Impact Research, in an interview for
Der Spiegel *(2011)*

We live in remarkable times. Amidst high-quality and well-funded scientific research into the causes and consequences of climate change, conversations about climate change in our lives – and climate communications – are stuck. Consciously or unconsciously, a feeling of complacency has often weighed on our collective and on our individual selves.

Those of us who have waded into these choppy waters of climate discussions have often found turbulent, polarized and partisan exchanges. Too often, when many of us feel those instabilities – amid daily challenges of putting food on the table, staying healthy, caring for loved ones – we choose to not rock the boat. Instead, we have sensed that the most viable alternative to avoid these rough waters is to stay in the proverbial shallows and to choose to remain silent. Over time, such individual choices have contributed to patterns of "climate silence." This social norming of silence on climate change (Marshall, 2014) has limited our abilities to coherently and adequately address one of the most looming challenges in the twenty-first century.

In addition, over the past decades many of us who have devoted our professional lives to working on climate change have been saddled with a recognition – perhaps most acutely within the climate science communities – that more information about climate change has not adequately addressed the chronic challenges of climate literacy, public awareness and engagement on its own. We have then sensed that more creative approaches are needed to more effectively meet people where they are on climate change.

Responding to these emergent needs, in recent years has been a blossoming of valuable research in the peer-reviewed literature addressing various elements of this larger challenge. More research groups, organizations, institutions and practitioners around the world have increasingly explored creative spaces of climate communications to better understand what works where, with whom (what audiences), when and why.

This book seeks to more comprehensively make sense of the developments, movements and key challenges therein. Within these chapters, I draw out varying modes, methods, audiences and cultural contexts while analyzing larger considerations of awareness, inspiration and engagement (see the Preface for more). As I move through these elements, I work to pivot from a limited place of convincing people of the facts, of winning arguments, of mere naming and shaming into more creative spaces in communications about climate change. In other words, to address this collective action problem I encourage a creative shift from "turning on each other" to "turning to each other" for support and collaboration. By drawing out trends, patterns, experiments, findings and key successes as well as challenges associated with creative (climate) communications through all this research and experimentation, I then provide some guidance on effective and successful communications in the face of today's climate challenges (see Chapter 7 for more).

There is a time and place for just about everything. However, to find common ground and to work collectively to address climate change, there is a burning need to consider mindfully and methodically how we communicate about it. To be clear, finding common ground does not mean violating one's own commitments, concerns and aspirations. It is important to be authentic in these interactions (see Chapter 7 for more). It is not productive to act as an apologist for positions with which one disagrees in order to just get along. Finding common ground means listening carefully to other points of view and entering into open, respectful and honest dialogue about both different perspectives and shared values. Creative pathways through these processes can then be seen as ones that "smarten up" communications and thereby facilitate more effective connections through issues, people and things that everyday citizens care about.

Creative approaches involve the deployment of multimodal communications. A mode is "a system of choices used to communicate meaning. What might count as a mode is an open-ended set, ranging across a number of systems, including but not limited to language, image, color, typography, music, voice, quality, dress, gesture, special resources, perfume, and cuisine" (Page, 2010, p. 6).

Amid many elements seeping into these environments, I consider dynamics that shape creative and potentially effective messages as well as messengers of those climate change communications. Over time, broad references to communications through media platforms have generally pointed to television, films, books, flyers, newspapers, magazines, radio and internet as pathways for large-scale communication. These processes have typically involved publishers, editors, journalists, professional content producers and members of the communications industry who produce, interpret and communicate texts, images, information and imaginaries.

But clearly, modes of communication are not limited to speeches, textbooks, video interviews, advertisements or news media pieces. While there are many great texts to guide science communication (e.g., Bucchi and Trench, 2008; Bennett and Jennings, 2011; Leshner et al., 2017), environmental communication (e.g., Perrault, 2013; Pezzullo and Cox, 2017) and analyses of news media that influence public discourse (e.g., Anderson, 1997; Boyce and Lewis, 2009; Boykoff, 2011; Painter, 2013), these works take us only partway down a road that must be traveled more extensively on these topics (Blanding, 2017) (see Chapter 7 for more). Along with those important contributions, we must also take into account how creative (climate) communications shape perspectives, attitudes, intentions, beliefs and behaviors among public citizens around the world. In addition, we must recognize the significant expansion into new, more creative and interactive webs of democratized, peer-to-peer communications (van Dijk, 2006; O'Neill and Boykoff, 2010).

Additional modes and manifestations of communication also include (analyses of) documentary films about dystopian futures, stand-up comedy about climate and cultures, podcasts about climate science and policy interactions, lawn sculptures made from reusable water bottles and choreographed human glacial melt. Kathryn Cooper and Eric Nisbet (2017) have commented, "Influencing audiences about climate change is a challenging task due to the diversity of the media landscape, audience predispositions and selective exposure, and psychological biases such as affect. Documentaries, both those made to inform as well as those made to influence audiences to action, have the potential to overcome these challenges."

Participatory and experiential activities (Osnes, 2014) have been considered as a powerful way to consider resonant climate challenges (Smith and Joffe, 2009). Moreover, extensions into entertainment media and interactive platforms have been increasingly recognized as important facets of making climate change meaningful (Boykoff, 2011; Dudo et al., 2017). Therefore, multimodal techniques draw on many systems of communication.

Meeting people where they are takes carefully planned and methodical work. It does not mean "dumbing things down" for different audiences. Through this process of assessment of research and practice in these areas, conversations can more capably seek answers to a provocative question that Mike Hulme posed in his seminal 2009 book *Why We Disagree about Climate Change*. He asked, "How does the idea of climate change alter the way we arrive at and achieve our personal aspirations and our collective social goals?" (Hulme, 2009, p. 56). Pursuing answers to that fundamental question necessitates dialogue, deliberation, active listening to other points of view and consideration of one's place in the collective. These then become productive yet more manageable forays into dynamic, immense and complex systems of meaning-making at the interface of climate science, policy and society.

As such, these climate change conversations are not contained solely in the province of science or environmental communication. They involve politics, economics, culture, ideology, environment and society. These expanded considerations help to more comprehensively make sense of ways in which meaning and knowledge are derived from communications, interactions, listening, exchanges and dialogues.

Philip Smith and Nicolas Howe (2015) have alluded to climate change as "social drama," writing, "we believe there is a real possibility for climate change to emerge as a truly compelling social drama – a cultural form that will change history for *us* before climate-change-the-natural-event changes it radically *for* us" (p. 209, emphasis in original). As I stated at the beginning of this chapter, we are living through momentous times as we fundamentally grapple with issues that cut to the heart of how we live, work, play and relax in society. However, the "we" here in my blanket statement, and the "us" in Smith and Howe's exhortation are in fact very differentiated groupings.

There are cruel realities in gaps in opportunity and access to natural resources and to meeting livelihood needs for large segments of our global population (Agyeman et al., 2007; Pezzullo, 2009). Many grim paradoxes are associated with people and places at the forefront of climate impacts (see Chapter 4 for more). Among them, (1) those at the forefront of impacts are those with the least capacity to address them; (2) those most impacted are often those with the least influential voices in decision-making; and (3) mechanisms to confront associated problems are often weak, under-resourced and fragmented across scales.

In his book *Earth Odyssey*, Mark Hertsgaard (1998) wrote, "On my way to Brazil from Asia, I had stopped off in San Francisco . . . After a year of travel, much of it in Africa and Asia, seeing my old hometown again was more than a little disorienting . . . the sheer wealth of the place was staggering. With their leather jackets, designer eyeglasses, and stylish haircuts, many San Franciscans were *wearing* more money than African and Chinese peasants would earn in a lifetime" (p. 195, emphasis added). Therefore, treatments of "we" and of "us" must be approached carefully and mindfully. Doing so helps to better understand how assorted players – from competent citizens and audacious activists to willing ignoramuses or cunning obstructionists – shape the theater of contemporary emotional, rational and intellectual conversations.

Know Thy Audience[1]

These creative (climate) communication endeavors must start with considerations of audience. These may be imagined, target, (un)intended or actual audiences. Researchers and practitioners have increasingly paid attention to differentiated audiences as key components to deliberate development of effective communications.

Anders Hansen (2015) has stressed the importance of integrating conceptions of perceived audiences into the production of environmental communications. Long-time journalist Richard Black concurred that "effective communication always begins with the audience" (p. 283). Dietram Scheufele (2018) has pointed out that empirical social science research has helped "enable more effective communication with publics whose demographic, socio-structural, or value-based characteristics position them squarely outside of the proverbial choir that science communication is often preaching to" (p. 1123). Sheila Jasanoff (2014) has pushed for consideration of "a more robust conception of *publics* – not treating them as natural collectives (e.g., housewives or teenage women) but as dynamically constituted by changes in social contexts" (p. 23). John Besley and Matt Nisbet (2013) have examined surveys of scientists' perceptions of imagined "public audiences" and motivations to participate in public life. They found that the strongest predictors of participation were attributed, among other things, to a view that "a lack of public knowledge is harmful" and therefore communicating their work represented a "commitment to the public good" (p. 971).

Perceived audiences vary. At times, one may intend to rally supporters and those with a common perspective; at other times, one may endeavor to reach audiences with other points of view. Effectively reaching these different audiences necessarily requires different communication strategies. Moreover, clearly we do not all think the same; we do not all interpret a given meme or message equally to the friend or family member next to us. Even those in tight epistemic communities, families, or marriages have different ways of knowing

---

[1] This is a reference to a part of an enduring adage from Stephen Schneider, "Know thy audience, know thyself, know thy stuff." It is discussed in more detail in Chapter 7.

about climate change, as well as different perspectives on how to communicate effectively about it. Therefore, we need to tailor messaging to meet unique people where they uniquely may be.

Audience segmentation and consequent message alteration has been a part of marketing and associated communication strategies since the 1950s (Smith, 1956; Slater, 1996). Audience segmentation endeavors, as they relate to climate change communications, have proliferated over the past decade (Leal Filho, 2019). For example, Julia Metag and Mike Schäfer (2018) have mapped out a schematic representing processes of audience segmentation in relation to scientific and environmental issues (see Figure 1.1). Going forward, they called for segmentation work to enable more detailed accounting of "how people belonging to a specific segment get in contact with information about science or environmental issues in their everyday life, how they evaluate this information, and how this relates to their attitudes" (Metag and Schäfer, 2018, p. 1001).

Most prominently among audience segmentation work resides the "Global Warming's Six Americas" project on climate communication, the results of which were first published in 2009. The project has been a latent class analysis of the US public to create perspective segmentation based on responses to a survey about climate change (Maibach et al., 2009).[2] Six categories of responses emerged from these survey questions, defined as "dismissive," "doubtful," "disengaged," "cautious," "concerned" and "alarmed." Since its inception, Six Americas has tracked public perspectives on concern, belief and motivation in regard to climate change or global warming. This categorization was first applied to the US context but has since been tested in the Chinese (Wang et al., 2017; Wang and Zhou, in press), Indian (Thaker and Leiserowitz, 2014) and German (Metag et al., 2017) contexts as well.

Similarly, in the US context John Besley (2018) has examined US National Science Foundation survey responses on views of science and technology. He categorized respondents into six groups: "disengaged," "moderate," "optimists," "worried," "liberal friends of science," "cautious conservative" and "conservative friends of science." Besley argued that this categorization makes sense "to help understand views about science and technology and communicate more effectively" (Besley, 2018, p. 14). In other audience segmentation research, Megan Brenan and Lydia Saad (2018) sorted US adults into three perspectives on climate change: "concerned believers," "cool skeptics" and "mixed middle" based on Gallup survey data[3] showing deep divisions on climate change concern depending on political ideology. They found that

---

[2] In 2018, this was whittled down to four key questions in order to segment audiences (Zhang et al., 2018).

[3] This was a survey conducted March 1–8, 2018 of 1,041 US adults.

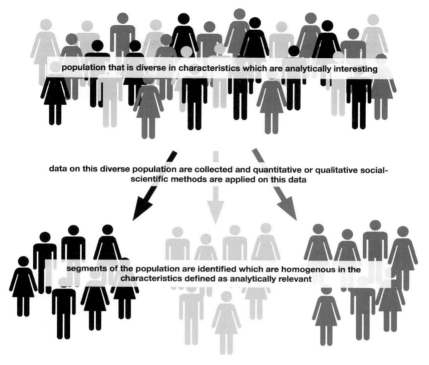

Figure 1.1 A schematic depicting the logic of segmentation analyses (from Metag and Schäfer, 2018). This is a process in which a more generalized audience is subdivided into groups based on factors such as socioeconomic characteristics, political worldviews, psychological traits, ideological preferences, demonstrated attitudes, or behaviors relating to environmental issues. These are not necessarily real-world communities but are clusters of people with these similar characteristics.

these categories were represented by 48%, 19% and 32% of the overall population but just 17%, 45% and 38% for self-described Republicans (Brenan and Saad, 2018). Furthermore, Donald Wine, Wendy Phillips, Aaron Driver and Mark Morrison (2017) have explored segmentation through socio-demographic factors and across attitude–behavior dimensions. Together, audience segmentation research endeavors have sought to better understand how to tailor climate change communications to engage effectively with a targeted subgroup (e.g., Flora et al., 2014; Monroe et al., 2015).

Understanding that a population is diverse and its members therefore respond to stimuli differently is important. It then helps to facilitate processes of more effective communications about climate change. For example, Jordan

Harold, Irene Lorenzoni, Thomas Shipley and Kenny Coventry (2016) have studied how to tailor climate change graphics and images as visual representations for different audiences. Others have examined how color graphics can be helpful for some and alienating for those who face color blindness challenges (Light and Bartlein, 2004). Moreover, one's native spoken language has been found to influence color discrimination (Thierry et al., 2009). Therefore, something mundane like color choices (e.g., achromatic versus chromatic color scales) needs to be thoughtfully selected (Harold et al., 2016).

Moreover, placing value on how different audiences have different ways of knowing about climate change helps to produce more effective communication strategies. Shane Gunster (2017) has commented, "How one conceives of an audience shapes how one communicates with it" (p. 54). He continued, "in the case of climate change, assumptions about audience needs, interests, scientific and political literacy, beliefs and values inform key decisions about . . . content, including areas of emphasis (and avoidance), framing, sourcing, diversity of argument, level of complexity and analysis, tone and style" (p. 54). Carefully evaluating people's conceptions and expressions of themselves in relation to climate change demonstrates respect that then effectively opens up discussions rather than closing them down. As a result, opening up ways of knowing (see Chapter 4 for more) and of communicating about climate change serve as democratizing forces for new voices and perspectives.

To illustrate, Texas Tech University professor Katharine Hayhoe has shown deftness in effectively tailoring ways of discussing climate change with varied audiences. She has also been someone who has drawn strength by speaking clearly from her identities as (1) an atmospheric scientist studying climate change, (2) a political science professor and (3) an Evangelical Christian. As a few examples of her communication work, Katharine has been the host of the "Global Weirding" series of short YouTube videos, answering questions about global warming and climate change (see Chapter 7 for more). She was also in the first episode of the documentary TV series *Years of Living Dangerously* featuring her work and communication with religious audiences in Texas. For a number of years now, Katharine has effectively communicated about dimensions of climate change through a three-step approach she has called "bond/connect, explain, and inspire." By this she means that effective communication first must involve a conscious exploration of what the speaker and audience have in common, what they both may care about at the human–environment interface. Second, it must entail why those involved in the communication might care about what is happening (e.g., that climate change is exacerbating drought in certain regions). Third, it must contain considerations about how those involved can help confront the problem in ways that are compatible with shared values, inspiring action. According to Katharine Hayhoe, these steps involve a sense of collective struggle and positive actions.

However, her approaches are more anomalous than representative of wider communications engagements to date by researchers, scholars and practitioners in the public sphere. Candice Howarth and Richard Black (2015) have pointed out that "the communication of climate change historically has been generic, untailored and untargeted" (p. 506). As such, more is needed to carefully tailor facilitated communications and dialogue that values different perspectives on climate change in order to increase concern and engagement. From these more systematic and methodical approaches, evidence-based communications can be designed to creatively resonate with target audiences.

Trust Us, We're the Experts[4]

Considerations of audience lead into important considerations of whom to trust and who the "experts" or "authorized speakers" are in the context of relationships between science and society (Engdahl and Lidskog, 2014). Trust in communications has been defined as "willingness to make oneself vulnerable to another based on a judgment of similarity of intentions or values" (Siegrist et al., 2005, p. 147). More broadly, claims makers who are perceived to be within a person's perceived affinity or identity groups have higher legitimacy, credibility and trustworthiness (Makri, 2017). Consequently, public citizens more willingly adhere to the claims that these trusted sources make (Dearing et al., 1994).

At the interface of science, policy and society, many have routinely relied on "expert" perspectives and advice to make sense of the complexities of climate change. However, developments of user-generated, peer-to-peer, democratized and interactive communications have led to substantive changes in how people access and interact with information as well as in who they consider authorized definers of the various dimensions of climate issues. Andrew Hoffman (2015) has posited that "the *messenger* is as important as the message" and therefore communicators must "address the *process* by which the message was created; choose *messages* that are accessible; and present *solutions* that represent a commonly desired future" (p. 53, italics in the original).

Social science research into the relationships between trust and environmental issues has found that trust can lead to higher acceptance of risks and greater

---

[4] This section title is adapted from the Sheldon Rampton and John Stauber book *Trust Us, We're Experts: How Industry Manipulates Science and Gambles with Your Future*, published in 2002.

support for recommended policy actions (e.g., Poortinga and Pidgeon, 2003). Trust can also boost conceptions of reliability. In research on wildlife management policy support, Hwanseok Song, Katherine McComas and Krysten Schuler (2018) found that sources or messengers that were viewed as more trustworthy led to more support for the policy measure they advocated. This is also consistent with work by Michael Manfredo, Tara Teel and Harry Zinn (2009), who mapped conceptions of trust onto notions of "care." With respect to younger people, research by Matthew Motta found that interest in science in early teen years (ages 12 to 14) is associated with greater trust in climate scientists and in climate science in adulthood. He also found these trends occur independently of political ideology (Motta, 2018).[5] Therefore, cultivating interest in science among young people can be a pathway to trust in scientists and in science, as well as to a greater likelihood to advocate for support of sciences as well as policy interventions based on science (Gauchat, 2018) (see Chapter 8 for more about youth and climate communications). In conjunction with this, Ishani Mukherjee and Michael Howlett (2017) have found that "the ebb and flow of ideas gaining government attention is heavily dependent on the actions and interactions of not one, but rather several identifiable groups of actors involved in defining problems, articulating solutions to them, and gaining and retaining political support for specific matches of problems and solutions" (p. 70; see also Mukherjee and Howlett, 2015; Béland and Howlett, 2016).

Over time, the enlargement of scientific ways of knowing about climate change into aesthetic, affective,[6] emotional, visceral, tactile and experiential ways of knowing (see Chapter 4 for more) has prompted a democratization of who are considered authorized, legitimate expert voices in the public sphere. Increasingly, researchers and practitioners have studied how these various ways of knowing and learning about climate change influence awareness and engagement (e.g., Baum and Groeling, 2008; Fahy and Nisbet, 2011; Jacobson, 2012; Horan, 2013; Zhu and Dukes, 2015). For instance, through a rich ethnographic account of various perspectives on climate change including religious groups, business associations and Inuit leaders, Candis Callison (2014) has commented, "Even though climate change may have begun as a scientific concept, it has flourished and it's been adopted, torqued, politicized, paired.

---

[5] Through these findings, he offered an insight that "attention-grabbing and engaging content of board and video games offers an intriguing medium" for more successful communication and engagement (Motta, 2018, p. 487).

[6] Anthony Leiserowitz and Nicholas Smith (2017) have defined affect as "the specific quality of 'goodness' or 'badness' experienced as a feeling state (with or without conscious awareness) or the positive or negative quality of a stimulus" (p. 2).

In short, it's been filled with meaning through its interaction with belief systems, practices, and other systems of knowledge" (p. 247; see also Callison, 2017). This democratization has then promoted more sophisticated and textured understandings of what works, how, when, why and under what conditions in creative (climate) communications.

These more democratized spaces of meaning-making have taken form, for example, as citizen science projects, where professional and amateur scientists work together in the construction of research through observations and data gathering. With the technical tools of smartphones, GPS, and the internet, these collaborations have become more prolific and powerful than ever. Initially met with some skepticism in the scientific community (and in the peer review process), improvements in verification of data have led to widespread partnerships, from bird research (Silvertown, 2009) to cooperation to developing ways to confront key challenges associated with climate change. For examples, Climate CoLab, climateprediction.net and weather@home have drawn on citizen volunteerism and computing power to complete a range of tasks from helping small businesses go low carbon to more efficiently running more robust climate model ensembles. This process of getting together has helped improve the power of research as well as enhancing engagement at various stages of discovery and action.

Diverse voices effectively have thus become new "knowledge brokers" on climate change (Priest, 2016; Lam, 2017). Andreas Sieber (2017) has described them as "new influencers with heterogeneous networks." James Shanahan (2017) has noted that the narratives set by these leaders are critical to setting agendas on climate change. Jack Stilgoe, Simon Lock and James Wilsdon (2014) have pointed out the added value of grappling with varied perspectives and voices when deliberating science in society.

For example, this cooperation and brokering is evident in the approaches of activists in the Appalachian region of the United States, where there is a heavy presence of coal extraction and coal-related jobs. Coal River Mountain Watch adopted the cheeky slogan "Save the Endangered Hillbilly" as the group partnered with United Mine Workers to campaign for safer working conditions, and with "Appalachia Rising" activists campaigning to ban mountaintop removal and other environmentally damaging practices by the coal industry (Skibell, 2017a). Incidentally, this approach through trusted messengers within coal communities stands in stark contrast to a March 2016 town hall comment by then-US presidential candidate Hillary Clinton, referring to needed transitions to clean energy. In a tone-deaf moment, Clinton commented, "We're going to put a lot of coal miners and coal companies out of business" (Strauss, 2016). Working to meet West Virginians and residents from nearby states where they are on issues of environmental degradation, public health and safety through the "Save the Endangered Hillbilly" campaign has led to successful communications and engagement on these issues.

As another example, religious leaders can be effective climate communicators (Wilkinson, 2012). Survey research by Christopher Scheitle, David Johnson and Elaine Ecklund (2018) has found that members of the general public often trust science information from religious leaders as much as they do from the scientists themselves. What's more, they found that more religious citizens turned more readily to religious leaders as authorities on science (Scheitle et al., 2018). They concluded that by relying on non-scientists as sources for information about science "the risks are indeed high, but so too may be the reward for expanding our conception of where people turn to learn about science" (p. 15). In particular, Pope Francis has effectively ushered in messages of radical confrontation of environmental degradation and human exploitation through his 2015 encyclical "Laudato Si" (Brulle and Antonio, 2015). The Pope was credited for Roman Catholics' concern for climate change as the highest among US Christian groups (Maza, 2016).[7] Moreover, in a 2018 editorial Katharine Hayhoe (2018) commented, "As uncomfortable as this is for a scientist in today's world, the most effective thing I've done is to let people know that I am a Christian. Why? Because it is essential to connect the impacts of a changing climate directly to what's already meaningful in one's life, and for many people, faith is central to who they are" (p. 943). Based on his research, Randolph Haluza-DeLay (2017) has argued, "evangelical Christians are more likely to listen to an evangelical such as climate scientist Katharine Hayhoe than the encyclical of Pope Francis or a leader of another religious tradition" (p. 91). He added, "Each is an élite person who is valued differently by different social groups . . . this does not mean that there cannot be inter-religious cross communication, but that the salient narratives, values, images, institutions and even solutions may not resonate with different religious backgrounds, even within the same religious tradition" (Haluza-DeLay, 2017, p. 91).

As a third example, a "Climate Matters" project in a group called Climate Central (led by Bernadette Woods Placky, Chief Meteorologist and Climate Matters Program Director) began in 2010, working with Jim Gandy of WXLT in Columbia, South Carolina. Since then, Climate Matters has worked with weathercasters in partnership with George Mason University (led by Ed Maibach), Climate Communication Science and Outreach (led by Susan Joy Hassol), NASA, NOAA and the American

---

[7] See http://nineteensixty-four.blogspot.com/2016/06/laudato-si-catholic-attitudes-about.html for more details.

Meteorological Society (AMS). They have worked from the premise that weathercasters can be local celebrities-as-public-intellectuals in a region, and they have therefore been seen as trusted and highly credible sources of information over time. Through his research, Kris Wilson (2007) has noted "There isn't a politician, entertainer, or athlete in the world who wouldn't kill for a fraction of the power that television weathercasters command from the public's attention" (p. 84). The Climate Matters project has therefore sought to capitalize on that trust and recognition in order to help meteorologists report on climate impacts as they relate to weather. This then has helped to also make these issues local, immediate and personal (see page 15 for more about the importance of bringing climate change "home"). Over these past years, the Climate Matters project has co-developed open-source visual representations – graphics, charts, videos – to help interested weathercasters bridge complexities of climate change to the everyday forecasting of changing weather. As such it has gained funding support and reach, and now provides more than 500 TV weathercasters in the USA and other 100 TV weathercasters internationally with weekly materials in English and in Spanish.

Adam Corner and Jamie Clarke (2017) have argued that "Nurturing and supporting representatives of diverse social groups is crucial – people who can speak with authenticity and integrity, using language and themes that life climate change out of the 'green ghetto' ... " (p. 89). The foregoing three examples of self-described "hillbillies," religious leaders, weathercasters and political conservatives show that these issues are no longer merely in the domain of those stereotypical "environmentalist" concerns. These instances instead demonstrate the power of authentically reaching new audiences through trusted messengers and accessible messages. These approaches then help break out of a pattern or convention that climate and environmental concern is the province of tree-hugging, Birkenstock-wearing environmental- ists. While there are those humans-as-environmentalists too, concerns and engagements clearly emanate beyond that caricatured collective.

Nonetheless, a perception of a "green ghetto" was on display in a televised debate between Barack Obama and Mitt Romney in the lead up to the 2012 US presidential election. Coming out of that event, moderator (and *CNN* journalist) Candy Crowley referred to "all of you climate change people" as some kind of an isolated "environmentalist" pressure group (Downie, 2012). In so doing, Crowley revealed her twentieth-century mindset while showing what may be a fairly typical residual perception of those who have climate and environment concerns in our current communications ecosystem. This perspective has pervaded many discussions and conceptions since that time. For example, it was embodied by Stewart Easterby's utterances in the *Wall Street Journal* in

2018. In his opinion piece, he dismissed threats of climate change while he cynically recommended to those who are concerned about the dangers of climate change that they "will attract more supporters to your cause if you can pick a name and stick with it . . . " (Easterby, 2018). While more of a benign dig than malignant incitement, Easterby framed these concerns those of a kind of special interest group while prodding "environmentalists" to smarten up their communications.

As such, "environmentalism" is a twentieth-century construct that may have lost its efficacy and luster in a twenty-first-century communications environment. Stewardship, pollution concerns and health have been considerations at the human–environment interface since the emergence of humans (*Homo sapiens*) from other hominids about 300,000 years ago. However, "environmentalism" and its "environmentalist" advocates have been seen to have emerged as a contrivance that addressed concerns of human quality-of-life issues amid pressures of industrialization, urbanization and consumption in the twentieth century (Gottlieb, 1993; Shabecoff, 2003).

Part of the lost potency of "environmentalism" can be attributed to the fact that Millennials or Generation Y (born between the mid-1980s and the mid-1990s) and those in Generation Z (born in the mid-1990s through the early 2000s) were born into a world after many social movements that led to the formulation of Earth Day, to the founding of the Environmental Protection Agency, to the Intergovernmental Panel on Climate Change and to the establishment of the US Clean Air Act, Clean Water Act and Endangered Species Act, among others (Gottlieb, 2002). As such they did not bear witness to the many struggles associated with their founding that have built the associated archetypal "environmentalist." However, Millennials and Generation Z members were born into a world when environmental and climate issues were already on the public agenda, and already a part of many school curricula (see Chapter 8 for more). In this way, "environmentalism" has effectively overcome that stereotypically long-haired hippy subject and has now permeated many perspectives, identities, characters, social groups and subcultures.

Yet, reinterpretations of what "environmentalism" means in present-day conditions have not been limited to young, potentially sincere, members of the public citizenry. Some claims are dubious, reckless or arguably nefarious. For example, in a 2017 meeting with CEOs of GM, Ford and Fiat Chrysler, US President Trump claimed, "I am to a very large extent an environmentalist, I believe in it" (Yachnin, 2017). This claim immediately followed a pronouncement that his administration would reduce "unnecessary regulations" on the automotive industry (Chang, 2017). Similarly, in 2017 then-EPA administrator Scott Pruitt pondered aloud (to those gathered at a right-wing

Federalist Society event), "I've been asking the question lately, what is true environmentalism? What do you consider true environmentalism? And from my perspective, it's environmental stewardship, not prohibition ... we have been blessed, as a country, with tremendous natural resources. . . . I believe that we have an obligation to feed the world and power the world, with a sensitivity, as far as environmental stewardship, for future generations" (Cama, 2017).

In a 2018 commentary entitled "Environmentalism Is a Long-Term Investment," Austan Goolsbee – former economic policy advisor to US president Obama – challenged common framings of "business versus government" on the subject of conservation and the environment. He challenged this as "zero–sum" thinking and opined, "Making rules more favorable to business can lead markets to fail and destroy private sector value, while cleaning up pollution or protecting public spaces can unlock value in the private sector and allow it to grow" (Goolsebee, 2018, p. SB4). This commentary, while useful in expanding notions of "environmentalism" to smart business and longer-term thinking, still served to further an entrenchment and naturalization of capitalist logics of human–environmental relationships.

Along with developments like the "Green Issues" Getty Images collection (Hansen and Machin, 2008), these have appropriated ongoing trends of differentiated consumption that perpetuate rather than alleviate challenges associated with climate change. This consumption can be cast as not just material consumption but also consumption of narratives and storylines (see Chapters 4 and 5 for more). These arenas of consumption have further strengthened atomized rather than collective movements for change, and have favored adherence rather than resistance to powerful interests that have buttressed current climate challenges still safely within an "environmentalist" construct.

In a November 2018 interview former California Governor (and movie star) Arnold Schwarzenegger commented that "environmentalists were doing a terrible job selling climate change concerns" (Concha, 2018). While he made cogent points about moving communications to the here and now and to focus on public health concerns to bring climate change "home" (see Chapter 2 for more), his comments also revealed two additional insights: first, he casually referred to communication about climate change with the public as "selling" it; second, he characterized communicators as mere "environmentalists." Neither of these items was challenged, and this indicated the transgressions of stalled out discourses about climate change. Owing in part to examples and instances like these, some have now refuted the "environmentalist" moniker (e.g., Landsburg, 2011) while others have asserted that it is an outdated term (e.g., Curtis, 2012).

Maybe we are all environmentalists now (Tesch and Kempton, 2004)? Maybe this has helped break out of that "green ghetto"? Maybe that is not a good thing. Despite that creative climate communications do not amount to Faustian bargains (exchanging knowledge about climate change for "selling out"), their perceptions as such have revealed vulnerabilities in treatments of "environmentalism" in present-day communications environments. In their provocative 2004 article "The Death of Environmentalism," Michael Shellenberger and Ted Nordhaus (2004) argued that "Environmentalism must die, they concluded, so that something new can be born" (p. 121). While "environmentalism" may be in need of an update in these spaces of creative climate communications, "new knowledge brokers" and "influencers" help to catalyze these spaces of engagement.

In these contemporary communications environments, social science and humanities researchers have analyzed how celebrities have increasingly stepped in as spokespeople for climate concern. These knowledge brokers have shaped cultural dimensions of climate change in contemporary times (Boykoff et al., 2010; Hoffman, 2015). For examples, research by Jason Carmichael and Robert Brulle (2017) as well as Michael Tesler (2018) has pointed to the influence of celebrity politicians (or elite political actors) on public attention to climate change. Their involvement has been dubbed an emergent "climate science–policy–celebrity complex" (Boykoff and Goodman, 2009).

By way of their power in the cultural and social (and sometimes political) arenas, their interventions have led to greater awareness of climate issues (Doyle et al., 2017; Hammond, 2017; Stever, 2019). By extension, in the USA most people still see often-celebrity-fueled science-related entertainment in a positive light while also drawing a diverse audience (Funk et al., 2017; Kneas, 2017). Yet, research by Matthew Atkinson and Darin DeWitt found that celebrity advocacy has more nuanced resonance. Through examinations of celebrity witness testimony at US congressional hearings, Atkinson and DeWitt (2018) found that celebrity witnesses were three times more likely to be mentioned in legacy media coverage. However, they found that this did not similarly permeate nonpolitical social network infrastructures. They concluded, "engaging the public's interest in political issues is a challenge for the famous, admired, and wealthy ... there are severe limits on the extent to which the Internet can facilitate the vibrant participatory democracy many optimists anticipate" (Atkinson and DeWitt, 2018, p. 13). In other words, they found that (through political engagements such as congressional hearing testimony) celebrities were successful in hooking news in certain venues – such as the *New York Times* – but they still came up short in engaging other audiences through nonpolitical social network arenas.

Paris Hilton ✔

@ParisHilton

This is earth
It's hot
Don't pollute

1:26 PM - 25 Sep 2017

50,657 Retweets **106,390** Likes

Figure 1.2 @ParisHilton tweet September 25, 2017.

Mike Goodman and I have cautioned how these interventions have faced potential perils. Among them, we warned of dangers where their involvement might reduce concern to the domain of fashion and fad, where their involvement may fuel ad hominem attacks on their individualized heroism,[8] and where their efforts might trivialize behavioral change rather than mobilizing certain audiences and fan bases (Boykoff and Goodman, 2009). In other words, on the highly politicized and highly contentious terrain of climate discourse and action in the contemporary public sphere, we warned that these extraordinary celebrities provided relatively easy opportunities to isolate or dismiss efforts as "leftist campaign communications" (Ereaut and Segnit, 2006, p. 23) or those of the "liberal Hollywood elites" (Boykoff and Goodman, 2009).

For example, a Paris Hilton tweet about climate change garnered considerable traction (noted through retweets and likes) but also considerable consternation and mocking along these lines (noted through replies to her Tweet) (see Figure 1.2). Phaedra Pezzullo (2016, p. 805) has warned that the spectacle of "sexy" or celebrity – "with bright lights, loud noises, and big egos" – runs risks of obscuring and even displacing attention that needs to be paid to structural causes of climate challenges.

Among the many persistent challenges for effective climate communication, Libby Lester has commented that "A public sphere is both legitimate and effective when it (1) provides an opportunity for all those affected to

---

[8] We commented that "by the very act of making it personal, it makes it personal" (Boykoff and Goodman, 2009, p. 404).

participate in the public debate, (2) provides a space for a diverse range of views to be put and importantly heard, and (3) holds decision-makers accountable through the processes of publicity and the pressures of public opinion" (2015, p. 3; see also Fraser, 2007). Addressing these elements shaping an effective public sphere for discussions and deliberations about (climate) science, Eric Kennedy, Eric Jensen and Monae Verbeke studied participation in science festivals. They found "such events are disproportionately reaching economically privileged and educated audiences already invested in science, as opposed to diverse and broadly representative sample of the general public ... [therefore] science festivals are falling short of their aims to make science accessible to a broad audience" (Kennedy et al., 2018, p. 14). They then recommended ways to more effectively reach new audiences and promote social inclusion through varied promotions as well as science festival activities.

Increasing participation through involvement clearly is important. However, with participation there remain challenges about deepening engagements through inclusion. Racial and gender dimensions of climate challenges have often been overlooked, and this has served to weaken sustained, inclusive and thus effective engagement. Charles Ellison (2017) has commented that many current collective efforts are "leaving people of color out of the discussion, even though climate and environmental disasters disproportionately affect our communities" Moreover, female climate researchers have reported numerous insults, taunts and degrading comments – mostly through social media and email – as they have carried out their work in the public arena (Waldman and Heikkinen, 2018). These considerations of inclusion in climate communications have woven into ongoing deliberations of what constitute "just sustainabilities," striving for a better and more equitable quality of life within the context of sustainable development and sustainability (Agyeman, 2008). William Gibson once said, "The future is already here, it is just not evenly distributed."

Going forward, Lloyd Davis and colleagues have called for continued work in environmental and science communication in order to enhance public legitimacy of science as evidence shaping public pressures for policy action (Davis et al., 2018, p. 5). Meanwhile, Aysha Fleming, Frank Vanclay, Claire Hiller and Stephen Wilson (2014) have argued that "The key contributor to positive change is to recognise the power of language in shaping perceptions and actions, and to proactively and reflexively use language to create new conditions" (p. 416). The power of creation and creatively therefore shape these spaces of possibility for effective communication and engagement.

The Post-Truth Anthropocene Era

Considering audience(s) and notions of trust in messengers and messaging are critical. In addition, conditions of communications ecosystems as they relate to the quality of messages matter. In the Anthropocene enter "fake news" and "post truth."

At present, these communications ecosystems are functioning in an epoch now often referred to as the "Anthropocene." The term "Anthropocene" comes from the Greek root *anthropos*, meaning "human" and *cene*, meaning "recent." This does not mean that this is all about humans; for example, flora and fauna are significant in their own rights. Nonetheless, the Anthropocene has developed as a term to signify the most recent period of geological time when humans have had dominant and unprecedented influence on the climate and environment on planet Earth.

For billions of years, radiative forcing – producing changes in the climate – has occurred in response to energy imbalances. While this has contributed to a range of environmental impacts, there has been a focus on atmospheric temperature change, as it is the most evident effect of biophysical work done to rebalance this energy distribution (Wigley, 1999). This notion of a new geological age has been discussed over the last several decades. It was a concept first coined by ecologist Eugene Stoermer (Steffen et al., 2011) and popularized by atmospheric chemist Paul Crutzen (2002) in the early 2000s.

A 2017 study by Owen Gaffney and Will Steffen mapped out an "Anthropocene Equation" to clearly articulate human influences on climate change. In it, they noted that humans are changing the climate about 170 times faster than natural forces would do alone. Moreover, they pointed out that natural forces would have driven a slight temperature decrease (of 0.01 degree Celsius over the last 100 years) due to the general conditions of the current glacial–interglacial late Quaternary period (and Holocene epoch) that we are in. That temperatures are rising and other indicators – such as warming of the troposphere and cooling of the stratosphere – all have in fact pointed to what Gaffney and Steffen (2017) surmised as global warming entirely attributed to human activities. They concluded that "the rate of change of the Earth system over the last 40 to 50 years is purely a function of industrialized societies" (Gaffney, 2017).

Anthropogenic sources contributing to this Anthropocene Equation – affecting the distribution of energy across the planet – include fossil fuel burning (primarily coal, gas and oil) and land use change. Considerations have then developed regarding how to develop policies to address these developments in the Anthropocene. These processes have been referred to as dimensions of environmental politics (Lövbrand et al., 2015) and have also been dubbed "Anthropocene Geopolitics" (Dalby, 2007).

Among creative endeavors bridging to the public sphere, a "Welcome to the Anthropocene" exhibition was on display in Munich, Germany at the Deutsches Museum from 2014 through 2016. Assembled through support from the Rachel Carson Center, this was seen to be the first major exhibition that grappled with the realities of the Anthropocene. The project addressed topics such as urbanization, mobility and food in the context of this unprecedented era. The exhibition then prompted visitors to consider historical antecedents as well as future possibilities based on our collective human behaviors as well as collective human responses to these contemporary environmental challenges.[9] Also, *Anthropocene Magazine* has been developed by the global science consortium Future Earth. *Anthropocene Magazine* has invested in digital and print stories as well as live events called "Anthropocene Dialogues" in order to realize the self-described mission "to curate a global conversation about data, technology, and innovation that lead to solutions to the persistent environmental challenges of our time."[10] These are examples of projects that have been working to foster discussions on sustainability and development within a communications ecosystem and that have been grappling with discursive pollution from "fake news" and "post truth" emissions.

"Post truth" was celebrated as the word (or phrase) of the year in both Macquarie and Oxford dictionaries. The *Oxford English Dictionary* (2016) defines "post truth" as "relating to or denoting circumstances in which objective facts are less influential in shaping public opinion that appeals to emotions and personal belief." This notion of post truth has increased in influence along with its close cousin "fake news." Roots of fake news have been traced back to the "swiftboating" of then US senator John Kerry in the lead up to the 2004 US presidential election. At that time, he was the Democratic nominee for US president. In the lead up to Election Day, a right-wing Political Action Committee (PAC) called "Swift Boat Veterans for Peace" developed a strategy to call into question Kerry's honorable military service. The PAC allegations were deployed through a set of television advertisements that leveled charges about his efforts, particularly during his time in Vietnam as a Swift Boat commander. While those allegations were later determined to be unsubstantiated, unwarranted and slander, the ability to call into question John Kerry's suitability to be commander-in-chief were nonetheless on display (Kinsley, 2008).

[9] More about this exhibit and project can be found here: www.deutsches-museum.de/en/exhibitions/special-exhibitions/archive/2015/anthropocene/

[10] More information can be found here: www.anthropocenemagazine.org/

As fake news developed on the landscapes of politics and ideology, the term has been routinely been co-opted and elaborated as "left-wing fake news" by purveyors such as Rush Limbaugh and Breitbart News (Peters, 2016). Along with the increasing power of super-PACs through the 2010 *Citizen's United v. Federal Election Commission* ruling in the US Supreme Court, in this contemporary milieu, fact-defenders and fact-checkers now face an uphill battle against the agenda-setting power of fake news purveyors (Vargo et al., 2018). Andrew Maynard and Dietram Scheufele (2016) have observed that "truth seems to be an increasingly flexible concept."

These considerations of post truth, fake news, the Anthropocene and climate change have become intertwined our contemporary communications environment. For example, from the bully pulpit in 2018 US president Donald J. Trump hosted a "Fake News Awards" (Team GOP, 2018). Among the winners, he named a *New York Times* story by Lisa Friedman (2017) about a federal climate change study stating that climate impacts in the USA were being experienced now. Team GOP stated that the reason for this was that "The New York Times FALSELY claimed on the front page that the Trump administration had hidden a climate report" (caps in the original).

The article, entitled "Scientists Fear Trump Will Dismiss Blunt Climate Report," did issue two corrections in the days following initial reporting, but they were in reference to the fact that (1) the report was already available (and therefore not leaked) and (2) that coauthor Katharine Hayhoe is a Texas Tech University professor and not a government scientist (Friedman, 2017). However, much like the "swiftboating" of John Kerry nearly a decade and a half earlier, the surface-level spectre of fake news was enough, for Friedman's piece to make Trump's "Fake News Awards" list of honors.

In an extension of the Six Americas work mentioned earlier, Sander van der Lindon, Anthony Leiserowitz, Seth Rosenthal and Ed Maibach have posited that one effective way to counteract fake news and misinformation regarding climate change is to communicate the high-level convergent agreement (or "consensus") among relevant experts about climate change. They called this "inoculation," "pre-bunking," or "pre-emptive protection" against the real-world repression of information across the political spectrum (van der Linden et al., 2017; see also Cook, 2017; Cook et al., 2017) (see Chapter 5 for more).

This process of inoculation can effectively come from communications by climate science researchers and policy actors themselves. Survey research by James Beebe et al. (2018) found that among these science–policy actors, "climate science is more settled than ideological pundits would have us believe and settled enough to base public policy on it" (p. 14). But John Corner has warned that this is not just a solely US-based concern; rather this is a global

Global Temperatures Plunge. Icy Silence from Climate Alarmists
Land temperatures have plummeted by 1 degree - the biggest and steepest fall on record. But the news has been greeted with an eerie silence.
breitbart.com

11:12 AM - 1 Dec 2016

Figure 1.3 @HouseScience tweet December 1, 2016.

challenge.[11] He has commented on how changing socioeconomic, cultural, political and environmental conditions in various country contexts has seeded "suspicions towards perceived forms of the 'official' and the 'established' which are thereby being generated" (Corner, 2017, p. 1106). He continued, "Such suspicions can drive democratic change but they can also present an opening for further, novel modes of the unreliable and the deliberately untrue" (Corner, 2017, p. 1106).

Stephan Lewandowsky, Ulrich Ecker and John Cook (2017) have pointed out that as post truth and fake news "have become increasingly prevalent in public discourse," this then has devastating impacts on the dissemination of quality information vis-à-vis misinformation on climate change (p. 1). And this repression of good information threatens a well-functioning democracy (Cook et al., 2017). Furthermore, in the spaces of science and the environment John Foley (2017) has argued that "the war on facts is a war on democracy."

To illustrate, in late 2016 a Breitbart News Network story was tweeted by the US House Committee on Science, Space and Technology (see Figure 1.3). Stating "Global Temperatures Plunge. Icy Silence from Climate Alarmist," this politicized jab was irksome to many relevant expert climate researchers as an exercise in cherry picking data for political ends. But the story and backlash

[11] As an example of the proliferation of fake news elsewhere, a study in Singapore found challenges to successfully authenticating information encountered in social media (Tandoc et al., 2018).

then led to back-and-forth exchanges where accusations of "alarmist" scientists overhyping data then created an impression of vigorous debate in the public arena (Balaraman, 2016). Ultimately, it was deeply troubling that this factually deficient piece from Brietbart was then promoted through one of the most influential Congressional committees addressing climate science policy. Effectively it was one more signal that in a fake-news and post-truth communications environment, facts, information quality and democracy were put on notice.

While the antidote to post truth might seem to be injections of facts and truth itself, the realities here are more complex when considering who speaks on behalf of climate change (Boykoff, 2011) and how we each interpret facts and truths and integrate them into our lives differently (de Certeau, 1984; Rutherford, 2007). As such, "post truth" can meld into notions of "selective truth." Steve Rayner (2006) has observed, "For good or ill, we live in an era when science is culturally privileged as the ultimate source of authority in relation to decision-making ... [but] we know that science is not capable of delivering the kinds of final authority that is often ascribed to it" (p. 6). Moreover, Stephen H. Schneider has commented,

> On the one hand, as scientists we are ethically bound to the scientific method, in effect promising to tell the truth, the whole truth and nothing but – which means that we must include all doubts, the caveats, the ifs, ands and buts. On the other hand, we are not just scientists but human beings as well. And like most people we'd like to see the world a better place, which in this context translates into our working to reduce the risk of potentially disastrous climate change. To do that we need to get some broad based support, to capture the public's imagination. That, of course, means getting loads of media coverage. So we have to offer up scary scenarios, make simplified, dramatic statements, and make little mention of any doubts we might have. This 'double ethical bind' we frequently find ourselves in cannot be solved by any formula. Each of us has to decide what the right balance is between being effective and being honest. I hope that means being both. (quoted in Schell, 1989, p. 47)

While we often rely on climate scientists to tell us the truth and give us the facts about climate change, Stephen Duncombe (2012) has pointed out, "The truth does not reveal itself by virtue of being the truth: it must be told, and we need to learn how to tell the truth more effectively. It must have stories woven around it, works of art made about it; it must be communicated in new ways" (p. 20). As such, authorized voices make choices about how to represent truths.

But this approach does not then facilitate an apologetic stance for comments like that from Rudy Giuliani in 2018 that "truth is not truth" (*Meet the Press*, 2018); rather this enriched view of perspective and interpretation actually helps to more capably dissect and expose these statements as both damaging and foolish. It therefore takes additional critical work to unpack and interrogate *how* meanings are made and maintained, and to analyze what historical and biophysical contingencies shape our perceptions of truths about opportunities for climate engagement. A context-sensitive lens of cultural politics can effectively help consider claims and claims makers representing "truth" and "truths" (see Chapter 2 for more).

Conclusions: Five Ws

Former US Senator Daniel Patrick Moynihan (D-NY) famously quipped, "Everyone is entitled to his own opinion, but not to his own facts" (Moynihan, 2010). But this adage has been under attack as much in the contemporary public sphere as ever. These attacks on fundamental considerations of truth have flowed through the massive issue of climate change as they together have then pumped up deep currents of ideologies, values, culture and worldviews. As such, Jason Carmichael, Robert Brulle and Joanna Huxster (2017) have called for more systematic analyses of fake news as they relate to climate change concern, "while there is ample anecdotal evidence that fake news is being released to mislead the public" (p. 611).

In our everyday experience, it can be very challenging to make sense of these intersections, in part due to the historically infused complexities of these issues. But this is also due in part to how we as people communicate and interact with one another. These challenges can lead to psychological duress and anxiety. Robert Pérez and Amy Simon (2017) have noted that human perspectives and decision-making are influenced by "an elaborate psycho-social circuit board that connects people's emotions, values, beliefs, identity and lived experiences" (p. 9). Moreover, Matthew Lieberman (2013) has researched how we humans are social beings with a need to communicate and connect with each other. He has found that our brain activities are similar for both physical pain and threats to social connections. Physical and mental health are connected; furthermore, human connections are an important part of our mental health. Mélodie Trolliet, Thibaut Barbier and Julie Jacquet (2019) have argued that a better understanding of and accounting for underlying brain mechanisms (disposed to emotions and cognitive processing) then can create opportunities for effective communication about climate change.

In contemporary communication ecosystems, these are far from straightforward challenges. Philip Smith and Nicolas Howe (2015) have argued, "Although the social drama of climate change has come into being, it is by no means potent or very well organized as a cultural system. It is somewhat disorganized drama marked by incoherence, disengagement, and proliferation as much as compression, commitment and consensus" (p. 53). Moreover, the quantity of exposure, not even the framing or messenger simply determines knowledge or engagement (Boykoff, 2011; Shi et al., 2016). At best, it shapes their possibilities (Carvalho and Burgess, 2005). As such, there are dynamic and complex challenges we all face when communicating scientific content while also attending to associated cultural meanings (Kahan et al., 2011).

There is urgency derived from the recognition that climate change is "here and now" as the chapter title suggests (see Chapter 7 for more). In her book *Reclaiming Conversation*, Sherry Turkle has pointed out that while there is never any perfect public square, there is a need to promote public conversations as pathways for vital democratic deliberation. She concluded with a recommendation "to remember who we are – creatures of history, of deep psychology, or complex relationships. Of conversations artless, risky, and face-to-face" (Turkle, 2016, p. 362). With this in mind, the next chapter confronts how perceptions that climate change is a distant threat in space and in time may dampen urgency for engagement. So after considering in this chapter dimensions of these challenges such as **who** might be creative and effective messengers, **what** are some creative messages about climate change, **where** are intended or perceived audiences, **when** can these communication strategies effectively be deployed and **why** are current conditions of communications ecosystems constructed as they are, it is worth also considering **how** we have come to know what we know about climate change.

2

How We Know What We Know

People come to know about climate and climate change through a variety of pathways, including through science, aesthetics, emotions and experiences (see Chapter 4 for more). In reality, there is not a simple distinction between science as an evidence-seeking enterprise and science as a process interpreted through various political, social, cultural and ideological lenses. In *Connecting on Climate*, Ezra Markowitz, Caroline Hodge and Gabriel Harp (2014) have pointed out that "people interpret new information through the lens of their past experiences, knowledge and context" (p. 6). Knowledge then can feed into deeper understanding (Su et al., 2014; Huxster et al., 2017). This knowledge and understanding sometimes, but not always, leads to concern (Malka et al., 2009; Kahan et al., 2012).

Moreover, social science research has helped make sense of ways in which spatial reasoning and mental modeling differ between groups, and that impacts cognition/interpretation (e.g., Sterman and Sweeney, 2007; Shipley et al., 2013). Examples of mental models can help enhance understanding of climate change issues such as stratospheric ozone depletion ("the ozone hole") and sources and sinks relating to greenhouse gases in the atmosphere, biosphere, hydrosphere and lithosphere ("like water filling and draining in a bathtub") (Bostrom, 2017). In research into how residents in the Rocky Mountains come to know about climate and climate change, Katherine Clifford and William Travis (2018) conducted interviews with people who work closely with the land and environment (e.g., ranchers and public land managers). This work helped them gain insights that in fact "climate is a social-ecological-atmospheric construct," noting "climate processes are imbricated with ecological and human processes . . . [where] people do not separate human from atmosphere in the same way that models or climate scientists do" (p. 8).

Researchers have also examined and experimented with how visuals and images can be useful pathways for climate change knowledge and understanding (see Chapter 4 for more). They can provide a "cognitive short cut compressing a complex argument into one that is easily comprehensible and ethically stimulating" (Hannigan, 2014, p. 77). For example, in 2018 Mike Willis and colleagues published an article in *Earth and Planetary Science Letters* capturing abrupt climate change through the massive destabilization of the Vavilov ice cap. This was significant in its documentation of a higher rate of change among glaciers and ice caps that collectively hold about a foot of sea level rise. The researchers used remote sensing techniques to show how weak sediments can rapidly give way to land ice flows into marine waters (Willis et al., 2018). While this article dramatically analyzed what had transpired in the Russian High Arctic, its description of movements of more than 80 feet a day (when previously it was two inches a day) was locked in a veritable peer-review vault where few outside of specialists and relevant-expert researchers had a key. However, these revelations were made visual and accessible through an additional 35-second representation posted on YouTube.[1] Within days of its posting, it earned hundreds of thousands of views, including repostings in outlets like the Weather Channel. This added step of translation then effectively opened up and made accessible to a wider public this critical set of findings about newfound vulnerabilities of glaciers and ice caps. Through the animation of high-resolution topographic maps, viewers could watch a time-lapse video of how the ice caps slowly crept forward for many years but then surged in 2015. It effectively opened up new avenues for ways of knowing about these phenomena impacting 300,000 square miles of the Earth's surface. Later, Mike Willis commented, "We've never seen anything like this before, this study has raised as many questions as it has answered. And we're now working on modelling the whole situation to get a better handle of the physics involved" (Weeman, 2018).

As another example, Gisela Böhm and colleagues experimented with imagery and affect through public perception of energy transitions, as they call it "individual behaviors, political strategies, and technologies that aim to foster a shift toward a low-carbon and sustainable society" (Böhm et al., 2018, p. 6). Experimenting with Norwegian and German university students, among their results they found that frequently the respondents expressed that affective images helped to concretize linkages to multiscale decarbonization pathways, from national policies to personal choices. These were also found to be mainly positive associations, rather than negative or fear-inducing considerations and connections.

---

[1] This can be seen here: https://youtu.be/jeC47jxiuuA

Making It/Keeping It Real

Andrew Hoffman (2018) has noted that "one of the great challenges of tackling climate change is making it real for people without a scientific background. That's because the threat it poses can be so hard to see or feel." Ezra Markowitz, Caroline Hodge and Gabriel Harp (2014) have also commented, "Climate communicators should appeal to values held by their target audiences to make it easier for audience members to recognize climate change as a personally meaningful issue" (p. 7).

Creative communications therefore entail work to stir feelings and conjure emotional ways of knowing about climate change. While emotion – by extension, "being emotional" – is typically associated with anger, and fear (Moser, 2017), creative communications can also whip up positive feelings and emotions (see Chapter 5 for more). These are most usefully considered as spectral rather than binary distinctions (Anderson, 2017). In some audiences and individuals, fear can induce feelings of paralysis through powerlessness and disbelief. Andrew Hoffman has said, "Typically, if you really want to mobilize people to act, you don't scare the hell out of them and convince them that the situation is hopeless" (Ryzik, 2017). But with other audiences and people, fear can inspire motivation and a willingness to take action in the face of climate threats. Dan Chapman, Brian Lickel and Ezra Markowitz (2017) have observed, "The bifurcation between 'go positive' and 'go negative' simultaneously oversimplifies the rich base of research on emotion while overcomplicating the very real communications challenges advocates face by demanding that each message have the right 'emotional recipe' to maximize effectiveness" (p. 848).

Over the past years, numerous social science studies have examined the efficacy of emotions in climate communications. In particular, they have examined how fear can raise or lower awareness and inspire action on climate change. For example, through their experimental research about emotional messaging in environmental communication campaigns, Jennifer Hoewe and Lee Ahern (2017) found that emotional messages impact participants more personally. They also found that "emotional appeals may be a way to make pro-environmental messages more effective with traditionally unreceptive audiences" (p. 817). Kari Marie Norgaard (2006) has also examined how fear may influence people to resist information about anthropogenic climate change in the name of psychological self-protection.

Furthermore, research by Saffron O'Neill and Sophie Nicholson-Cole (2009) tested how various images raised awareness as well as inspired engagement in the public citizenry. They found that dramatic and fearful representations raised awareness and concern, but that these kinds of images were "also likely to distance or disengage individuals from climate change, tending to render them feeling helpless and overwhelmed when they try to comprehend their own relationship with the issue" (p. 375). They concluded that sensational, shocking and fear-inducing images "have a place, given their power to hook audiences and their attention. However, they must at least be used selectively, with caution, and in combination with other kinds of representations in order to avoid causing denial, apathy, avoidance, and negative associations that may come as a result of coping with any unpleasant feelings evoked" (O'Neill and Nicholson-Cole, 2009, p. 376).

These findings helped to uncover more textured understandings of the role of emotional, specifically fear-based, ways of learning about climate change. This research provided early insights into more nuanced ways to meet people where they are on climate change. John Meyer (2015) encouraged ongoing endeavors that are "attuned to the concerns, frustrations, pleasures, and fears that animate us every day" (p. 172).

Nonetheless, researchers and practitioners have often communicated fear in their communications about climate change. For example, in 2018 social scientist Mayer Hillman declared to Patrick Barkham of the *Guardian*, "We are doomed. The outcome is death, and it's the end of most life on the planet because we're so dependent on the burning of fossil fuels. There are no means of reversing the process which is melting the polar ice caps. And very few appear to be prepared to say so." Hillman argued that optimism is mere wishful thinking and "accepting that our civilisation is doomed could make humanity rather like an individual who recognises he is terminally ill. Such people rarely go on a disastrous binge; instead, they do all they can to prolong their lives" (Barkham, 2018).

As another example, Amanda Reilly (2016b) from *Energy & Environment Daily* reported on the results of a three-year study by the US Global Change Research Program with the top line "Every American is vulnerable to global warming." And ten years following the influential "Stern Review" in 2006 outlining the economic dangers of climate change, Nicholas Stern commented, "with hindsight, I now realize that I underestimated the risks. I should have been much stronger in what I said in the report about the costs of inaction. I underplayed the dangers" (McKie, 2016).

As a fourth example, in 2018 a study dubbed "Hothouse Earth" was published in the *Proceedings of the National Academy of Sciences* (PNAS) describing how runaway global warming could result from positive climate feedback loop in the climate system (Steffen et al., 2018). The authors wrote, "The present dominant socioeconomic system is based on high-carbon economic growth and exploitative resource use. Attempts to modify this system have met with some success locally but little success globally in reducing greenhouse gas emissions or building more effective stewardship of the biosphere. Incremental linear changes to the present socioeconomic system are not enough to stabilize the Earth system; these include changes in behaviour, technology, innovation, governance and values" (Steffen et al., 2018). In an interview with Kate Aronoff of the *Intercept*, coauthor Will Steffen commented that to avoid planetary collapse there can be "absolutely no new fossil fuel developments . . . [and] you need to have a rapid phase-out plan for existing fossil fuels" (Aronoff, 2018).

Debates persist about how to "make it real" or "keep it real" through contemporary communications about climate change. A prominent example of these vibrant discussions and disagreements can be seen through an influential 2017 *New York Times Magazine* article titled "The Uninhabitable Earth" Written by David Wallace-Wells, it catalogued worst case scenarios on impacts and futures in a warming world. This piece generated instant attention and scrutiny. While some were quick to call its content "the Silent Spring of our Time" (Matthews, 2017), others argued that its value was in its traction it quickly gained, as the most read story in the magazine's history. David Roberts from *Vox* commented, "By any sane accounting, the ranks [of] the under-alarmed outnumber the over-alarmed by many multiples. The vast majority of people do not have an accurate understanding of how bad climate change has already gotten or how bad it is likely to get, much less how bad it *could* get if we keep electing crazy people" (Roberts, 2017a).

Yet, numerous researchers and practitioners voiced discontent with fear-inducing framings of climate challenges in "The Uninhabitable Earth" article. For example, Michael Mann, Susan Joy Hassol and Tom Toles (2017) argued that the *New York Times Magazine*'s "The Uninhabitable Earth" story was prone to overstatements like an exaggeration of the near-term threats of climate feedbacks, such as the release of methane clathrates locked in the permafrost. They posited that the realities of global warming were bad enough so there was no need for hyperbole (Mann et al., 2017). Journalist Chris Mooney reported that many felt "this particular story has gone too far," quoting tweets from John Foley, who called it "a deeply irresponsible article," and Eric Steig, who commented "what's written's actually beyond worst possible case. THIS is the 'alarmism' we get accused of. It's important to speak out against it" (Mooney, 2017, capitalization in the original).

Nonetheless, Bill McKibben (2018) followed with a similarly doom-and-gloom piece titled "Life on a Shrinking Planet" for the *New Yorker* magazine. He argued that carbon-based industry interests have unyieldingly continued to pollute the atmosphere and contribute to climate change amidst evidence of exacerbated wildfires, heat waves and sea level rise that are making swaths of planet Earth now uninhabitable (McKibben, 2018). Published just after devastating California wildfires and before the Fourth US National Climate Assessment documenting climate risks and impacts in the USA (USGCRP, 2018), the temporal context may have tempered similar critiques coming forward from climate researchers, social scientists and media pundits.

While these cases can draw out different vantage points and communication choices, they also can help distinguish between notions of being "alarmist" and being "alarmed." In a 2008 journal article titled "The new climate discourse: Alarmist or alarming?" James Risbey (2008) systematically reviewed claims that climate change is "'catastrophic', 'rapid', 'urgent', 'irreversible', 'chaotic', and 'worse than previously thought'" (p. 26). He found that the use of these terms in public discourse was subjectively yet accurately invoked based on interpretations of evidence from climate science and was therefore "alarming" rather than "alarmist" (Risbey, 2008). Elsewhere, Keynyn Brysse, Naomi Oreskes, Jessica O'Reilly and Michael Oppenheimer (2013) interrogated accusations of "alarmism" on climate change in the public sphere. Through their analyses of UN Intergovernmental Panel on Climate Change (IPCC) reports, they found that climate scientists have actually been *overly cautious* in their projections of the impacts of climate change (emphasis added). They found a preponderance of relatively conservative estimates with respect to the evidence of potential disintegration of the West Antarctic ice sheet and Arctic ozone depletion they assessed. The authors named this trend "Erring on the Side of Least Drama (ESLD)" (Brysse et al., 2013, p. 179). ESLD is also relevant in the context of an alarming October 2018 IPCC Special Report on 1.5°C (Allen et al., 2018) (see Chapter 3 for more).

Dan Chapman and colleagues (2017) have stressed that context as well as varied human responses matters when examining how emotional ways of knowing shape awareness and engagement. These examples and associated debates show that understanding and engagement is not merely a cognitive process; it also involves affective and behavioral dimensions (Lorenzoni et al., 2007). Emotional learning (also referred to as "hot cognition") can provide pathways for long-term retention that mere academic learning cannot facilitate (Minsky, 2007; Forgas, 2008).

These dynamics are often at play when having conversations and communicating about links between everyday weather and longer-term climate patterns. Weather patterns and weather extremes have often provided visceral and experiential entry points for understanding longer-term changes in the climate. Connections through weather can provide meaningful and logical bridges for thinking and talking about climate change. These are distinct but not mutually exclusive categories. For instance, Jon Krosnick and colleagues found that beliefs about climate change were a function of factors that prominently included relevant personal experiences (e.g., exposure to weather disasters) (Krosnik et al., 2006). As examples of communications of distinctions between weather and climate, Stephen H. Schneider (2009) has commented, "Weather is what you get; climate is what you expect . . . weather is day-to-day fluctuations; climate is the long-term averages, the patterns and probability of extremes" (p. 91). Journalist Bud Ward has added, "Weather helps us decide what to wear each day; climate influences the wardrobe we buy" (in Russell, 2008), while Marshall Shepherd (2018) has said (crediting John Knox), "weather is mood, climate is personality."

History has shown that weather disasters such as hurricanes making landfall in the USA have not proven to be a silver bullet that generates significant action to address climate threats. Despite widespread damages from Hurricane Katrina, which made landfall in August 2005 in the Gulf Coast of the USA, from Superstorm Sandy making landfall in October 2012 in the Northeastern USA near centers of influence and power in Washington, DC and New York, and from Hurricane Michael that made landfall in the Florida panhandle in October 2018, discussions and actions making connections between weather and climate change have remained relatively stuck. For example, soon after the devastating 2017 Atlantic hurricane season featuring Harvey, Irma, Nate and Ophelia, UN Secretary-General António Guterres said, "I have not yet lost my hope that what is happening will be making those that are still sceptical about climate change to be more and more realizing that this, indeed, is a major threat for the international community at the present moment" (Nichols, 2017). Michelle Nichols (2017) from *Reuters* reported that Guterres went on to comment that he hoped the 2017 Atlantic hurricane season "will convince Trump and others that climate change is a 'major threat'" and that "the threat is real." But these continued hopes for a sea change have not materialized.

When pondering how to meaningfully address connections between weather and climate change for rural and often conservative agriculturalists, sixth-generation US farmer Casey Cox commented, "We have got to find a way to talk to people about what's happening. I'm not sure how to do that. The best I can come up with is to call it 'climate variability'. That's an expression people seem to accept" (Thrush, 2018, p. A10).

While vague allusions linking weather and climate have been detected in media coverage dating back to the late-1700s, 1988 was seen as a breakthrough year when discussions about climate change grew beyond mere scientific pursuits and surfaced in the wider public sphere. Political

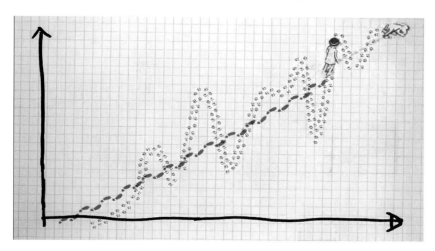

Figure 2.1 This video analogy between a man walking his dog and trends in climate along with variations in weather was produced by TeddyTV for NRK Animation by Ole Christoffer Haga (2012).[2]

pronouncements, science–policy developments and ecological events contributed to this emergence (Boykoff, 2011). Prominently contributing to public discussions was a series of heat waves along with associated drought conditions and wildfires throughout North America that summer. Commenting on this period of time, David Demeritt (2001) wrote that these ecological and meteorological events "were arguably as influential in fostering public concern as any of the more formal scientific advice" (p. 307).

The University Corporation for Atmospheric Research (UCAR) Center for Science Education has capitalized on these entry points and has assembled some creative and resonant short videos to explain wider climate trends. For example, a one-minute video produced in 2012 depicts an animated man walking his dog. While the man follows a relatively straight line, the dog takes a more circuitous route, sniffing and checking things out while still ultimately walking in the same direction as its owner. The narrator then connects these patterns of the footsteps of the man with climate trends, and the behaviors of the dog with weather patterns, while their movements are superimposed on a figure charting increasing temperature over time (see Figure 2.1).

It is worth pointing that that there are differences in strict definitions of "climate change" and "global warming." "Climate change" is a broader term

---

[2] The video can be viewed here: https://scied.ucar.edu/dog-walking-weather-and-climate

that accounts for changes in many climate characteristics, such as rainfall, ice extent and sea levels. "Global warming" refers to a more specific facet of climate change: the increase in temperature over time. Clearly, temperature increases do not occur in isolation from other climate characteristics; rather, many other sources and feedback processes contribute to changes across time and space. Temperature (particularly atmospheric temperature increases) is seen as a clear and distinguishable climate characteristic that indicates more general climate change, and has been called the "fingerprint" for climate change (Wigley, 1999).

The ways in which various aspects of climate change or global warming are communicated – in the news, through interactive and participatory media or by way of discussions with friends – provide critical links between people's perspectives, experiences and possible actions in various cultural, political, social and environmental contexts (e.g., Adams and Gwynnald, 2013; Schuurman, 2013). Over time, various research teams and strategists have conducted experiments and convened focus groups to find out what the terms "climate change" and "global warming" evoked in citizens across the political spectrum (e.g., Schuldt and Roh, 2014; Wiest et al., 2015). These have been pursued with the motivations to learn more about how awareness, concern and possible actions may be shaped by what the phenomena may be called, or how they are described.

For an example, Lorraine Whitmarsh (2008) surveyed UK citizens and found that "global warming" was frequently associated with heat-related impacts, human causes, ozone depletion and the greenhouse effect, while "climate change" was often associated with observed weather and climate impacts, and natural variation in the climate. She concluded that citizens considered "global warming" as "a more emotive term, in part because it suggests a clear direction of change towards *increasing* temperatures" and found that "implications of 'climate change' are more ambiguous" (p. 16, emphasis in the original).In the US context, a 2003 memo revealed culturally and politically contingent uses of "climate change" and "global warming." Memo author Frank Luntz has been a well-known US political strategist and advisor to the George W. Bush administration. The memo was titled "Winning the global warming debate – an overview" and Luntz (2003) made the case, through focus group findings, that, "It's time for us to start talking about 'climate change' instead of global warming ... 'Climate change' is less frightening than 'global warming'" (p. 142). Two years later, ActionMedia (2005) produced a report called "Naming Global Warming." In efforts to deliberately shape perceptions, they recommended, "DO NOT call the problem 'climate change'. 'Climate change' is understood as the natural process the earth's climate has undergone in the past'. DO call the problem 'global warming'. 'Global warming' is the result of human activity'" (p. 6, emphasis in the original). However, as these terms populate policy arenas and more widely in the public sphere, their meanings are increasingly interchangeable (e.g., Houghton et al., 1995; see Boykoff, 2011 for more).[3]

The Narratives and Stories We Tell Ourselves (and Each Other)

Clifford Geertz has defined culture as "the ensemble of stories we tell ourselves about ourselves" (Geertz, 1973, p. 448). Similarly, Fred Inglis has suggested that identity is formed through a process whereby we observe and "inhabit the narratives of how to live' available within a given culture" (Inglis, 2005, pp. 1–2).

There is great psychological, social, political, ideological and cultural power to stories we tell ourselves (and each other). Stories have a beginning, middle and an end. Narratives are representations of stories. Narratives can juggle the temporal dimension of storytelling to maximize resonance with selected audiences. Dawn Lerman, Robert Morais and David Luna (2018) have written, "A narrative is a story and is built from a series of connected events" (p. 15). In reality, stories and narratives are what make the world go around. Information, facts and data only get us so far; stories and narratives have been inextricable elements of culture and society for millennia.

People have little trouble turning information into narratives (Arnold, 2018). Moreover, narratives are primary drivers of human cognition (Lyotard, 1984). Research into narratives as it relates to policy action and public engagement suggests that four structural elements must be present for successful communication: setting, plot, characters and a moral of the story (Jones, 2010). Setting provides context; plot signifies interactions and relations over time; characters are animated by heroes, villains and victims; while a moral of the story can be considered solutions, gains or benefits from action.[4] Stories and narratives are generally populated by actors (protagonists, antagonists, witnesses); rhetoric (*logos, pathos* and *ethos*); and motivations, locations, plots and genres (Smith and Howe, 2015). Nonetheless, challenges lurk about how to construct these stories and narratives in coherent

---

[3] Therefore, in this book, I use the terms "climate change" and "global warming" interchangeably.
[4] This has been called the Narrative Policy Framework (NPF).

and coordinated ways in order to induce engagement and action (Goldstein and Keohane, 1993; Miskimmon et al., 2014).

For example, Bruce Goldstein and colleagues examined how resilience of urban systems could be enhanced through narrative storytelling (Goldstein et al., 2015). As another example, in *The Strategic Storyteller* author Alexander Jutkowitz (2017) explored keys to effectively decoding complexity for compelling storytelling in the context of content marketing. He reasoned, "Stories command our attention and open our minds to receive new ideas. They aren't effective because they force ideas, but because they awaken our vital needs for wisdom, wonder and delight" (p. 5). Storytelling can also open up vulnerabilities as the storyteller reveals what he or she cares about and values. This same storytelling can build reliance and trust through authenticity.

(Astoundingly), when it has come to communications about climate change many of those who have been deemed legitimate spokespeople and authorized voices have chosen to cling to "more facts" and "more information" in hopes and expectations that intended audiences and members of the public will then make the "right" choices (see Chapter 3 for more). Alexander Jutkowitz (2017) has commented, "The world is in dire need of stories. Information is abundant, but stories are rare . . . storytelling is also an inherently disruptive activity" (pp. ix–xi). He has therefore argued, "We need more storytellers. We need more people with the tools and the desire to dig into the world's information and build their own stories out of it" (p. x). Simon Bushell, Thomas Colley and Mark Workman (2015) have opined, "Narratives are always present, but a 'strategic' narrative is one that is consciously developed to achieve certain aims" (p. 971). Strategic narratives in practice must be multifarious and multifaceted.

These considerations regarding how to more effectively talk about climate change through narratives and storytelling have increasingly pervaded journalistic reporting and opinion writing. These approaches of storytelling and attention to audience are consistent with what has been found to be effective communication approaches (see Chapter 7 for more).[5] For example, into these spaces has ventured *The Story Group* led by Dan Glick and Ted Wood. They are an independent, multimedia journalism company that is committed to the power of storytelling through their work on climate and environment issues in the USA. Through in-depth reporting and short videos, they have worked

---

[5] For more, see Marisa Beck's (2017) research on the power of narrative in shaping policy outcomes through a case study exploration of integrated assessment models of global climate change.

with groups such as the National Climate Assessment to depict stories of "Americans on the Front Lines of Climate Change" like Texas ranchers, Iowa agriculturalists and Washington State oyster farmers.[6] As another example, Randy Olson has developed the "ABT method"(And, But, Therefore) to help de-jargonize language and clearly as well as compellingly get to the point in particular communications.[7] Through the building of tension, interest and attention are garnered. Through this structure of narrative, information becomes story and story becomes more specific, more fascinating and therefore more powerful and resonant for target audiences. Randy has used this method to help academic researchers (his former job) to communicate more creatively and effectively.

Similarly, in *Energy and Environment Daily* Erika Bolstad (2016) has written about challenges for climate researchers to talk about climate in plain and not jargon-laden language. And, Amelia Urry (2016) from *Grist* has provided "do's and don'ts" that include "tell a story rather than reciting facts" and "tailor that story to your audience." Meanwhile, Andrew Thaler (2017) from *Southern Fried Science* has made the case that "Data is the map, storytelling is the journey." He argued, "to teach, to inform, and to connect with people, we need to identify the narrative threads in our own research and tell a better story." Furthermore, a 2017 editorial in *Nature Climate Change* argued, "Protecting science-based policymaking requires engaging the public, not politicians. Cultural institutions and the arts provide non-partisan platforms for communication that can connect scientific climate change data to people's lives." In these ways, the content producers have emphasized that resonant and effective communication involves a "smartening up" of framing using narrative structure and expository writing (e.g., Green, 2018).

Why Creative (Climate) Communications

This book is about creative climate communications. However, the parentheses signify that while this book is about climate, there are many transferable guidelines, rules and lessons that can apply to many analogous scientific, political, cultural and societal issues coursing through the veins of collective society. The parentheses surrounding climate in the title also convey that when one is mindful of audience and context, creative – and ultimately effective –

---

[6] Selected works from "The Story Group" can be viewed here: http://thestorygroup.org/category/videos/nationalclimateassessment/aflcc/

[7] More about this "And, But, Therefore" (ABT) tool can be viewed through Randy Olson's TEDtalk here: www.youtube.com/watch?v=ERB7ITvabA4

communications about climate change sometimes may importantly involve *not* invoking the term "climate" or "climate change" explicitly. For instance, Andrew Thaler (2017) has made the case that for certain audiences he does not talk about science when talking about climate change. He wrote, "When I talk about climate change, I talk about fishing . . . I talk about flooding . . . I talk about farming . . . I talk about the future."

Of course, this does not mean to suggest a shying away from the issue; rather it means that communications must be creative in order to effectively find common ground on climate change. It makes the argument that there are times and places to effectively make climate communications explicit, and other times to embed climate communications implicitly in other ways of discussing climate change. Researchers and practitioners have increasingly paid attention to this more discerning approach to (climate) communications, noting and analyzing how these choices influence engagement among different audiences.

For example, dimensions of climate change may more effectively be communicated to some farming communities by way of conversations about environmental stewardship and land ethics. In fact, journalist Hiroko Tabuchi (2017b) has observed, "in breadbasket [US] states, where warming directly affects bottom lines, farmers are discussing the climate – without saying 'climate change'" (p. B1). The view as that while farmers seek to confront a range of pertinent issues such as soil erosion, carbon sequestration and water availability inside the farm gate, explicit framing of the issues through "climate change" in the politicized and polarized US context are counterproductive. To illustrate this perspective, she interviewed fourth-generation Kansas grain farmer Doug Palen, who commented, "if politicians want to exhaust themselves debating the climate, that's their choice . . . I have a farm to run" (Tabuchi, 2017b, p. B1). Tabuchi noted, "while climate change is part of daily conversation, it gets disguised as something else" (p. B1).

As another example, facets of climate change can be more effectively discussed with some fiscally conservative business communities through communications about the benefits of deploying renewable energy technologies at scale. In the USA, self-identified Republicans have been found to support clean and renewable energy because of the economic benefits, not the climate change benefits (Bain et al., 2012; Bolderdijk et al., 2013; Roser-Renouf et al., 2015). Citizens' Climate Lobby in the USA has enlisted efforts to communicate in resonant ways across the political spectrum. In successfully engaging with Republicans, they have argued that "stewardship" and "responsibility" are resonant terms (Brugger, 2018b).

Furthermore, research by Lauren Feldman and P. Sol Hart (2018) has found that support for low-carbon energy policies – such as a carbon tax or renewable energy subsidies – increased among self-identified US Republicans when survey questions were framed as "air pollution" or "energy security" instead of "climate change." They concluded, "Our findings suggest that strategic messages designed to promote support for clean energy and climate mitigation policies would reduce opinion polarization and more effectively engage Republican audiences if the messages did not overtly mention climate change" (p. 63). However, this is not to suggest that discussing climate change through renewables and clean energy is a silver bullet strategy. In research on the influence of extractive activities on public support for renewable energy, Shawn Olson-Hazboun, Peter Howe and Anthony Leiserowitz (2018) found that "Individuals living in both mining-dependent counties and counties with natural gas production are somewhat less likely to support renewable energy policies than individuals living outside such places" (p. 117).

Together, a great deal of evidence from social science research as well as practitioner experience supports the notion that being explicit about climate change may alienate rather than appeal to target audiences and constituencies (e.g., Vezirgiannidou, 2013). With these considerations in mind, the editors of *Nature Climate Change* (2017) have commented, "this begs the question of whether the success of climate change communication is actually hampered by mentioning climate change at all" (p. 1). In short, smart communication discretion is advised.

In this contemporary environment, it is a miscalculation to always choose communications approaches about climate change that explicitly mention climate change. As the foregoing findings point out, leading with decarbonization and sustainability considerations with political conservatives in the US explicitly through the language of "climate challenges" and "climate crises" can erect more barriers than bridges. There are tensions, of course. For example, Adam Corner has discussed these challenges in the context of connections between weather and climate change. In the context of the warm summer of 2018 in Britain, he challenged readers to be "prepared to call out climate change when we see it." He concluded, "If we don't learn to call a heat wave like this what it is, we are doomed to keep repeating the same mistakes over and over again – equivocating, avoiding and getting hotter all the time (Corner, 2018).

Conversely, it is a mistake to succumb to worries and antagonisms to therefore make "climate change" unmentionable, where communicators do not dare to utter the words for fear of backlash. In J. K. Rowling's *Harry Potter* series, Lord Voldemort is a chief antagonist. As an imposing character, most wizards (according to my son Elijah) do not dare say his name. They instead refer to him as "you know who" and "he who must not be named." It would be a mistake to treat climate change like Voldemort. Doing so would amount to "self-silencing" about climate change (Geiger and Swim 2016) (see Chapter 5 for more). Nonetheless, Anthony Leiserowitz, Ed Maibach, Connie Roser-Renouf, Seth Rosenthal and Matthew Cutler (2017) found that in the US context few people regularly discussed climate change. Through their polling work they found that one in three discussed climate change with family and friends "often" or "occasionally"; the rest said they discussed it "rarely" or "never" (Leiserowitz et al., 2017; see also Maibach et al., 2016).

There are clear costs to self-silencing in terms of marginalizing climate change discourse in the public arena as well as reducing salience of the issue for policy actors and everyday people (Walker et al., 2017). Moreover, silence can provide social cues that climate change may not be such a big issue or threat (Geiger et al., 2017). It also squanders opportunities to confront misinformation and skepticism in the public view (Leombruni, 2015).

Yet self-silencing and self-censorship can be the result of more formal external pressures and threats. In recent years, due to the contentious US cultural and political milieu, there have been reports that scientists – particularly embedded in agencies in the US federal government – have avoided using the phrase "climate change" for fear of blowback of various sorts (Waldman, 2016; Hersher, 2017). For example, in 2018, many US EPA staffers reevaluated the effectiveness of explicitly invoking the term "climate change" when moving forward with climate-related agency work. Robin Bravender reported that "given the sensitivity of the topic, some EPA employees have sought guidance about how to approach climate change without jeopardizing their jobs and their programs" (Bravender, 2018). Such tactics were then evident in a 2018 EPA statement outlining automobile efficiency standards where the term "climate change" did not appear at all in the thirty-eight-page document (Beene, 2018). Also in 2018, a survey of US government employees found that more than 50% of scientists across sixteen agencies agreed that "consideration of political interests" curtailed engagement in science-based decisions (Marshall and Reilly, 2018). These views were expressed in the context of Trump administration actions to remove references to climate change across multiple government websites (Davenport, 2017, 2018). These have been perspectives also embedded in current conditions of partisan divides among policy actors (Helmuth et al., 2016; Guber, 2017) and the general public (Carmichael et al., 2017).

Political polarization can play out through the media (e.g., Boykoff, 2011; Bolsen and Shapiro, 2018) and through informal communication arenas like

Twitter (e.g., Tandoc and Eng, 2017; Fownes et al., 2018). For example, after conducting a study of climate communications of Twitter accounts followed by US senators, Brian Helmuth et al. (2016) found "distinct and semi-isolated sub-networks" (p. 10).[8] They concluded that, sadly, "our results strongly suggest that overt interest in science may now primarily be a 'Democrat' value" (p. 10).

Trends in carrying these creative communications through new and social media unfold in the context of a wider and fundamental set of questions involving how these mediatized communications may take place in echo chambers or whether they open up novel discussions, considerations and behaviors (e.g., Anderson, 2017; Segerberg, 2017; Tandoc and Eng, 2017). Michael Shank (2017) from the Carbon Neutral Cities Alliance has argued that social media memes are key to successful climate communications. He stated, "if we can't translate a meaty message for the myriad social media vehicles out there, we haven't tried hard enough."

According to social amplification of risk theory, evaluations of risk are shaped in part by the extent to which they are part of a public discourse (Kasperson et al., 1988; Renn, 2011). Yet, researchers – across the sciences and humanities – have struggled to effectively communicate risks. A former US National Security official has said, "We need to be more willing to talk . . . because if you talk, if it's completely consensus, are we at risk for understating the risk?" (Reilly, 2016a). Kristie Ebi has also noted the reality that "Any of us who have worked on this field [of risk management and climate change] for a long time have been accused of scaremongering when we try to talk about risk" (Reilly, 2016a). Elsewhere, Gernot Wagner and Martin Weitzman (2015) have referred to low-probability and high-consequence events that are under-reported as the "fat tail of climate risk."

Andrew Hoffman (2015) has characterized the cultural and political dimensions of current conditions as a "schism" where "opposing sides do not debate the same issues, seek only information that supports their positions or discon-firms the other's, and begin to demonize those who disagree with them" (p. 6). In other words, this can be considered as a "Don't confuse me with the facts; I have my mind made up already" stance.

This milieu has also been explained through theories of cultural cognition as it relates to climate awareness and engagement (Kahan, 2015a). Cultural cognition broadly explains why groups with different cultural perspectives disagree about critical social and political issues. The thesis explains that

---

[8] Senate Democrats were three times more likely than Republicans to follow science-related Twitter handles "suggesting overt interest in science may partly define party identity." From their findings, they stated, "Hence, instead of being viewed as a neutral source of objective informa-tion, science may now be considered a special interest in US politics" (Helmuth et al., 2016, p. 1).

these differences are not attributed to low literacy or misunderstanding, but rather they are about reinforcing identity and belonging among desired groups.

By extension, people engage in motivated reasoning – emotion-biased decision-making – as a way to combat and reduce cognitive dissonance (Festinger, 1957; Kunda, 1990) on controversial science issues such as climate science (Stenhouse et al., 2018). As an example of motivated reasoning one can think of the notion held by supporters of a country's World Cup soccer team that their team will win the cup in 2022 despite the fact that they did not make the tournament in 2018 and have not been in the top twenty FIFA world rankings for many years.[9] In other words, motivated reasoning "refers to the tendency of people to conform their assessments of all sorts of evidence to some goal unrelated to accuracy" (Kahan, 2015a, p. 3). Research by Nathan Walter, Sandra Ball-Rokeach, Yu Xu and Garrett Broad (2018) found that ideological perspectives fueled motivated reasoning that then manifested in adoption of false beliefs on both climate change, evolution and vaccines, even in the face of abundant information to the contrary. Motivated reasoning may manifest both consciously and unconsciously. It can also cut across the ideological spectrum (Porter, 2016).

In the case of climate change, social science and humanities research has increasingly mapped out how people may seek out information that conforms to the beliefs of their perceived ideological group (or tribe) rather than prioritizing information itself. In this arena, values can be powerful predictors of belief about climate change (e.g., Feinberg and Willer, 2013). The amplifying or diminishing levels of environmental concern enforced or challenged by social group belonging along with motivated reasoning runs strong. A thirty-eight-country study by Steven Brieger (2018) found that social identity is closely intertwined with people's active willingness to engage in environmental protection.

Rusi Jaspal, Brigitte Nerlich and Marco Cinnirella have explored considerations of resilience of identity under threat from climate change information (Jaspal et al., 2014). Moreover, through research into cultural cognition – where "certain types of group affinities are integral to the mental processes ordinary members of the public use to assess risk" (Kahan et al., 2010, p. 502) – Dan Kahan (2015b) has recommended several "disentanglement strategies." These strategies involve separating ways of knowing and knowledge acquisition from identity when communicating (and by extension receiving) scientific information on topics such as climate change (Kahan, 2015b).

---

[9] USA (!?!?) USA (!?!?)

However, Johannes Persson, Nils-Eric Sahlin and Annika Wallin (2015) have drawn in ethical considerations regarding the valuation of any beliefs even if they are in direct conflict with scientific evidence, as is done by the Cultural Cognition thesis. They critiqued "over-enthusiastic adoption" of cultural cognition and questioned whether this ultimately degrades trust in science. They concluded, "It does not seem far-fetched to suggest that if we conclude that values have to be respected ... we shall only have ourselves to blame when distrust in science deepens" (Persson et al., 2015, p. 5). In other words, differences between perspectives can be seen in some ways in terms of right and wrong as well as of differing assessments of evidence and risk.

Matthew Hornsey, Emily Harris and Kelly Fielding (2018) have pointed out that while conspiratorial beliefs and climate contrarianism are most strong in the USA, "there is little inherent to conspiratorial ideation or conservative ideologies that predisposes people to reject climate science" (p. 614). When motivated reasoning meets consistent (and deliberate) misinformation, it can lead to deep cultural divisions that must be carefully overcome through fundamental reconsiderations regarding how we creatively communicate about climate change (see Chapter 3 for more). But, there are ways to creatively find common ground on climate change through smart and effective communications approaches.

On a practical level, part of a smartening up process can involve a more effective deployment of analogies and metaphors (Nerlich et al., 2010; Larson, 2011; Ford, 2015; Hassol et al., 2016; Flusberg et al., 2017). Beth Osnes (2017) has commented, "Dramatic metaphor has the capacity to tease out a more nuanced understanding of both problems and solutions. A movement, action, property, or piece of dialogue can make a comparison to something dissimilar in order to enhance its meaning and to reveal what they might have in common. Metaphor extends beyond one thing merely serving as a symbol for the other" (p. 143). For example, Sandra van der Hel, Iina Hellsten and Gerard Steen (2018) found that the "tipping points" metaphor in climate communications provided a "multi-purpose bridge" and therefore an effective communication tool (p. 1).

Another part of this smartening up can be found through "speaking the same language" as the audiences that communicators intend to reach. For example, in a 2018 meeting of experts at the International Weather and Climate Forum in Paris, speakers asserted that "efforts to adapt to climate change would fall on deaf ears unless scientists did a better job of explaining climate issues" (Tabary, 2018). This approach can be seen to cohere with what John Dryzek and Alex Lo (2015) called for in terms of bridging rhetoric than then "reaches those who do not share the speaker's perspective" on climate change (p. 15).[10]

Therefore, this "smartening up" is a process of listening and adapting rather than one of winning an argument or talking people into something. Through providing space and perspective these creative approaches can lead to effective connections (see Chapter 7 for more). Considering communications in this way, we can more capably then return to productive discussions and responses to Mike Hulme's (2009) aforementioned question presented in Chapter 1: "How does the idea of climate change alter the way we arrive at and achieve our personal aspirations and our collective social goals?" (p. 56).

Cultural Politics and Creative Interdisciplinary Communications

Creativity essentially is applied imagination. Creativity can involve experimentation, risk-taking, openness to other points of view, suspension of stigmatism and a willingness to possibly make mistakes as well as keep your self-consciousness in check. Wider conceptions of creativity and creative industries, along with new tools and platforms for expression help provide context. Ken Robinson has argued, "If you're not prepared to be wrong, you won't come up with anything original ... we don't grow into creativity, we grow out of it."[11]

Studies of creativity, associated mainly with the discipline of psychology, began in the late 1940s (Guilford, 1950) and have been a burgeoning area of research since then (Sternberg and Dess, 2001). In those early days, research focused on how creativity could unlock many different answers or solutions to various problems (Sawyer, 2017). Educational programs then emerged in the 1960s to bring creativity into the formal learning environment (Wallach, 1988).

For example, Jane Piirto (2017) has studied creativity and has arrived at "seven I's" of the creative process that can help guide these considerations going forward: inspiration, insight, intuition, incubation, improvisation, imagery and imagination (p. 133). Among them, incubation involves the value of rest during the creative process, to unlock creativity (think of "the shower thought"). Improvisation means experimentation and play as key elements of

---

[10] In contrast, "bonding rhetoric" is seen to be that which appeals only to one's own perspective.
[11] Ken Robinson's October 2014 Ted Talk on Creativity can be viewed here: www.ted.com/talks/ken_robinson_says_schools_kill_creativity?language=en

creative processes. In terms of intuition, Piirto recommends tapping into multiple ways of knowing (including "gut feelings") in order to optimally access creativity (p. 149).

The wider spaces of cultural politics further inform these realms of creative (climate) communication. It helps us to understand why things matter. Cultural politics refer to processes of how meaning is constructed and negotiated across space/place and nested questions of how people make sense of, value and act in the world. Through this lens, we can consider how the ways that we act in this world are shaped by the ways that we view and describe it. Cultural politics also involve analyses as to how asymmetrical power processes are contested in the spaces of "subpolitics" that manage the conditions of our lives. Subpolitics are places where power circulates via networks to produce discourses, institutions and practices (Hall, 1997). Such an approach helps link how discourses (the ways we talk about climate change) then contribute to social and material practices (the ways we respond to climate challenges) at the human–environment interface. To explore communication about climate change in the spaces of cultural politics is then to interrogate complex entanglements that shape resonant ways of learning and knowing as well as construction of meaning in our lives.

Over the years, creativity has manifested in many interdisciplinary engagements with the arts, through a variety of projects and endeavors (Moser, 2017). One example is "conservation theatre" where young participants in a project in Mexico engaged in participatory theatre to draw out considerations of transformative processes to support community-based natural resource management (Heras and Tàbara, 2016). María Heras and J. David Tàbara (2016) found that this approach helped enhance learning about local conservation issues while catalyzing "social interaction, diversity recognition and empathic dialogues" (p. 948).

As another example, adaptation professionals, public officials and other stakeholders engaged in role-play simulation exercises in the Northeastern US. These helped to improve their considerations of adaptive capacity, social learning and understanding of local risks (Rumore et al., 2016). Researchers Danya Rumore, Todd Schenk and Lawrence Susskind (2016) concluded that "serious games" "should be more widely embraced as part of adaptation professionals' education and engagement toolkits" (p. 745) (see Chapter 7 for more).

Endeavors like these have been found to boost creativity and collective imagination by opening new pathways to learning and knowing through experience, affect, emotion and aesthetics (Ormrod, 2017). While Paul Slovic (1987) has characterized climate change as something not readily

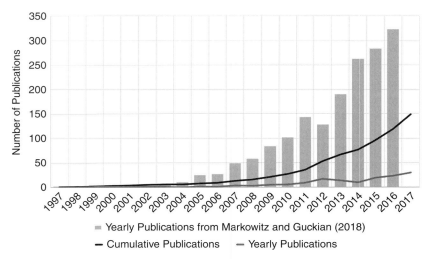

Figure 2.2 This Web of Science search was conducted to track the abundance (or scarcity) of "creative" invoked in climate or science research from 1997 through 2017 (noted as line graphs). This was a topical search with terms climat* NEAR creativ* or scien* NEAR creativ*. For comparison, this is then contrasted with the Markowitz and Guckian's (2018) search of climate science communication research over the period 1997 through 2016 (noted as bar charts). This approach was consistent with Markowitz and Guckian (2018), garnering an expanded database including relevant articles beyond just titles or keyword searches. Like in Markowitz and Guckian's investigation, after an initial search a manual screening was done to remove unrelated material. $N = 149$ articles were then found over this twenty-one-year period.

imaginable, Peter Kalmus (2017) has countered that "global warming is the result of the greatest failure of imagination the world has ever seen" (p. 293).

However, despite this multidecade history into creativity through the arts, comparatively there has been limited engagement explicitly with creativity in the sciences. Through a Web of Science search of peer-reviewed research publications, Susanne Moser tracked the rise in publications on climate change communication over a decade (2005–2015). During this time, the number of these publications has increased roughly tenfold (Moser, 2017). Similarly, Ezra Markowitz and Meaghan Guckian searched Web of Science for climate science communication research over the period 1997 through 2016. They found a steep rise in publications as well (see Figure 2.2). By comparison, my additional search in Web of Science sought to detect creative climate research or creative science research articles. While this search documented increases, a maximum of thirty publications in 2017 demonstrated that there remains unfulfilled potential for further research into these areas of creative climate, environment and science communication (see the Appendix for more details).

R. Keith Sawyer (2017) has pointed out that the sciences have "for the most part ignored research on creativity" (p. 281). While Suzannah Evans Comfort and Young Eun Park (2018) have pointed out that "attention to environmental communication has exploded in recent years" (p. 1), their assessment narrowly omits creative environmental communications. Yet Pam Matson, William Clark and Krister Andersson (2016) have noted, "To understand social-environmental systems, and to pursue sustainability within them, researchers, decision makers, and concerned citizens have to draw on many kinds of knowledge and know-how ... different kinds of expertise and knowledge would fall far short if they operated alone. Researchers and practicing professionals need to understand each other's languages, respect and appreciate each other's methods and approaches, and recognize where and how their limited knowledge can make a difference" (p. 81). Richard Lachman (2018), among others, has therefore called for scientists to get further training in the arts and for STEM (science, technology, engineering and math) to expand to STEAM with explicit inclusion of the arts in order to most effectively confront society's grand challenges.

Various scientific institutions are increasingly supporting a shift toward trainings to improve climate and science communications. Organizations and societies like the American Association for the Advancement of Science (AAAS), American Meteorological Society (AMS), American Geophysical Union (AGU), National Academy of Sciences (NAS) and governmental agencies such as the National Oceanic and Atmospheric Administration (NOAA) and the National Aeronautics and Space Administration (NASA) are offering communication trainings for climate researchers. Also, the group COMPASS has emerged as a good resource to help train scientists to more effectively communicate their research to journalists, policy actors and members of the public citizenry. COMPASS is housed at the National Center for Ecological Analysis and Synthesis (NCEAS) at the University of California Santa Barbara and is led by Nancy Baron. They have specialized in marine conservation work, but they have also worked more widely to train scientists in other programs such as the Aldo Leopold Leadership Program.[12]

In addition, the Alan Alda Center for Communicating Science at Stony Brook University has offered training workshops for the past ten years. Stemming from his interest through hosting a US Public Broadcasting

---

[12] I have benefitted from a COMPASS training through this Aldo Leopold Leadership Program.

System (PBS) series *Scientific American Frontiers*, Alan Alda melded his talent for making conversation with improvisation and acting to help improve science communication through effective public engagements. The Climate Advocacy Lab has also stepped into these spaces, to help with translations into the public sphere and consequent action in the face of climate challenges. This is a group that was founded in 2015 through support from a constellation of foundations including the Skoll Foundation, the William and Flora Hewlett Foundation and the McKnight Foundation. In the past years, Carina Barnett-Loro, Sean Kosofsky and others have worked to provide resources, tools, tactics and innovations to help catalyze engagement (focused in the US context) from a range of actors on climate change.[13] In the UK and across Europe, Climate Outreach is an organization that deliberately has worked (since 2004) to find common ground in communications about climate change. Led by George Marshall, Jamie Clarke and Adam Corner among others, this team has worked diligently to develop advice and practical tools for people, collectives and organizations across age groups, religious affiliations and political perspectives to understand and engage meaningfully with climate change.[14] Through endeavors like these as well as across many other institutions and universities, new training programs focused mainly on science and environmental communication have been emerging. As these programs have grown, still only a smaller subset has engaged explicitly with creativity.

In this volume creativity is not pursued merely for the sake of creativity. Creativity may at times work against effectiveness. This is a book about creativity directed toward effective climate communications. Therefore, by "smartening up" creative (climate) communication strategies, one can then more effectively recapture what may be seen to be a "missing middle ground" (Kirk, 2018) or common ground on climate change in the public arena.

Regarding interdisciplinarity, Daniel Abbasi (2006) has advocated for natural and social science connections in order to "close the gap between science and action." Susan Clayton et al. (2015) have pointed out, "Given the complexity of global climate change and the many factors involved, interdisciplinary collaboration is needed to research human interactions with climate" going forward (p. 643). Moreover, Alison Anderson has commented, "We need to direct more energy into outreach and interdisciplinary collaborations . . . we need to think more creatively about how we can collaborate in creative ways to bridge traditional divides between arts-based subjects and the natural sciences.

---

[13] More information about Climate Advocacy Lab can be found here: https://climateadvocacylab .org/

[14] More information about Climate Outreach can be found here: https://climateoutreach.org/#

In my view, the most significant challenges are to overcome disciplinary focused ways of thinking, target research findings for different types of audiences, and wield more influence in policy arenas" (p. 4).

As such, more needs to be done to support and lead on creative projects in these contemporary spaces. This can range from fostering spaces of collaboration in research and practice to changing reward systems for innovative scholarship and practice to increased funding of research and experimental forays into creative (climate) communications (see Chapter 7 for more). These endeavors can then further advance the value of entanglements between the arts and sciences on the topic of climate change in these critical areas at this important time. Consequently, interdisciplinary groups in the sciences and arts/humanities can more capably accelerate collective learning and collective action (Corbett and Clark, 2017; Moser, 2017).

Co-production of creative works on climate change – between arts and sciences, between research and practice – have produced many useful results to date. Co-production is defined as the interplay between science and society, where conditions are shaped by the cognitive processes of what we know, the social processes of how we come to know it and the normative processes of incorporating or resisting them in our everyday lives (Jasanoff, 2004) This can be expanded to help understand how knowledge connects with action in non-linear ways in the sciences and elsewhere (Cash et al., 2006; Dilling and Lemos, 2011).

Matt Nisbet et al. (2010) have mapped out a new communication infrastructure needed for meaningful interdisciplinary advances in climate change engagement where people are "empowered to learn about the scientific and social dimensions of climate change, inspired to take personal responsibility, able to constructively deliberate and meaningfully participate and emotionally and creatively engage in personal change and collective action" (p. 329). Creative communications can then attend to Susanne Moser's (2017) call "to improve the interaction between climate communication research and practice . . . [and] dedicated science-practice boundary work" (p. 345).

Conclusions: Finding Common Ground

To establish common ground literally means to share an immediate space or location. Figuratively this means to connect in meaningful ways, to relate to others with different perspectives (DeMarco, 2017). Finding common ground is vital to finding productive pathways to confront climate threats, and to carving out possibilities for solutions to pressing challenges.

This often entails a process of finding common values and shared concerns. Adam Corner and Jamie Clarke (2017) have reasoned, "The facts of climate science are like a dictionary: they provide the basic vocabulary. The real challenge is in weaving poetry and prose to inspire people to care about the problem. Fundamentally, this means engaging with people's values" (p. 48).

There are certainly many ways to address climate change, including calling out culprits involved in perpetuating associated problems. In fact, it can be tempting to villainize chief contributors to climate change, and to disparage obstructionists to climate action. Bob Johnson (2014) has commented that amidst "a thickening in the web of our dependencies on prehistoric carbons ... today we only cling more tightly to them despite mounting evidence that the costs of doing so are rising and that we are on the verge of an existential threat to the ecological experiment we have been conducting for a century and a half" (pp. 163–164). Those who cling to their beliefs despite overwhelming evidence can be annoying to say the least.

At times it may feel cathartic or palliative to vent about frustrations associated with bad actors; it may even serve certain purposes to be forthright and straightforward and to expose deliberate strategies and campaigns spreading misinformation (see Chapter 3 for more). At times it is easier to yell or lament than to listen and consider other points of view. However, unyielding antagonistic approaches can sometimes give contrarians and doubters the outsized attention that many of them seek. Also, not all these manifestations of contrarianism ought to be attributed to nefarious or attention-seeking objectives. It could merely be laziness and habit more than it is deliberate bias (Pennycook and Rand, 2018).

Ridiculing what are often outlier positions can also then privilege static characterizations of positions over a dynamism of learning and knowing about climate change. It can also contribute to a caricaturing of science–policy deliberations on climate change, thereby lending credence to silliness and senselessness over seriousness and sincerity. Moreover, such approaches can then be tone deaf about what may be most effective in certain contexts, and may then blunt and undermine communication goals of engaging and inspiring citizens to take action on climate change (see Chapter 3 for more).

There are certainly researchers and practitioners who have argued that this effort to find common ground across the political spectrum is futile. For example, Ben Adler (2015) has argued that "an environmentalist calling for the climate movement to incorporate conservatives is just as nonsensical as a fundamentalist pastor calling for the anti-gay marriage movement to incorporate liberals."[15] Nonetheless, from a politically conservative perspective, Stewart Easterby (2018) has pled "don't lambaste them," reasoning that this leads to destructive rather than constructive conversations. In other words, consistently castigating other points of view ends up proving counterproductive when working to address a common challenge and collective action problem of climate change. Instead, more consistent processes of establishing common ground involves commitments to respectful, open and considered dialogue about common concerns. These bridge-building approaches serve to open up spaces for productive exchanges, rather than put up defenses that impede the finding of common ground (see Chapter 7 for more).

An important part of this process of finding common ground is overcoming what is often a sense of psychological distance from climate change. Social science research over the past years has found that far-off future impacts or distant physical threats from climate change contribute to a diminished sense of urgency to address associated climate challenges. Through their assessment of research on psychological, social and cultural barriers to climate action, Nathaniel Geiger, Brianna Middlewood and Janet Swim (2017) have posited that perceptions of distant impacts "can lead individuals to disengage from the phenomenon" (p. 323).

Rachel McDonald (2017) has commented that "perceptions of climate change as distant in both space and time have important implications for public opinion on climate change" (p. 240). Furthermore, Julie Doyle (2016) has commented, "The reason why climate change has failed to engaged the public in any meaningful way is not simply due to the complexity of the issue, although this is certainly an important factor ... how climate change is made relevant to people's everyday lives, how it engages people through their existing social values and norms, *matters*. Climate change needs to be understood as a concern for the "here and now," rather than a distant future "out there somewhere." This involves making climate change temporally, spatially and socially meaningful and relevant" (p. 8, emphasis in the original).

As such, finding common ground on climate change also involves processes of "bringing climate change home" (Slocum, 2004). Sander Van der Linden and colleagues have posited that effective communication involves "emphasiz[ing] climate change as a present, local, and personal risk" (van der Linden et al., 2015, p. 758). Many science–policy actors and communications practitioners have sensed this, while social science research also has substantiated these impressions (see Chapter 7 for more). For example (as was mentioned in

---

[15] It is worth noting that this argument preceded the election of Donald J. Trump as US president in fall of the following year.

Chapter 1), former California governor (and movie star) Arnold Schwarzenegger pointed out that climate communications needed to focus on the here and now rather than distant impacts in far-flung places (Concha, 2018). In terms of overcoming senses of distance, focus group research relating notions of care, compassion, pity, sympathy, empathy and denial in the context of suffering found that audiences "make [distant threats] relevant and real to themselves by imagining how they would react to the event based on pre-existing perceptions and their own experiences" (Huiberts and Joye, 2018, p. 12). Eline Huiberts and Stijn Joye concluded, "The complex and sometimes contradictory (lack of) feelings of empathy, sympathetic expression and strategies of denial show the moral struggle that participants go through, and there is no singular, easy, or 'right' answer to the way that people ought to, or in fact do, react toward mediated distant suffering" (Huiberts and Joye, 2018, p. 12).[16] They defined this as "experiential overlap" and attributed it to a sense of caring and concern (Huiberts and Joye, 2018).[17]

Yet even though a great deal of social science research has pointed to distance as a barrier to engagement, distant images and video such as that of an emaciated polar bear still strike a nerve. For example, in late 2017, Paul Nicklen from *National Geographic* shot a video of a polar bear scavenging for food in a garbage bin. This video went viral on Instagram and beyond. Nicklen's caption in the video read "It's a soul crushing scene that still haunts me, but I know we need to share both the beautiful and the heart-breaking if we are going to break down the walls of apathy. This is what starvations looks like. The muscles atrophy. No energy. It's a slow and painful death" (Rosenberg, 2017).

However, at this distance there is more room for counterclaims because distant suffering may not be verified through first-hand experiential, emotional or visceral ways of knowing. For example, outlier perspectives espoused by Susan Crockford (Canadian zoologist), that polar bears are not as threatened by climate change, as one may believe, have had outsized influence in these ongoing discussions and deliberations (Goode, 2018).

With these nuanced dynamics in mind, it is important to further emphasize the importance of tailored communications, along with the problematization of silver bullet logics. As such, as R. Kelley Garrett (2017) has pointed out,

---

[16] The character Creed Bratton from *The Office* television show (US version) once said, "No matter how you get there or where you end up, humans have a miraculous gift to make that place 'home'."

[17] Incidentally, this could make a case for steering reality television programming into these arenas of climate change content, owing to the compassion and caring that they may motivate in some audiences (Nikunen, 2016).

"There is no single 'solution' to climate change misperceptions" (p. 519). To dig into these considerations further, Chapter 3 interrogates how information deficit model logics have significantly contributed to stalled out communications about climate change, while also amplifying outliner perspectives (like Crockford's mentioned earlier).

Appendix

A Web of Science search was conducted to track the abundance (or scarcity) of "creative" invoked in climate or science research from 1997 through 2017. This was a topical search with the terms climat* NEAR creativ* or scien* NEAR creativ*. For comparison, this is then contrasted with Markowitz and Guckian's (2018) search of climate science communication research over the period 1997 through 2016. This approach was consistent with that of Markowitz and Guckian (2018), garnering an expanded inclusion of relevant articles beyond just titles or keyword searches. Like Markowitz and Guckian, after an initial search a manual screening was done to remove unrelated material. A total of $N = 149$ articles were then found over this twenty-one-year period.

Year	Cumulative publications	Yearly publication
1997	1	1
1998	1	0
1999	2	1
2000	3	1
2001	4	1
2002	5	1
2003	6	1
2004	6	0
2005	8	2
2006	9	1
2007	13	4
2008	16	3
2009	21	5
2010	27	6
2011	36	9
2012	53	17
2013	67	14
2014	77	10
2015	96	19
2016	119	23
2017	149	30

3

Do the Right Thing[*]

> Climate change is the defining issue of our time, and we are at a defining
> moment. Scientists have been telling us for decades. Over and over again.
> Far too many leaders have refused to listen.
>
> *— António Guterres, the United Nations Secretary General*
> *(in Sengupta, 2018)*

What is the "right" approach to communicating about climate change so that
people "get it"? How should one craft climate change messages so that folks do
something about this pressing challenge? Why can't people just listen and then
"do the right thing"? These are some logical yet elusive questions that many
practitioners, scholars and everyday people have asked many times over the
years. These questions often articulate a niggling frustration: why can't we find
the silver bullet in climate communications and consequently "solve" climate
change?

The fact of the matter is that this is not that simple. If it were, we would be
doing a more effective job of protecting the climate from us and protecting us
from the climate. From there we could then move on to another global
challenge. The grim reality is that climate change is a vexing, complicated
and difficult threat, requiring a fundamental rethink of how communication and
engagement work effectively between science and policy in the public sphere.

There are clear challenges to narrate and habitually embody common purpose amid very diverse ways of viewing and engaging at the human–environment interface. However, in the face of evidence of these complexities, many
have clung to information-deficit model logics when communicating about
climate change.

---

[*] This chapter title is drawn from the 1989 film title from director and writer Spike Lee. In that film,
while raising many critical social issues, he also wove in a more complicated picture about what
some may feel is the "right thing to do" regarding intersecting challenges of racial and socioeconomic inequality in the urban USA.

Despite this need for a fundamental rethink, over the past decades many offering relevant expertise from the scientific community, seeking to face up to this challenge, have sought to influence change by way of mainly prioritizing the throwing of more information, more data, more findings and more evidence into the mix in the public sphere. An atmospheric scientist once joked that this approach is likely solely relying on IKEA directions to build furniture: the information is there, yes, but there are other elements that stand between you and a finished bookshelf.[1]

By extension, communications about climate change – in particular, climate science – has followed a similar route. The goals of communications have effectively been the transmission of information that effectively overcomes this deficit and changes audience cognition, affect or behavior in order to bring it into alignment with scientific understanding (Suldovsky, 2017). Ongoing adherence to the tenets of an information-deficit view displaces and impedes more creative, comprehensive and effective climate communications. Together, these information-deficit model-laden approaches have amounted to failing views-cum-strategies that the facts speak for themselves, or that mere communications of the facts of climate change will detonate caring as it activates public citizens. As such, this "hitting people over the head with science" approach is a naïve and twentieth-century climate communication model.

Yet, scientific findings are often dominant ways through which climate change is articulated (Boykoff, 2011). This can be attributed in part to the ongoing influences of Enlightenment thinking and its dominant paradigm that scientific facts and techno-rationality guide optimal decision-making (e.g., Jasanoff and Kim, 2009). This can also be attributed to ways in which climate change was identified through science as "a pollution problem" when it entered public discussions in the 1980s. This can also be attributed in part to the difficulties of interpreting climate change signals from day-to-day observational trends.

Work to provide scientific ways of knowing is critically important and central to addressing climate change, no doubt. Exposure to more information via the sciences about climate change over the past century has helped provide pathways for enhanced understanding and improved decision-making on associated climate issues. The tireless, steadfast and committed efforts from many of the brightest minds of our time have helped to develop a strong diagnosis of the problem. Organizations such as the influential UN Intergovernmental Panel

---

[1] This was a comment from a participant at the University Corporation for Atmospheric Research (UCAR) Annual Members Meeting on October 10, 2018, www.ucar.edu/who-we-are/membership-governance/member-institutions/annual-members-meetings/2018-members-meeting

on Climate Change (IPCC) have worked very hard over the past thirty years to provide and communicate aggregated assessments for decision-makers and wider societal actors.

It is important to note that these approaches are perfectly logical. This is seemingly what science at its core has been meant to do: science is a systematic enterprise that identifies patterns; builds and organizes knowledge; and provides pathways to learning, understanding and explaining the world around us. Through these pursuits, science builds information and insights from testable questions through data and observations to provide pathways to knowing about climate change and other phenomena.

Yet, Susanne Moser (2009) has rightly warned that "providing information and filling knowledge gaps is at best necessary but rarely sufficient to create active behavioral engagement" (p. 165). To this I would add the need for "sustained" behavioral engagement as well. However, the mindset behind António Guterres' quote at the beginning of the chapter (and many like it) is that, despite sincere concern about the threats of climate change along with loads of scientific findings-as-warnings, leaders and others just are not listening!

It is critically important to recognize and remember that scientific "ways of knowing" are not the only pathways to knowing and learning about climate change. Andrew Hoffman (2015) has argued, "The debate over climate change . . . is not about carbon dioxide and greenhouse gas models; it is about opposing cultural values and worldviews through which that science is seen" (p. 4). Lynda Walsh (2015) has posited, "climate change, as an exigence, is a political program and not a scientific one" (p. 366). Moreover, scientific explanations about climate change are never wholly "objective" or independent of politics and culture; rather they are co-produced by scientists embedded in society (Jasanoff, 1996).[2]

The aggregate impacts of a dominant information-deficit mental model over time are manifold. Among them, there is the sense that (1) poor choices and actions are attributed to "deficits" of knowledge and information that impede one's ability to make "correct" choices, and (2) those with deficits are then seen to be ignorant or ill-informed (Priest, 2001). Consequently, ignorance is often blamed for science-inconsistent understanding and lack of support for action based on evidence from science. Further, any lack of public and policy engagement with climate issues is attributed to deficiencies in knowledge of science.

---

[2] While Declan Fahy (2018) has pointed out that by focusing on "objectivity" as "the application of trained judgment, the implementation of a transparent method, and the pluralistic search for consensus around areas of shared understanding," environmental journalists have effectively lit a path for effective communication in the twenty-first century (p. 860), though it remained a guiding light rather than a destination.

But confounding examples abound in contemporary society. For example, if the information-deficit model were an accurate representation of the way the world worked, the 2016 US Senate Environment and Public Works Committee hearing on "Ozone Depletion, the Greenhouse Effect, and Climate Change" (convened by Republican Senator John Chafee (Rhode Island) would have sparked a process to comprehensively address trans-boundary environmental threats to security in the United States. At that hearing Senator Chafee stated, "This is not a matter of Chicken Little telling us the sky is falling. The scientific evidence . . . is telling us we have a problem, a serious problem" (Mooney, 2016). If information-deficit model logics held in this case, public and policy-actor outrage, engagement and urgent action would have then surely followed . . . but it did not.

Here, I delve into the predicament that has resulted from ongoing and dominant reliance on what has been called an information-deficit model approach to climate change in the twenty-first century. I then relate this to the reality introduced in Chapter 2 that our ways of communicating about climate change remain relatively stuck. To examine impacts of this dominant approach to ongoing climate communications, I consider how this approach has limited communication goals while it has provided space for amplified influence by outlier perspectives on climate change, commonly referred to as "contrarians" or "deniers." I also make connections between these patterns of practice and social networks of climate contrarianism and climate countermovement (CCM) activities in the USA.

CCM perspectives thrive on ideological or evidentiary disagreement to the orthodox views of science, a drive to fulfill the perceived desires of special interests and exhilaration from self-perceived notoriety. However, dominant information-deficit model approaches to climate communication – even through naming, shaming and critiquing these perspectives – serve to also provide oxygen to invigorate rather than diminish the strength of outlier claims. Therefore, focusing on contrarians displaces sorely needed attention that should be paid to communication shortcomings from adherence to information-deficit model logics. Woven through elements of this chapter, I continue to build on the argument that we collectively need to get more creative and innovative in order to more effectively and meaningfully communicate and engage with climate change in the public sphere.

The Belly of the Beast

In this chapter and across the book, the focus is often on the influential US context. From these considerations, we can extract some explanatory power to explore these conditions in other Western democracies (Lewis et al., 2018). For example, in a study of beliefs, conservatism and climate contrarianism across twenty-five countries, Matthew Hornsey, Emily Harris and Kelly Fielding (2018) examined five indices of ideology: left–right political ideology, liberal–conservative political ideology, conspiracy beliefs, individualism and hierarchical values. They found that this relationship between contrarianism and ideology has played out most intensely across all these five measures in the USA, where "political cultures have emerged that encourage citizens to appraise climate science through the lens of their conservative ideologies" (Hornsey et al., 2018, p. 619). Interestingly, they also noted, "Nations that displayed the largest relationships (e.g., the United States, Australia, Canada and Brazil) tend to be those whose economies are relatively highly reliant on fossil fuel industries" (p. 618).

This coheres with previous analyses in Canada (e.g., Lachapelle et al., 2012), Brazil (e.g., Jylhä et al., 2016) and Australia (e.g., Fielding et al., 2012). However, Gregory Lewis, Risa Palm and Bo Feng (2018) have found that "members of left/liberal parties worry more about the effects of climate change than members of conservative parties in Western democracies, but not in the rest of the world. Women, young people, and the less religious express more concern about climate change in the English-speaking Western democracies, but in most of the world gender differences are small, and older and more religious people express more concern" (p. 14). Through their examinations of Pew Research Center's Global Attitudes Surveys they found that party identification and political ideology are much weaker drivers of concern for climate change outside of US borders. Moreover, their research revealed that concern was driven more by commitment to democratic values (manifested in preferences for gender equality, egalitarianism) and educational attainment than gender, religiosity or age demographics (Lewis et al., 2018).

In their research, Matthew Hornsey, Emily Harris and Kelly Fielding (2018) also found that, in contrast, "relationships between conservative ideologies and climate scepticism appear to be relatively weak and inconsistent in Europe and the United Kingdom" (p. 618). This finding is quite interesting, amid the realities that both Europe (EU) and the USA have many intertwined political economic and cultural identities for a better part of two centuries. In both contexts, commitments to economic growth and to carbon-based industry, and deeply entrenched technological optimism, have been forces influencing discussions of climate change in the public sphere (Boykoff and Rajan, 2007).

However, the extents to which these political economic and hence cultural factors influence each context differed. These divergent contexts have been shaped by varied origins and histories from the sixteenth century onward. Precapitalist roots, primacy of the nation-state, the power of traditional cultural institutions and a more constrained physical geography shaped EU inhabitant imaginaries and behaviors, while economic freedom via comparatively more liberal democracies and mindsets linked with "manifest destiny" for the more powerful and associated consumption patterns in the USA fueled citizen actions as well as expectations (Starr, 2004; Boykoff and Rajan, 2007). Attitudes to science and environment between the USA and EU has also differed (Rajan, 2006). Moreover, differentiated regulatory and societal networks and institutions have shaped varied carbon-based industry decision-making and practices, while divergent institutional arrangements designed to address climate change have emerged over time (Levy and Kolk, 2002; Pulver, 2007).

Contrarians in the USA, China, Germany, Brazil and in other national contexts around the world have been profiled as predominately older white men (e.g., McCright and Dunlap, 2011; Young and Coutinho, 2013; Anshelm and Hultman, 2014; Forchtner and Kølvraa, 2015; Liu, 2015; Jylhä et al., 2016). In research in the Norwegian context, Olve Krange, Bjørn Kaltenborn and Martin Hultman (2018) found associations with right-wing nationalism and xenoskepticm,[3] noting, "Climate change denial is but one facet of a more general complex of resistance to various societal issues such as economic growth, environmental conservation, globalization, governance and relationships to other social groups" (p. 9; see also Forchtner et al., 2018). Meanwhile, the relationship between CCM organizations and climate contrarians has been studied extensively in the US context, in part because of the influence of carbon-based industry at the interfaces of science, policy and society and because of intense political and ideological polarization in past decades (e.g., Dunlap and McCright, 2011).

The King Is Dead, Long Live the King!

In the original French the foregoing declaration is "Le Roi est mort, vive le Roi!" It was declared in 1422 when Charles VI died, and Charles VII acceded to the throne. It is a seemingly contradictory phrase that has been repurposed and modified over time for a variety of situations because it usefully points to a notion that despite that one thing/belief/concept/idea/notion/view is "dead," another immediately rises to take its place and continue with the same or similar ordering.

---

[3] They define xenoskepticism as "suspicion or dislike of immigrants combined with a belief that immigration rates are too high."

Similarly, Bryan Wynne (2008) once declared, "The information deficit model is dead, long live the information deficit model!" (p. 23). This was a somewhat tongue-in-cheek comment. But in this comment and larger treatment in the piece, Wynne pointed to decades of social science and humanities scholarship that consistently demonstrated that the information-deficit model was insufficient.

Over many years, research – emanating from Science and Technology Studies and Public Understanding of Science research in particular – consistently has provided evidence that the "informational deficit" model of communication is not an accurate representation of how communication in the world actually works (e.g., Hansen et al., 2003; Sturgis and Allum, 2004; Davies, 2008; Tøsse, 2013; Hetland, 2014). By extension, science and environmental communication is much more dynamic, nonlinear and complex (e.g., Davies, 2008; Nisbet and Scheufele, 2009). Despite all this work, it has been a concept that nonetheless continued to consequentially live on, after each "death" through social science and humanities research finding.

The information-deficit model approach is one that essentially posits that poor choices and actions are attributed to "deficits" of knowledge and information to make the "correct" choice. Further, any lack of public and policy engagement with climate issues is attributed to deficiencies in knowledge of science.

So what explains its persistence? Marin Bauer (2016) has pointed out, "This concept has an unusual staying power" and attributes it in part to "the persistence of a common way of thinking: the public understanding of science poses a problem, and this problem is to be attributed on the side of the public, and changing the public is the solution" (p. 398). Molly Simis, Haley Madden, Michael Cacciatore and Sara Yeo have offered reasons why the deficit model is hard to shake from public communications of science. Among them, "Scientists' training results in the belief that public audiences can and do process information in a rational manner" (p. 401). Furthermore, Brianne Suldovsky (2016) has argued that guarding of authoritative privilege is the most compelling reason why scientists continue to adhere to deficit model logic. She has commented, "The deficit model is particularly useful . . . when communicators concurrently assume the epistemic authority of science" (p. 422). Carina Cortassa (2016) has referred to this as "epistemic asymmetry" (p. 454). By engaging in a more dialogical or democratic form of engagement, many scientists may feel they are giving up their authority or sacrificing their legitimacy in the public sphere if they move away from a more comfortable information-deficit approach.

The information-deficit model may be an appealing way of viewing science in society because it can help "experts" cope. Carina Cortassa (2016) explained it by writing, "It is an intuitive and optimistic way to frame the gap between science and society" (p. 447). Such views are further supported through the development of institutional architectures and reward systems that help these sorts of approaches proliferate.[4] Molly Simis and colleagues have observed that "the persistence of this model may be a product of current institutional structures" (Simis et al., 2016, p. 401). In the eyes of everyday people with plenty demanding their daily attention, this model may be alluring because it implies that there are omniscient and reliable authorities looking out for their needs, and responsibilities to diagnose the problem and deal with it lies with the "experts."

It may suit policy actor perspectives too, as scientific information is seemingly neatly packaged up and delivered to them for their decision-making and problem solving. For example, preceding the December 2009 UN Conference of Parties meeting, there was a March 2009 Copenhagen Climate Congress. The closing session featured a panel including Professor Daniel Kamman, Professor Will Steffen, Lord Nicholas Stern and Dr. Stefan Ramsdorf. After their presentations, discussant Professor Katharine Richardson invited then-Danish prime minister Anders Fogh Rassmussen to comment. She prompted his remarks by saying, "Mr. Prime Minister I'm wondering after this round here as we've presented our findings and our heartfelt feelings if you'd like to comment how this can be used." In his remarks, Danish prime minister Rassmussen commented, "Your input has been very helpful" before pressing the panel by stating that "I need some concrete advice now . . . now I need to know from the panel . . . do we need to set the bar higher . . . I need to know and I need to know today . . . as a politician, I have to make a decision." His provocation set forth a fascinating display and discussion of the challenges of translation of scientific ways of knowing about climate risk to policy decision-making. In the discussion that ensued after the prime minister plainly asked for yes or no answers to whether a two-degree Celsius temperature target is enough, the panel did not simply answer the questions. Instead, they elaborated on scientific evidence revealing the complexities and uncertainties involved in a reasoned response.[5]

Influential and widely cited scholar Anthony Downs and his contemporaries in the 1950s and 1960s are often held responsible for the alluring notion that

[4] Reward systems (promotion and tenure) within academia have largely supported the value and pursuit of an authority and expert stance within the scientific community.
[5] An archived webcast of the session can be viewed here: https://video.ku.dk/climate-congress-2009-closing-session-12

humans are all rational actors. They argued that by being rational actors they engaged in predictable and rational decisions based on the best evidence available (Downs, 1957). This perspective was fueled by the work of sociologist Robert Merton (1942, 1973) a decade before. He postulated that well-functioning scientific norms featured four key elements: (1) disinterestedness: scientists should not think of one's personal position relative to science but rather they should consider the collective good; (2) universalism: all scientists should contribute to a growing body of scientific understanding about the world; (3) communalism: scientists should share scientific knowledge with each other and with the world in order to build successfully on it; and (4) organized skepticism: scientists should always question scientific claims in order to strengthen both the processes and products of science. Together, there was an appeal to a neatness, elegance and simplicity of the requisite engagements that necessarily followed: assemble good evidence-filled arguments to sway these rational people. Among these Mertonian norms, disinterestedness in particular has led to a reticence of relevant expert researchers to advocate for evidence-based discussions in society and this has been detrimental to more effective engagement in the public sphere (see Chapter 6 for more).

Yet our realities are much messier, where decisions and actions are routinely irrational and unpredictable. We violate these notions on a daily basis in just about all elements of our lives, as our theoretical selves depart from our practicing selves. Even in the face of a great deal of information about how our behaviors impact climate change, we nonetheless continue to cling to our carbon-based conveniences: as examples, we choose to drive gas-guzzling vehicles when we could bike or walk, and we eat red meat when protein from other sources would do. We engage in these behaviors and increase our carbon footprints on planet Earth even though we know better (Suldovsky et al., 2017). Research by Robert Gifford (2014) has actually found that more knowledge of the negative environmental impacts of one's actions does not effectively dissuade most people from continuing with those behaviors.

From a psychoanalytic perspective, Sally Weintrobe (2013) has written about anxiety that may derive from these daily contradictions. She has noted that anxiety is "the biggest psychic barrier to facing the reality of anthropogenic global warming" (p. 46), and has suggested that anxiety arising from our collective destructiveness can be attributed in part to "the narcissistic part of the self [that] dreads giving up our sense of entitlement to have whatever we want and entitlement to apply our magical 'quick fixes' to the problems of reality" (p. 42). Also from this perspective, Renee Lertzman (2015) has argued that these "'gaps' between what people profess to value and their actions" may be "outward expressions of ambivalence, anxiety, potentially unresolved mourning and unconscious defence mechanisms such as denial and projection" (p. 23). They and others have posited that such unresolved angst can translate to mental health challenges (Fritze et al., 2008).

More widely, studies of climate change and mental health have extended to considerations of "ecological grief" (grief from anticipated or experienced ecological loss) (Cunsolo and Ellis, 2018) as well as mental health risks faced by climate researchers because of their immersion in depressing information (Richardson, 2015; Yong, 2017; Clayton, 2018). While that may seem a little intense, Lertzman latches onto an important consideration of how the contradictions we live by we thereby carry with us.[6] In this way, apathy may be a manifestation of tucked away anxiety and an unconscious defence mechanism (Searles, 1972). While a little anxiety can catalyze action, too much can be paralyzing (Cassady, 2010).

Essentially, when working to more comprehensively attend to these multifarious dimensions regarding how climate change pervades our lives and livelihoods, facts derived from scientific findings are helpful but they are not enough. In describing his experiences working at the human–environment interface in Alaska, Alex Lee (2017) has observed, "We can't throw more facts at a problem that is not about facts. I *value* biodiversity, I think we *should* take care of vulnerable populations, I *like* glaciers. Let's talk more about what sort of world we want to live in, what options we have going forward and the moral commitments we have in a changing world" (italics in the original).

Yet, the information-deficit model persists through the ongoing and seductive notion that science (as process and as product) provides the more direct pathway to public understanding. In the policy and public spheres, there have been many notable manifestations of an information-deficit model view. These have included reliance on "sound" science for decision-making, and objectives to eliminate uncertainty as a precondition for action.

However, in reality we find that these pathways are much more circuitous and labyrinthine. For example, "post-truth" and "fake news" memes and practices persist despite our knowing better, despite copious fact-checking and record-correcting (Huertas, 2016; Borel, 2017) (see Chapter 1 for more). The powers of other ways of knowing that shape our perceptions and perspectives on climate change are therefore on display (see Chapter 4 for more).

Harry Collins and Robert Evans (2002) have delineated competing perspectives alongside this information-deficit model. Alternative models of engagement point to more dialogical and democratic forms, where informal actors

---

[6] This could in some ways be akin to the premise of soldiers carrying memories of war, as depicted in the classic collection of short stories *The Things They Carried* by Tim O'Brien.

influence ongoing processes at the science–policy and science–policy–society interfaces. One alternative model was principally advanced by Ulrich Beck (1992). Another model of engagement, called a "normative theory of expertise," was promoted principally by Bruno Latour (2004), among others. With similar democratizing commitments, these perspectives mapped more porous institutional boundaries between formalized spaces of science-governance and the informal public arena along with more variegated roles of expertise and authority. These approaches reconceptualized elements of dynamism and non-linearity of processes as well as how power, influence and scale operate within and between science–policy–public spheres. These approaches more fully accounted for complex relationships between information and action (Collins and Evans, 2002). These alternative models are useful for reconsidering how creative and effective communications can and should be pursued in our present-day public sphere.

Freedom to Choose (Ignorance)

Isaac Asimov (1980) has argued that "there is a cult of ignorance in the United States . . . the strain of anti-intellectualism has been a constant thread winding its way through our political and cultural life, nurtured by the false notion that democracy means that 'my ignorance is just as good as your knowledge'" (p. 19). Anti-intellectualism has also been attached to attacks on education. Reece Walters (2018) has defined ignorance in the context of climate contrarianism as "unjustifiable belief." He has argued that it "provides those in positions of power and entitlement to assert their own unquestioning expertise" (p. 163).

Nonetheless, authority derived from adherence to the information-deficit model can often be observed in health care fields (Ko, 2016). No doubt these trends, and consequent impacts, pervade the climate sciences as they relate to policy and society in current contexts. Yet, ongoing adherence to these approaches to communication about climate change exposes numerous short-comings and vulnerabilities in efforts to meet people where they are on climate change. In the contemporary polarized and partisan spaces of science–policy–society interactions, information-deficit model logics mixed with climate silence (see Chapter 1 for more) contribute to a communications environment where unproductive voices can more effectively contaminate productive discussions and engagement on climate change.

This also contributes to a "missing middle ground" (Kirk, 2018). These interactions are prominently on display when exploring the ways in which climate contrarians have essentially made squatters rights claims to climate change discussions in the public sphere. Through these conditions, over time there has been a proliferation of outlier perspectives that have become more animated and active from the oxygen put into these arenas.[7]

Here I focus considerations on contrarian voices – also called "climate skeptics," "dismissives," "doubters," "deniers," "rejectors,"[8] or "denialists"[9] – that have gained prominence and traction in the US public arena. These contrarian views are often situated in climate contrarian countermovement (CCM) organizations, with associated ideological and cultural inclinations. CCM organizations have been defined as groups that advocate against policies that seek action to mitigate climate change, especially mandatory restrictions and penalties on greenhouse gas emissions (Brulle, 2014).[10] These movements also advocate against substantive action to adapt to or mitigate climate change (McCright and Dunlap, 2000).

These labels are imperfect. In developing these labels there is a danger of excessively focusing on individual personalities at the expense of political economic, social and cultural forces. In other words, when focused on the movements of individual contrarians, such attention could displace deeper structures and architectures such as adherence to information-deficit model logics that give rise to the effectiveness of their claims in the public arena. Further complexity arises when drawing conclusions based solely on evident ties between carbon-based industry, contrarian lobbying and climate policy. In other words, it is not necessarily where the funding comes from, but whether these ties influenced the content of the claims made by funding recipients (Oreskes, 2004a).

Quick movements to labeling (and rejecting) risk dismissing legitimate and potentially useful critiques out of hand by way of dismissing the individual rather than the arguments put forward. In these ways, treatments of individuals through denigrating monikers do little to illuminate the contours of their arguments; they actually have the opposite obfuscating effect in the public sphere. In other words, placing blanket labels on claims makers overlooks the varied and context-dependent arguments they put forward (Boykoff, 2013).

---

[7] Previous work by Robert Brulle, Jason Carmichael and J. Craig Jenkins (2012) and Jay Hmielowski, Lauren Feldman, Teresa Myers, Anthony Leiserowitz and Ed Maibach (2014) has shown that politicized public discourse has fueled further polarization in the wider US publics.

[8] Robert Jay Lifton (2017a, 2017b) has chosen this adjective to describe those who have rejected scientific evidence linking extreme events such as hurricane activity to a changing climate.

[9] Matt Nisbet has tracked the proliferation of the suboptimal term "deniers" in articles on climate change at the *Guardian* and *Washington Post* from 2008 through 2017 and found a steep rise in its use over time (Nisbet, 2018a).

[10] I use "climate change countermovement organizations," "contrarian countermovement organizations," "climate change countermovement groups," "contrarian countermovement groups" and "think tanks" here interchangeably.

Instead, these movements erroneously communicate suggestions of homogeneity, when the realities are more nuanced. For example, US Science Advisor Kelvin Droegemeier has expressed strong allegiances with convergent scientific agreement that humans contribute to climate change, and "has expressed little tolerance for those who reject science . . . because that person is essentially a lost cause" (Waldman, 2018b). Scott Waldman quoted Droegemeier, saying, "Some people just get vilified if they even bring a topic up. If you're in a situation like that, somebody is so adamant that their view is right and that you're an idiot, walk away, walk away. Don't have the conservation with them; it's not going to be valuable" (Waldman, 2018b).

The nuances and distinctions between these labels have been picked apart over time. These are partly imperfect because they seek to describe a heterogeneous group of actors and organizations that counter many areas of convergent agreement in climate science and policy decision-making. In other words, they take up outlier perspectives. While many have pointed out that "skepticism" forms an integral and necessary element of scientific inquiry, its use when describing outlier views on climate change has been less positive. The term "skeptic" has been most commonly invoked to describe someone who (1) denies the seriousness of an environmental problem, (2) dismisses scientific evidence showing the problem, (3) questions the importance and wisdom of regulatory policies to address them and (4) considers environmental protection and progress to be competing goals (Jacques et al., 2008). While these authors discuss "environmental skepticism," the characterization holds for "climate skepticism" as well.

Candace Howarth and Amelia Sharman (2017) have developed "categories and subcategories" of skepticism, distinguishing among "(motivated) contrarianism," "policy-related skepticism" and "knowledge-related skepticism" (pp. 777–778).[11] They distinguish these labels from the category then of "denier," along with subcategories within (Howarth and Sharman, 2017).[12] Saffron O'Neill and I (2010) further developed a definition of "climate contrarianism" by disaggregating claims-making to include ideological motives behind critiques of climate science, and exclude individuals who are thus far unconvinced by the science or individuals who are unconvinced by proposed solutions, as these latter two elements can be more usefully captured through different terminology.

---

[11] See O'Neill and Boykoff (2010), Akter et al. (2012), Hobson and Niemeyer (2014) and Capstick and Pidgeon (2014), for more details regarding subcategories and descriptions of these labels.

[12] See Norgaard (2011) and Weintrobe (2013) for more details here. Also see Stern et al. (2016) for more regarding objections to urgent mitigation actions, dubbed "climate-change neoskepticism."

Moving between climate science, politics and policy, scholars such as Stephen Schneider (2009) and Riley Dunlap (2013) have pointed out differences between contrarianism derived from ideology and contrarianism derived from scientific evidence. Aaron McCright (2007) has defined "contrarians" as those who vocally challenge what they see as a false consensus of mainstream climate science through critical onslaughts on climate science and eminent climate scientists, often with substantial financial support from fossil fuels industry organizations and conservative think tanks. Those attacks have given rise to the Climate Science Legal Defense Fund, set up in 2011, to give legal advice and support to scientists who were facing legal issues (Schwartz, 2017).

Aaron Ley (2018) has examined how CCM groups have excessively used open records laws to call into question both messenger (asking questions about scientists' motives and legitimacy) and messages (raising the spectre of uncertainty and also accuracy of scientific findings from them). He found that the impact of these Freedom of Information Act suits on the activities of university researchers "has been overwhelmingly negative, has caused them to change their methods of communication, and has imposed a new work burden that draws them away from other work responsibilities" (p. 221). As a result, those most capable of speaking out about scientific evidence of climate change – "whose production of knowledge is a public good" (p. 221) – are significantly hampered by these suits instead. Many climate researchers have expressed fears of "McCarthyist attacks" in the wake of the election of US president Donald J. Trump. For example, Kerry Emmanuel from the Massachusetts Institute of Technology (MIT) has commented, "I think we're in a mild state of shock after the election. Politics has [sic] been turned upside down and all of these dark forces have erupted" (Milman, 2017).

Successes by CCM organizations in garnering attention in the public arena has often been attributed to the structural, political economic and cultural roots of why, in the face of overwhelming scientific consensus, fewer than half of US Americans believe that humans contribute to twenty-first-century climate change. In recent years, more attention has been paid to a mix of internal workings such as journalistic norms; institutional values and practices; and external political economic, cultural and social factors. However, the ways in which CCM organizations exploit the dominant information-deficit model has garnered less attention. Together, continued examination of these factors alongside communications efforts helps to identify, contextualize and tackle barriers to effective efforts to "meet people where they are."

Regarding political economic influences, Robert Brulle (2012) has commented, "Introducing new messages or information into an otherwise unchanged socioeconomic system will accomplish little" (p. 185). For instance, Robert Brulle conducted a study of lobbying spending on climate change in the US congress from 2000 to 2016. He found that among the over US$2 billion spent during this time period, spending by carbon-based industry entities and associated trade associations outnumbered spending by environmental organizations and renewable energy groups by ten to one (Brulle, 2018b). Brulle (2018b) concluded that this magnitude of differential spending – and by extension, influence – "illustrates the limitations of science advocacy efforts" (p. 13). Considering these factors, effective communications about climate needs to take into account meso- and macro-scale social change, political action and public mobilizations (see Chapter 5 for more). Furthermore, Justin Farrell (2016a, 2016b)has used computational social science methods, including large-scale network science and machine learning, to map the structure of organizations, companies and individuals involved in promulgating misinformation and disinformation about climate change. In so doing, we have been able to examine central messaging strategies as well as how messages in some cases are promoted through funding mechanisms. These political-economic investigations have revealed that ExxonMobil and Koch Family Foundation have been particularly influential backers of CCM groups. The Koch Family Foundation and its connected organizations have provided funding for the creation of a number of conservative organizations, including the Cato Institute and Americans for Prosperity. This family foundation has generated funds from the success of Koch Industries, which is the largest privately owned US-based energy company. Koch Industries generates energy from fossil fuels, and has a large stake in oil refining processes (Fifeld, 2009; Mayer, 2010).

It is also important to recognize that ideology and our culturally infused cognition pervade all of us as we seek to understand climate change. As such, nobody can effectively put themselves outside these filters of understanding (Maeseele and Peppermans, 2017). Ideology can be defined as a system of ideas that often form the basis of political or economic thought and policy.

Connected to this (as discussed in Chapter 2), cultural cognition explains why groups with different cultural perspectives disagree about critical social and political issues. The thesis, advanced prominently by Dan Kahan, explains that these differences are about reinforcing identity and belonging among desired groups. In fact, Dan Kahan has found that as science literacy increases among members of the public, cultural polarization rather than cultural consensus also increases (Kahan et al., 2012). In other words, our views are largely formed by our consideration of what referent group we belong to and therefore align with. Furthermore, more informed people dig in their heels on their views rather than expressing openness to changing them. Political elites are very influential in these processes as well (Converse, 1964; Zaller, 1992; Carmichael and Brulle, 2017).

Survey research by Josh Pasek (2017) further complicates the view that wrong answers to scientific knowledge questions means people are ignorant. Pasek pursued a more nuanced understanding of individual attributes associated with a willingness to take stances counter to convergent scientific agreement (or consensus) and how this willingness to disagree is distinct from ignorance or misperception (see also Clifford and Travis, 2018). His paper, entitled "Not my consensus," revealed that motivated reasoning may instead contribute to a questioning of scientific consensus, emanating from both religiosity and partisanship. By extension, his findings complicated a commonly held notion that questioning the veracity of scientific consensus indicates low literacy on the issues. He concluded, "We ignore an important part of the story if we assume that motivated assessments of science are limited either to identification of a consensus or to holding beliefs in line with that consensus" (p. 16).

Nonetheless, climate literacy is an important and pervasive dimension shaping the complex dynamics of awareness and action. Many groups like the Alliance for Climate Education led by Matt Lappé and the National Center for Science Education led by Ann Reid work to increase literacy as pathways for more informed and better decision-making about climate change. While scrutiny of climate literacy can lead to a slippery slope of information-deficit model logics in communications, there is some tension here: after all, some semblance of basic scientific literacy is critical to well-functioning societies (McNally, 2018). Through cross-country survey research by Jing Shi, Vivianne Visschers, Michael Siegrist and Joseph Arvai (2016), they have argued that "public education and risk communication efforts regarding climate change may not be the lost cause that some researchers (and some policymakers) assume they are" (p. 762). Moreover, in their book *Unscientific America: How Scientific Illiteracy Threatens Our Future*, Chris Mooney and Sheril Kirshenbaum (2010) have argued that recasting scientists as partners rather than elites can alleviate deficits in awareness and engagement challenges in society. But people who get science facts wrong are not necessarily ignorant or dismissive of scientists as geeks. These folks could instead be more accurately characterized as alienated, ostracized "knowledgeable disbelievers" (Roos, 2014).

In relying too heavily in consensus we run many risks (see Chapter 5 for more). Among them, we risk devaluing dialogue itself. Kyle Powys Whyte has commented that disagreement may be done within respectful discourse;

therefore "dissent is not the problem, it is an indicator of a problem."[13] In fact, this perspective helps to enrich our understanding of these complexities as half the US public are found to believe in at least one conspiracy theory of some sort (Oliver and Wood, 2014). An information-deficit model framing of these issues catalyzes rather than stymies productive discourse. Consequently, in an assessment of partisan cueing and polarization of climate change discourse, Deborah Lynn Guber (2017) has posited, "In the end, beliefs about climate change are as complex as the issue itself" (p. 217).

Bamboozled[14]

To illustrate these intertwined influences along with relations to information-deficit model logics, it is useful to take a look at the outsized influences of US-based conservative think tanks that feed into CCM activities and actions.

Among them, we can examine recent movements of a think tank called the Heartland Institute, for example. The Illinois-based Heartland Institute was founded by ideological elites in 1984 and is motivated by free-market policy approaches on issues including climate change. Over the past decade the aforementioned Heartland Institute has emerged as a leading contrarian countermovement organization that questions both diagnoses that humans contribute to climate change and a range of prognoses for mitigation policy action. Since 2008 the Heartland Institute has organized nearly annual meetings and called them "International Conferences on Climate Change." Attended by prominent political figures such as Congressional Representative Lamar Smith (Republican from Texas), the conferences have been geared to amplify outlier views at the science–policy interface (Mervis, 2017). For example, the Heartland Institute held their twelfth mostly annual conference on climate change in Washington, DC. At the meeting, panelist (and House Republican from Louisiana) Clay Higgins declared, "Environmentalists who believe the United States can be less reliant on coal, oil and natural gas are counting on rainbow dust and unicorn milk to power the nation" (Waldman, 2018a). This statement garnered cheers from the attendees and press coverage that followed. In recent years, the Heartland Institute has achieved high-profile status as a "primary American organization pushing climate change scepticism" (Gillis, 2012).

---

[13] Kyle's comment was made at workshop at University of Illinois, September 14, 2018.
[14] This is the title of Spike Lee's 2000 film, in part about success and notoriety gained through deliberate chicanery and (self-)deceit.

Heather Cann and Leigh Raymond have examined Heartland Institute reports and policy briefs to better understand the content of oppositional frames. They found that common science frames deployed by the group include the assertion that "climate change is a myth or scare tactic perpetuated by environmentalists, bureaucrats and political leaders" (appearing in 57.1% of the documents) and "mainstream climate science research is 'junk' science" (50%) while common policy frames circulated included "policy would economically harm consumers" (29.7%) and "policy would harm the economy overall, at the state or national level" (23.8%) (Cann and Raymond, 2018, p. 14).

How influential has the right-wing think tank Heartland Institute been in shaping the US Environmental Protection Agency (EPA) agenda – and climate storylines in the public arena – in the first years of the Trump administration? That was the main question that motivated a March 2018 lawsuit by the Environmental Defense Fund and the Southern Environmental Law Center. The legal suit claimed that the US EPA failed to respond to a Freedom of Information Act request from six months earlier that demanded correspondence between the Heartland Institute and the EPA specifically about a "red team–blue team" proposal for evaluating scientific evidence of climate change (Reilly, 2018).

Red team–blue team approaches have been fashioned from military discourse, where one group (a "red team") seeks to challenge another group (a "blue team") by taking up an antagonistic, contrarian or opposite point of view. This was originally seen as an effective way to rethink perspectives, strategies and approaches to complex problems. From this adversarial approach, more enriched perspectives and actions were thought to be enabled.

In its first year in power in the USA, the Trump administration proposed to form an adversarial "red team" to debate and debunk the science and climate change (seen as a "blue team" perspective) (Siciliano, 2017). In so doing, this approach effectively sought to restructure the peer review process and elevate outlier and contrarian views in the public arena. Then-EPA administrator Scott Pruitt introduced this military-strategy-style approach to evaluating climate research for policy applications, by proposing television debates to "advance science" in the public arena (Volcovici, 2017). Through this "red team–blue team" proposal (enlisting the help of the Heartland Institute), Pruitt began to identify potential contrarian scientists and economists as participants (Waldman, 2017). With Pruitt's resignation from the EPA in July 2018, the "red team–blue team" approach also lost momentum both inside and outside the administration. Nonetheless, before departing the EPA, Pruitt told another influential CCM – the Heritage Foundation – that there were still ongoing plans to constrict climate science under the guise of "reform" (Waldman and Bravender, 2018).

These approaches join with other ways the Heartland Institute continues to influence US federal administration considerations. For example, in late 2018 it was reported that senior White House aides reached out to the Heartland Institute to request a presentation on climate change. Heartland Institute Senior Fellow James Taylor reasoned, "[President Trump] is an open-minded and intelligent man, so of course he wanted the best information and arguments that both sides had to offer" (Waldman, 2018d). These examples, along with other possible efforts to constrict climate science influence in the public arena, thrives as an idea because of ongoing reliance on information-deficit model logics.

Polarization resulting from these strategies and tactics from US-based climate change countermovement organizations and think tanks has led to fundamentally different interpretations of scientific evidence, highly varied public perceptions of uncertainty and consequent policy confrontations and stalemates. CCM organizations have exploited vulnerabilities in climate communication systems that are driven by logics that lack of public engagement with climate issues is attributed to deficiencies in knowledge of science. Phil Williamson (2016) has called on relevant experts to take time to correct misinformation in the public arena. But this approach of reaction and correction is frankly feeble in the face of the larger influences that CCM organizations exploit in the public arena. In short, more smart and creative communications about climate change are sorely needed here.

Numerous events in recent years such as these have capitalized on information-deficit model thinking, and have thus recalibrated "contrarian" considerations in the public arena. CCM organizations now enjoy unparalleled access in the halls of the US federal government. Trump's nominations for key posts in the administration sparked worry among those who care about climate and environmental protection, justice and human well-being. These appointments included Secretary of the Department of Interior (Montana Congressman Ryan Zinke) and Secretary of the Department of Energy (former Texas Governor Rick Perry), who maintained ties to carbon-based industry interests. Moreover, influential contrarian actors have also moved from these CCM groups into posts in the Trump administration. For example, Myron Ebell has been the chair of the Cooler Heads Coalition as well as the Director of Global Warming and International Environmental Policy at the Competitive Enterprise Institute. In 2017 he was selected by President Trump to lead the EPA transition team. Ebell has been quoted acknowledging that his advocacy from these positions "does bleed into political persuasion and lobbying" for particular policy outcomes but countered that these activities are both commonplace and legal (O'Harrow, 2017, p. A1).

Developments such as these have contributed to a reality that ideological polarization around climate change issues – particularly in the USA – has increased in the past thirty years (Dunlap et al., 2016). These kinds of actions have also marked novel approaches to climate change countermovement or think tank strategies to oppose various forms of science and policy engagement from the local to national and international scales (Cann and Raymond, 2018) or to shape public impressions of climate change (McKeown, 2012; Young and Coutinho, 2013). In the USA, a patterned network of actors – individuals enmeshed in CCM organizations – bolstered by élite corporate benefactors have therefore demonstrably muddied the waters of discourse and action on climate change in part by preying on a vulnerable communications environment (Oreskes and Conway, 2011; Supran and Oreskes, 2017).

New York Times columnist Paul Krugman (2018) has written, "Climate denial is a deeply cynical enterprise" (p. A20). How and why has this enterprise garnered such disproportionate visibility in the public sphere through its claims and communications? Why does dissent through contrarian or "denier" voices persist and how do they find traction? Part of the answer resides in their successful communications strategies in their messaging and misinformation campaigns through US media outlets.

In the case of climate change, one can consider the overwhelming convergent agreement within relevant expert communities in science that humans play a significant role in today's changing climate (amidst an ongoing background of natural climate forcing) (see Chapter 5 for more). However, movements from this diagnosis to prognoses for action are contentious. In other words, the path from appraising the "way things are" to "the way they are thought they 'ought to be'" are fraught with discussions, debates and disagreements. It is within these spaces that one finds clashes as well as confluences of culture, politics, economics and society over time (Hoffman, 2015).

As such – in part by taking advantage of an information-deficit model communications environment – CCM organizations and associated climate contrarians have achieved veritable "celebrity status." They have drawn on this celebrity as a way to exploit networked access to decision-making within the dynamic architectures of contemporary climate science, politics and policy in the USA (see Chapter 1 for more). As examples, Tim Phillips (Americans for Prosperity) and the aforementioned Myron Ebell (Competitive Enterprise Institute) have shown themselves to thrive on recognition gained from these stances in the public arena as contrarian celebrity public intellectuals (see Chapter 6 for more on public intellectualism).

Shawn Olson-Hazboun and I have explored this phenomenon regarding how these individuals and groups have gained distinction primarily by way of activities associated with antiregulatory, antienvironmental, "skeptical" and neoliberal environmental movements aligned with those in the ideological right. We explored how these stances effectively dismiss, denigrate and demonize the climate science community and how these movements map onto the tenets of a US-based historical Wise Use movement from decades earlier (McCarthy 2002).

The Wise Use movement arose in the American West in the late 1980s, later spreading across the country as a national antienvironmental effort. Wise Use fought for private property rights; decreased environmental regulation; and unrestricted access to public land for mining, logging, grazing, drilling and motorized recreation. It was a coalition of individuals, movement leaders, nongovernmental organizations (NGOs), and corporations who aligned behind an "environment or economy" dichotomy. The birth of Wise Use marked the rise of the modern neoliberal, antiregulatory, antienvironmental movement prevalent today in which individual rights, private property and free enterprise are prioritized over environmental protection. Wise Use postured as advocating for rural residents' and resource laborers' rights, conceiving of environmentalists as distant, urban elites who remained out of touch with the needs of those who were on the ground and engaged in the production of natural resources. However, Wise Use had many corporate ties and simultaneously served the interests of extractive corporations whose profits could be affected by implementation of environmental regulations. Peter Jacques has theorized that an "organized deflection of accountability" is also inspired by the drive to defend the notion of an "American" ideology (Jacques, 2012). Thus, protecting corporate freedom and profits is another motivator.

Wise Use has been understood to be an expansion of the earlier Sagebrush Rebellion land revolts that spanned the 1960s and 1970s, which were a reaction to the advent of the new, vigorous environmentalism of the 1960s and 1970s. Wise Use proponents made use of "common sense" appeals to the conservative middle class, speaking to the ideology of "freedom" and "liberty," above all, which connected dually with the individual-centric doctrine as well as that of neoliberalism and the free-market (Boykoff and Olson, 2013). We pointed out how both Wise Use and CCM groups self-label ambiguously (e.g., The Cooler Heads Coalition) and in ways that invoke "environment-economy" and "regulation-freedom" dichotomies (e.g., Americans for Prosperity, Competitive Enterprise Institute) (see Bohr, 2016 for more). The slogan of Cooler Heads Coalition in fact is "May Cooler Heads Prevail," and this rhetoric of "common sense" is prevalent among both the Wise Use and climate contrarian movements.

Perceived academic martyrdom, exclusion from the institution and the unraveling of the scientific method are also complaints leveled against climate science and activism. They can also be elements that appear to generate psychological exhilaration within climate contrarian communities, particularly within the celebrity members themselves (Stever, 2019). Some – largely from within the movement itself – argue that contrarian stances have staked out part of a contemporary Copernican revolution in climate science and policy, helping the general public to overcome the collective delusion that humans play a role in modern climate change and that migration and adaptation actions need to be taken to address associated challenges.

Motivations are part cultural and psychological: research by Lahsen (2013) has found that many contrarian scientists in the 1960s and 1970s tended to view climate modeling and other developments in climate science that followed in the 1980s and beyond as inaccurate, variable and ungrounded. Meanwhile, these articulations are part politics and economics. For instance, at the time of the release of the UN Intergovernmental Panel on Climate Change (IPCC) Fourth Assessment Report, it was revealed that the American Enterprise Institute (AEI) – a group receiving funding from Exxon Mobil Corporation – was soliciting contrarian voices. At that time AEI was reportedly offering $10,000 "for articles that emphasized the shortcomings of [the Fourth Assessment] report from the UN IPCC" (Sample, 2007, p. 1). In this way, the group was calling out for particular and dissenting inputs and therefore undermining the integrity of the scientific process. As such, it has often been the case that funding-driven influences cohere with ideologically driven motivations by way of contrarian arguments questioning a range of relevant expert views across the climate science and policy spectrum (Barringer, 2012).

Amidst these highly contested, highly politicized and high-profile cases of climate science and policy, answers to questions of commitments, motivations and actions are complex, dynamic and varied across claims makers and the claims they make. Yet, the amplification of these arguments in the public sphere through US media accounts influences public understanding and engagement as these antiregulatory, antienvironmental and neoliberal environmental arguments coalesce in these CCM epistemic communities.

In recent research, Justin Farrell and I sought to further map CCM organization and contrarian voices and perspectives in US discourse, by way of analyses of US media since 1988. We worked to help unravel how contrarian CCM organizations in the USA demonstrate themselves to be (at times deliberately) detrimental to efforts seeking to enlarge rather than constrict the spectrum of

possibility for varied forms of climate action in this high-stakes, high-profile
and highly charged public arena (Boykoff and Farrell, 2019). We traced eleven
influential and US-based CCM organizations through media attention to their
movements and activities since 1988 in eleven prominent television and newspaper outlets.[15] The eleven CCM groups are the Cooler Heads Coalition, the
Global Climate Coalition, the Science and Environmental Policy Project,
Americans for Prosperity, Cato Institute, the American Enterprise Institute,
the Heartland Institute, the Heritage Foundation, Committee for a Constructive
Tomorrow, the George C. Marshall Institute and the Competitive Enterprise
Institute. The eleven US outlets are *ABC News*, *CBS News*, *CNN News*, *Fox
News*, *MSNBC*, *NBC News*, *Washington Post*, *Wall Street Journal*, *New York
Times*, *USA Today* and *Los Angeles Times*.

We were interested in their visibility through these sources because media
often connect formal science–policy and informal spaces of the everyday
together. Media representations then become powerful conduits of climate
science and policy (mis)information. There are many factors that shape how
members of the public consider possible responses to and engagements with
climate change. Analyses of media representations can help to see how decisions about portrayals (quantity and quality) shape how public citizens consider
possible responses and how they play into climate governance at multiple
scales in the USA (Brulle et al., 2012; Fisher, 2013).

Figure 3.1 depicts media attention for each of these CCM groups year
to year over these three decades across US television and US newspapers.
Figure 3.2 shows CCM organizations' media presence in US television
news segments and US newspaper articles, both from 1988 to 2017. In
these figures, one can see the CCM presence increased greatly after 2006.
With the exception of a spike in CCM visibility in the media around the
time of the 1997 Kyoto Protocol, coverage was considerably lower in the
past. The average year-to-year coverage of these CCM organizations from
1997 to 2006 was about a third (33%) of the average amount of their
visibility in the US media over the subsequent decade 2007–2016. The
years 1997–2006 saw an average of 63.1 stories or segments per year while
2007–2016 saw an average of 189 stories/segments per year.In particular,
there was a significant increase in the media presence of CCM organizations at the end of 2009 and through 2010, following the November 2009
so-named email hacking scandal emanating from the University of East
Anglia (also known as "Climategate"). There was also a notable uptick in

[15] This search used a Boolean string to search for the organization's name *and* "climate change" *or*
"global warming."

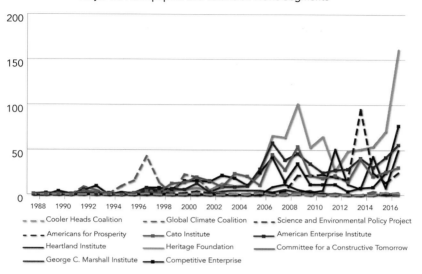

Figure 3.1 Media coverage year-to-year 1988–2017 of the Cooler Heads Coalition, the Global Climate Coalition, the Science and Environmental Policy Project, Americans for Prosperity, Cato Institute, the American Enterprise Institute, the Heartland Institute, the Heritage Foundation, Committee for a Constructive Tomorrow, the George C. Marshall Institute and the Competitive Enterprise Institute.

the Heartland Institute's media presence at the end of 2012, due in part to fallout from the release of its May 2012 billboard ad comparing climate "believers" with the notorious Ted Kaczynski (the "Unabomber"). In 2014, Americans for Prosperity (AFP) – a conservative think tank based in Washington DC – received a bump in media attention.

In 2014, AFP's anticlimate legislation campaigns were given a boost through a tripling of funding from the Koch Family Foundation (Mayer, 2016). AFP president Tim Phillips, along with others from AFP, effectively garnered attention in the media to shape public discourse surrounding the 2014 midterm elections in the USA, particularly stating how AFP were working to aggressively sink the election hopes of any candidate who supported a carbon tax or other climate regulations. Furthermore, there has been increased visibility of these eleven CCM groups in US media since the election, inauguration and establishment of the Trump administration.

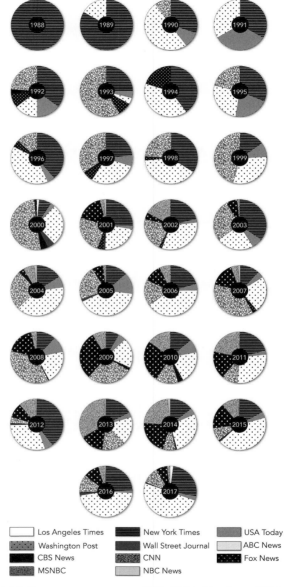

1988–2017 Counter-Movement Presence in Major US Newspapers and Television News Segments

Figure 3.2 Media coverage year-to-year 1988–2017 of the eleven CCM organizations on *ABC News*, *CBS News*, *CNN News*, *Fox News*, *MSNBC and NBC News* as well as the *Washington Post*, *Wall Street Journal*, *New York Times*, *USA Today* and the *Los Angeles Times*.

Total coverage in 2017 (403 stories/segments) was about double that of the average coverage over the previous decade of coverage of these groups (189 stories/year from 2007 to 2016). Specifically, the Heritage Foundation, Competitive Enterprise Institute, American Enterprise Institute and Heartland Institute gained increased visibility in 2017. US media accounts noted, for example, that the Trump administration embraced numerous Heritage Foundation policy recommendations articulated in their "Mandate for Leadership" series of publications. Among these recommendations was a strong stance on leaving the Paris climate change accord. By the Heritage Foundation's own boastful accounts, 64% of their policy prescriptions from that series were then included in Trump budget proposals (Bedard, 2018).

Through these analyses of media influence by these CCM organizations in the eleven US outlets, we found that influence in public discourse – indicated through media coverage – is shaped by founding and funding (e.g., the Global Climate Coalition was heavily supported in the 1990s, and the Heritage Foundation has been heavily supported in more recent years).

Examining the proportions of coverage from year to year in each outlet across these eleven organizations overall,[16] we found that the *Washington Post* and *New York Times* contributed most significantly to coverage of these prominent CCM organizations. In an era of "naming and shaming" of the *Fox News* network by many from the left, these findings ran counter to common perceptions that attention paid to these CCM groups in US media may be attributed to outlets with right-of-center ideologies, stances and reputations.

While I have referred to a "Rupert Murdoch effect" via *Fox News* in terms of how Fox shapes the content of climate change coverage in the US press in some of my previous work (2013), these findings did not support the notion that Fox or the *Wall Street Journal* are primarily responsible for the amplification of these particular outlier perspectives in climate change stories by way of the quantity of coverage. That said, it can be argued that small quantities may sufficiently sway media consumers. Jason Carmichael and Robert Brulle (2018) have found that "those who watch Fox News, read the Wall Street Journal or listen to Rush Limbaugh are about half as likely to accept anthropogenic climate change than those who do not consume these media" (8pp. 247–248).

---

[16] *Fox News* and *MSNBC* began coverage in 1996.

Expanding out across the twenty-one-year period from the inceptions of *Fox News* and *MSNBC* (1996–2017), we found that more than half the coverage (51%) of these CCM organizations was carried through the *Washington Post* and *New York Times*. Meanwhile, just 15% of the volume of coverage of these CCM organizations appeared on *Fox News* and in the *Wall Street Journal*, seen typically as bastions of right-of-center voices on climate change and other issues.

Coverage was found here to be 32% in the *Washington Post* and 19% was in the *New York Times* while 14% was on *Fox News* and less than 1% appeared in the *Wall Street Journal*. Over this same time period, 19% of coverage was on *CNN*, with 9% on *MSNBC*, 5% in *USA Today* and about 1% each on *ABC News*, *CBS News*, *NBC News*, and in the *Los Angeles Times*. Our findings showed, in part, that naming and shaming of right-wing outlets for amplifying the volume of CCM organizations and their associated contrarians was misplaced.

As we moved our analyses from the amount of coverage of CCM groups to the impact of this coverage, we interrogated how CCM voices worked through the media to stymy efforts seeking to enlarge rather than constrict the spectrum of possibility for mobilizing the public to appropriately address ongoing climate challenges. We explained that institutional features of media representational practices partially explained this state of affairs. Elsewhere, Justin Farrell (2016b) found that funding of CCM groups influenced the actual language and thematic content of media stories, and the polarization of discourse, while well-funded CCM organizations were more likely to have written and disseminated contrarian texts through the media.[17] Contrarian claims feed journalistic pressures to serve up attention-getting, dramatic personal conflicts, thereby drawing attention toward decontextualized individual claims making, and away from critical institutional and societal challenges regarding carbon consumption that calls collective behaviors, actions and decisions to account.

Our findings showed that greater funding for the Cato Institute, Heartland Institute and Heritage Foundation in recent years appear to translate to media visibility (see Figures 3.1 and 3.2). These most powerful organizations at the center of CCM organizational networks have received high levels of funding from powerful donors like ExxonMobil and the Koch family foundations.

---

[17] As examples, I have argued elsewhere that in the name of efficiency, reporters increasingly cover a vast range of beats, making it ever more difficult to satisfactorily portray the complexities of climate change. Meanwhile, media institutions and practices have produced content by seeking refuge in journalistic tendencies of personalization and drama, privileging incompatibility and contentiousness among messengers over treatment of arguments and assertions (Boykoff, 2011).

Together, our analyses sought to provide more textured understandings of how and why outlier views in climate science and governance are provided gratuitous media visibility.

The Fossils among Us

In 2005, the animated US television comedy *South Park* aired an episode called "Two Days Before the Day After Tomorrow"[18] parodying the doomsday action thriller film *The Day After Tomorrow* that had been released the previous year. Written by Trey Parker and Kenny Hotz, the episode portrayed how Stan and Cartman accidentally destroyed the beaver dam above town, causing Beaverton to face imminent flooding. Stan and Cartman kept their transgression quiet, and meanwhile the residents of Beaverton were whipped into hysteria as "top Colorado scientists and government officials" announced that the disaster was due to global warming.

In one scene, as cars lined the highway in standstill traffic from town, Randy Marsh is portrayed in a state of anxiety, rolling down his window to shout "We didn't listen!" to others trying to flee the flooding town. Upon layer and layer of jokes and cultural references to concurrent real-life plot lines such as Hurricane Katrina and then-US president George W. Bush's administration response, the story also poked fun at a penchant to blame any disaster event on climate change while also pointing out an overreliance on scientific ways of knowing.

Science alone – bolstered by information-deficit model communications logic – may provide necessary but not sufficient ways of knowing and acting on today's climate challenges. The traction of the "I'm not a scientist but . . . "contrarian meme has at times shown the limits of information-deficit commitments through scientific ways of knowing. Even though one may effectively be declaring his or her scientific ignorance, the resonance of this comment among contrarians could reveal a latent understanding of the value of experiential, emotional and aesthetic ways of knowing about climate change, among others.

The reality is that the fossils of the science community have shown themselves to be those who feel they have sole authority on learning and knowing about climate change. Part of this fossilized stance is due to ongoing adherence to information-deficit model communications logic.

---

[18] This was the season 9, episode 8 of *South Park* (#908), first airing on October 19, 2005.

Also, the fossils of the science community are those who feel their work is done once they have done the field research, and have written up and published their findings. It is soothing to take that view. Yet more is needed to discharge substantive engagement. Through dialogue and creativity via additional ways of knowing and learning about climate change, we can enlarge possibilities for effective (climate) communication (see Chapter 4 for more). Similarly, information-deficit model approaches can be overcome through a sustained commitment to meeting people where they are through more creative and dialogical processes.

One need not necessarily overthrow this model and replace it with another (Trench, 2008). However, greater recognition of its deployment, along with associated limits and constraints, will enrich opportunities for greater meaning-making (see Chapter 8 for more). The deficit model logic will undoubtedly persist in many relevant expert knowledge brokers you encounter. Henry Ko (2016) has described this condition as "many stakeholders dancing together in a multidirectional tango of communication" (p. 427). It can be overcome by reconsidering the role of "authority" among relevant experts engaged in creative (climate) communications. This more capably takes place when scientists and decision-makers commit themselves to new modes of interaction leading to social change. It can be overcome by considering the scientist as a citizen rather than separate from a perceived public citizenry (Meyer, 2016) (see Chapter 6 for more).

In other words, telling human stories about humans (e.g., about people affected by climate change and how they're responding to it), not necessarily scientific ones, can engage with new audiences and enhance possibilities for collective action. For example, the October 2018 IPCC Special Report on 1.5°C provided many opportunities for creative communications about climate change (Allen et al., 2018). This report sought to articulate the need for more aggressive efforts to meet representative concentration pathways that keep global average temperatures below 1.5°C in the next decades. Part of the backstory here is that in the UN meeting in Bonn, Germany – six months before the Paris meeting in December 2015 – a report was released stating that 2°C was not adequate to avoid serious impacts of warming. This articulation was prompted by pressure from many parties to the UN such as the Alliance of Small Island States (AOSIS) and fishing and ranching communities, who warned that 2°C of warming threatened to inundate and potentially eliminate their lives, livelihoods and place-based cultural histories. The Bonn report led then to the Paris Agreement statement to keep temperatures below 2°C but to "pursue efforts towards 1.5°C" (Allen et al., 2018). That statement then initiated a two-and-a-half-year process, incorporating more than 6,000 peer-reviewed studies, more than 42,000 comments on drafts and 91 authors from 44 countries that resulted in the Special Report.

The warning was stark: to avoid passing 1.5°C, emissions must drop 45% from 2010 levels by 2030, and must reach "net zero" by 2050 (incorporating "negative emissions technologies" involving carbon dioxide removal and solar radiation management). This all hands on deck pronouncement contained many stark warnings, such as no more coal by 2030 (in Chapter 2).

However, in the report acknowledgments of the role of many carbon-based industry groups as barriers to action were missing. These resistances have manifested through misinformation campaigns, often through communications via CCM organizations and contrarian voices (Oreskes and Conway, 2011; Farrell, 2016a; Supran and Oreskes, 2017; Brulle, 2018b). The report under-emphasized how asymmetrical power of carbon-based industry circulates through climate-related knowledge communities to produce not only discourses, but also particular institutional constellations, and (dominant) practices of knowledge production. This was a lost opportunity to connect some of the dots regarding challenges in staying below the articulated 1.5°C temperature target.

Overall, the blunt assessment generated media coverage but was largely met with mild public concern or uproar. On the heels of the IPCC Special Report on 1.5°C release, the satirical media outlet *The Onion* joked about this ongoing situation. It reported,

> Growing increasingly panicked as the American public remained oblivious to their efforts, numerous terrified climate scientists had resorted to frantically waving their arms while loudly begging to be acknowledged by throngs of passers-by who proceeded to walk straight through them, sources confirmed Tuesday. "Can anyone hear me? Hello?" shouted the discouraged climate scientists as they futilely jumped in front of pedestrians in desperate attempts to block their paths, only to have the people continue on down the block without so much as a glance in their direction. "Please, wait! Stop! Don't you see we're on a path to destruction? Hey! Come on, won't you please stop and listen to me for just a second?"[19]

With all the work put into the thorough and rigorous scientific report, the communications plan appeared to be more of an end-of-pipe fitting and thus had limited impact in the public arena. It then proved to be an opportunity lost for the IPCC to find more creative ways to reach public audiences through

---

[19] The article was titled "'Can Anyone Hear Me?' Shout Terrified Climate Scientists Frantically Waving Arms As Passersby Walk Straight Through Them" and can be found here: www.theonion.com/can-anyone-hear-me-shout-terrified-climate-scientist-1829652646

resonant and meaningful strategies where communication integrates insights gained from social sciences and humanities. For instance, a storytelling co-production could have told human stories from AOSIS members and others at the forefront of climate impacts. This could have animated warning such as that global annual catches for marine fisheries could decline by about 1.5 million metric tons for 1.5°C of global warming compared to a loss of more than 3 million metric tons for 2°C of global warming, and that livestock will be adversely affected with each tenth-of-a-degree of rising temperatures (Allen et al., 2018).

Staying within these "safe operating spaces" of communication merely through scientific information provided opportunities for contrarian voices to dismiss the report as ideologically infused and disingenuously motivated. For example, US Senator James Inhofe (R – OK) was quoted as dismissing the report as "the same old IPCC" (Sobczyk, 2018).

Instead, what was told through media accounts in the days following the report release were scientific ways of knowing about climate change by way of relevant-expert voices. Rather than resulting from a shortfall of creativity, this has been a function of limited capacity: despite having heard these critiques in the past, the IPCC communications team lacks the capacity to more ambitiously tackle these communications and engagement challenges. Consequently (as I opened Chapter 1), even alongside high-quality and well-funded scientific research into the causes and consequences of climate change, climate communications – and conversations about climate change in our lives – have remained stuck.

Richard Black (2015) has commented that "IPCC assessments present an unparalleled opportunity for climate science to speak directly to power" (p. 282). But he has also called for less jargon-laden summaries because the "summaries for policy makers" are still alienating for their intended audiences (Black, 2015). Silke Beck and Martin Mahony (2017) have called for IPCC reforms to fit the contemporary communications environment. In 2016, the IPCC organized a meeting in Oslo, Norway with the objective of improving efficacy and impact of communications (Mateu, 2016). However, in the words of long-time contributor Linda Mearns regarding readability of IPCC reports, "The IPCC is difficult to change. It's kind of like a very large ocean liner. It's not easy ... to steer and change course quickly" (Hijri, 2016). Perhaps as part of that slow change, Adam Corner and colleagues published a report in 2018 listing six principles for IPCC authors to use in public engagement. Among them, "be a confident communicator," "talk about real world, not abstract issues" and "connect with what matters to your audience" (Corner et al., 2018).

Going forward into the next Special Reports on oceans and the cryosphere (in 2019) and climate change and land (2020) as well as the IPCC Sixth Assessment Report (in 2021), IPCC communications could be fortified through co-produced and adequately funded, creative, smart and innovative ways to meet everyday people where they are on this critical (and existential) issue. Devoting more resources to sustained integration of social science research (along with findings from professionals, science communicators) can extend from "Summaries for Policy Makers" (SPMs) to Summaries for Everyday People (SEPs).[20] At present, "Carbon Brief" led by Leo Hickman most closely fits this call for an effective "science explainer" in the public sphere.[21] Based in the UK, "Carbon Brief" is an online news outlet covering climate and energy science and policy. Leo Hickman (2015) himself has noted that IPCC transmissions move too slowly for present-day rapid social media communications. He therefore has called for more authoritative boldness in order to remain influential. Yet the capacity for prima facie legitimacy and influence of the IPCC provides heightened potential for a SEPs endeavor. With a Sixth Assessment Report from the IPCC on the horizon in 2021, SEPs added to SPMs and Technical Reports could then provide path-ways for greater integration and deeper engagement with many different audiences beyond prototypical scientists and policy makers.[22]

In the meantime, information-deficit model approaches to communications about climate change amount to twentieth-century approaches to twenty-first-century challenges. Among the implications of such approaches are continued attention paid to embedded antienvironmental and neoliberal contrarian CCM groups. This therefore means continued influence in the US public arena. Adherence to the logic of the information-deficit model amidst the high-profile, highly contentious science, policy and politics around climate change in the USA has enabled contrarians to gain increased footholds in battles for public understanding and engagement regarding the causes and consequences of climate change.

There is the enduring notion (or hope) that these CCM groups will eventually be pushed to the fringes and become irrelevant along with their outlier views (Boykoff and Farrell, 2019). However, ongoing research in the social sciences

---

[20] Despite gestures/intentions/some substantive moves to improve IPCC communications, even the SPM texts remain jargon-laden and alienating texts even for policy makers themselves (e.g., members of the US Congress).

[21] More about Carbon Brief can be found here: www.carbonbrief.org/

[22] That said, these SPMs still likely remain relatively impenetrable for the typical "policy maker," so many policy makers could also benefit from SEPs. Recently, a friend asked me, "Where is the best place to get the cold hard facts about climate change?" It is an instructive example of questions that we all have as we try to make sense of these complex issues. SEPs could step into this role effectively, as SPMs remain relatively alienating for anyone but the most astute policy maker.

suggests otherwise. These amplified views are a reflection of contemporary cultural politics, and they will not disappear anytime soon (Lewandowsky et al., 2012). In particular, analyses through a lens of cultural politics helps to consider ways that claims and claims makers influence public discourse (see Chapter 2 for more). In these spaces, there are dynamic and real-time negotiations of boundaries between those who constitute "authorized" speakers (and who do not). Also, considerations of who are legitimate "claims-makers" are consistently being interrogated, and challenged (Gieryn, 1999).

Conclusions

In 2017, Reddit online commenters shared reasons why they changed their mind about climate change, moving effectively from the domain called climate contrarianism to acceptance of mainstream science. First-hand experience of abnormal weather patterns, questioning of contrarian credibility, affinity for stewardship ethics and scientific evidence appeared to all play prominent roles in shifts described in the 645 posts (Cho, 2017).[23] For example, one post read, "The past 3 or 4 years the weather has just totally been bizarre . . . it's gotten to the point where something just blatantly feels wrong about it." Another read, "I started thinking about it on my own and realized that everyone who was a 'denier' had a vested financial interest in ignoring the problems of fossil fuels." Among the nuggets of insights gained from this flurry of postings, Karin Kirk (2017) advised, "Be nice: none of the commenters lauded the effectiveness of flaming arguments, shaming or condescending treatment."

A feature by Livia Albeck-Ripka (2018) in the *New York Times* profiled how six people from various walks of life changed their minds about climate change. Among them she portrayed a meteorologist from Paducah, Kentucky – Jennifer Rukavina – who changed her mind after examining the scientific evidence herself and hearing directly from relevant expert scientists. She also portrayed evangelical leader Richard Cizik from Fredericksburg, Virginia who changed his mind after pressure from his congregants (Albeck-Ripka, 2018).

Along with these commenters, there have been a number of high-profile shifts from contrarianism for varying reasons. For example, Jerry Taylor – former Cato Institute Fellow and leading libertarian voice on energy issues – moved from contrarianism through his questions regarding fellow contrarian

---

[23] To read more, go to www.reddit.com/r/AskReddit/comments/5zvuxx/ former_climate_change_deniers_what_changed_your/

Figure 3.3 This flyer from Fueling U.S. Forward was circulated at 2016 gospel concerts in black churches; the group also co-sponsored the 2016 National Black Political Convention. Other Koch-funded organizations seeking to "meet people where they are" include the Libre Initiative focusing on Latino communities and Generation Opportunity focusing on millennials.

credibility along with his reexamination of the scientific evidence (Baptiste, 2017). As another example, renowned science and environmental filmmaker David Attenborough credited Ralph Ciccerone – then the National Academy of Sciences president – for delivering a rousing talk in Belgium in 2004 that convinced Attenborough of the case for anthropogenic climate change (Randerson, 2007).

Yet, these endeavors to "meet people where they are" have not been limited to those who seek to engage the more reticent among us about connections between human behavior and climate change. For example, the public relations group Fueling U.S. Forward – funded by petrochemical industry barons Charles and David Koch – has been working to "start winning hearts and minds" of minority groups, in the words of spokesman Alex Fitzsimmons (see Figure 3.3). The mission of Fueling U.S. Forward was a dedication "to educating the public about the value and potential of American energy, the vast majority of which comes from fossil fuels" (Tabuchi, 2017a, p. A3).

Matthias Lievens and Anneleen Kenis (2018) have commented, "If climate denialists have one merit, it is that they have forced us to re-think what it means to conduct (natural) science, and how science relates to the political, amongst other things" (p. 12). To this, I add a second: they have forced climate communicators to be creative and not to assume prima facie authority over (scientific) evidence as it meets public awareness and engagement; and they have shown how this focus on authority through science has squandered other opportunities for engagement.

In current times, where can we find spaces of dialogue and engagement that overcome the logic of the information-deficit model and sidestep political economic, cultural and ideological barriers to effective communication?

How about social media? Ashley Anderson and Heidi Huntington (2017) have actually argued that "social media discussions of science have the potential to detract from healthy public engagement with science" (p. 614). Moreover, Ashley Anderson (2017) has pointed out social media discussions can "provide space for framing climate change sceptically and activating those with a sceptical perspective of climate change" (p. 496). Studying Facebook, Emma Bloomfield and Denise Tillery (2018) found that climate contrarianism arguments espoused through "Watts Up with That" and "Global Warming Policy Forum" pages gained perceptions of legitimacy through systematic reposting and hyperlinking, "thus establishing a supportive, networked space among other skeptical sites, while distancing readers from original sources of scientific information" (p. 1).

Meanwhile, social science and humanities research into online behaviors and communication about climate change in the USA, the UK, Germany, India and Switzerland revealed useful insights. Among them, study authors Stefanie Walter, Michael Brüggeman and Sven Engesser (2017) found that users mainly stuck with their referent groups, forming insulated conversations rather than engagement across different social, political and cultural perspectives. They explained that these media comment sections thereby "serve as echo chambers rather than as corrective mechanisms" and consequently when "climate scep-tical readers find information that is consistent with their own beliefs . . . [it] hence gives them the impression that their opinion is the prevalent one in society" (pp. 213–214). Moreover, contrarian organizations have found that by financially bidding on search terms like "climate change" on Google then raises the profile of their content in search engine results (Tabuchi, 2017c). Regarding the toxicity of fake comments in internet comments sections and chat rooms, Brian Chen (2018) has commented that "there's not much you can do" and "the real leverage lies with the tech companies" (p. B7).

Can we look to traditional sources of media and news? In research into these arenas as potentially effective channels for climate communications, Todd Newman, Eric Nisbet and Matt Nisbet (2018) found that people "not only choose news outlets where they expect to find culturally congruent arguments about climate change, but they also selectively process the arguments they encounter" (p. 1). They concluded, "the partisan cleavage in climate perceptions are more likely a factor of whether individuals see some groups more or less equal than others (i.e., Hierarchical–Egalitarian values) than whether they are focused on what is better for the individual versus what is better for society as whole (i.e., Individualistic–Communitarian values)" (p. 13). They then determined that news media portrayals of climate change generally fortified rather than contested partisan identities along with underlying cultural world-views and conceptions of well-functioning societies (Newman et al., 2018).

In related research in the US context, Michael Tesler (2018) found that news outlet choices are "the strongest predictor of conservatives' climate change scepticism" (p. 306). He also found that, in turn, "The prevalence of ideological cues strongly affects public opinion about global warming" (p. 306). In other words, Tesler characterized media representations of climate change as a partisan issue that then played a part in cueing Republican policy actors to take up that stance; these together then influenced public perceptions of political polarization and partisanship on climate change in the US context.[24] Tesler (2018) concluded, "Ideological elites can dominate public opinion if and when scientific facts prove to be inconvenient to their interests . . . [and] it will be difficult to mobilize public opinion in an effort to curb the Earth's rising temperatures so long as ideological elites continue to cast doubts about climate science" (p. 323). These findings cohere with forerunning research by Jason Carmichael and Robert Brulle (2017). They found that in the US context elite cues have significantly influenced climate change discourse in the public arena over time, where media attention to partisan politics (particularly in the US Congress) increases public concern about climate change (Carmichael and Brulle, 2017).

How about longer-form media? For example, Seth Darling and Douglas Sisterson's book *How to Change Minds about Our Changing Climate* (2014) worked to overcome prevailing skepticism and end arguments "for good" as they "equip readers with the tools to distinguish fact from fiction, to see through the smoke and mirrors, and to understand what needs to be done to address climate change and why. You don't need a degree in science to understand the basic principles of climate change, but you do need to have some facts straight" (p. x). They also point out, "This isn't about pointing fingers. It's about identifying and understanding the problem and, more important [sic], taking action to do something about it" (p. xii). As another example, in 2018, Nathaniel Rich authored an influential long-form article called "Losing

---

[24] Tesler also examined how skepticism regarding evolution may have also followed these trends but he found no evidence that evolution beliefs were related to political discourse.

Earth" in the *New York Times Magazine*, describing how politicians, public actors and scientists sounded an alarm about climate change between 1979 and 1989. However, in his rich storytelling an absence of confronting cultures of consumption as well as the political economy of climate contrarianism drew strong critiques from natural and social scientists who have studied these influences (e.g., Colman, 2018; Leber, 2018a). Though this piece captured public attention through its scientific ways of knowing about climate change, such an approach can be enriched further through engagement with other ways of knowing as well.

Going forward, there are appropriate (and inappropriate) times and places for certain approaches to climate communication, amid varying audiences and contexts. Matt Nisbet (2018a) has cautioned that we have been blinded by an obsessive focus on climate contrarianism, writing, "as a scholarly community we have become obsessed with research intended to expose the faults in conservative psychology, the duplicitous nature of fossil fuel companies, and the many ways in which *Fox News* and right-wing think tanks seed 'denial,' and engage in a 'war on science.'" Sven Ove Hansson (2018) has similarly argued, "Climate science should primarily be presented to the public in ways that are independent of denialist activities, rather than reactively in response to those activities" (p. 5). Research by Adam Corner, Olga Roberts and Agathe Pellisier (2014) found that 18- to 25-year-old UK participants "were either unaware or uninterested in the idea of organised climate change scepticism, suggesting that campaigns to counteract science-based scepticism will not be particularly useful for this audience" (p. 5).

In fact, when media copiously cover and dissect President Trump's contrarian talking points (e.g., Waldman, 2018c), such attention can be argued to displace other more productive considerations, which then fall into a discursive moat around the bully pulpit. The Media and Climate Change Observatory (MeCCO) has called this a "Trump Dump," where media attention that would have focused on other climate-related events and issues instead was placed on Trump-related actions (leaving many other stories untold).[25]

When prompted to provide information in an interview, when sharing scientific facts in a public lecture, when delivering expert testimony in a hearing, filling knowledge gaps is important. Back to the IKEA example mentioned earlier: after all what are IKEA representatives supposed to do – go

---

[25] I am a part of the MeCCO research team as well; more about MeCCO can be found here: https:// sciencepolicy.colorado.edu/icecaps/research/media_coverage/index.html

to your house and dialogue with you while you consider the directions? No. The directions play an important role, but they do not lead consumers to always build that bookshelf perfectly.

As such, considering creative climate communications again, there is no "silver bullet" that creates active and sustained behavioral engagement; rather "silver buckshot" approaches to engagement through many pathways to learning and knowing about climate change are needed (see Chapter 7 for more). In this chapter I have interrogated the state of play of contrarian social networks and their effects – from individual attitudes to larger organizational and financial flows – in the US context, commonly referred to as the "belly of the beast" in terms of carbon-based industry power and political, societal and cultural polarization.

Without confrontation of the information-deficit model of communication as I have argued here, we are likely to remain encased in mere naming and shaming, in comfortable and ineffective blaming of the other. Efficacy of CCM organizations in the public arena can be attributed in part to their exploitation of the dominant information-deficit model. This model has effectively helped contrarians to find traction in battles for public understanding and engagement about the causes and consequences of climate change. While some point to the CCM organizations and connected individuals as the fossils of climate science, politics and policy, the fossils among us are those who fail to note that their strength is derived in part from our collective adherence to an information-deficit model view of communications.

Confronting this model (and meeting people where they are through awareness of audience perspectives) can diminish the power of contrarian outliers and can weaken the cultural schism mentioned in Chapter 2. Moving from deficit model logic and toward a more interactive, conversational and normative theory of expertise model forces a movement away from talking past one another along with cultures of naming and shaming and toward listening to one another through enhanced understanding of burden sharing. This movement pulls attention away from outlier perspectives and toward a "movable middle" (Moser and Berzonsky, 2015) (see Chapter 5 for more). It pushes away from a penchant for proving others wrong and "gotcha" mentalities and pulls toward a predilection for commonality and community mentalities.

Movement to a more dialogical model exploring shared values through open and respectful conversations involves a nimbleness and capacity to adapt to changing communications (and listening) demands and conditions. Authentically considering other points of view fosters meaningful exchanges and enhances possibilities for finding common ground. Facts established through scientific ways of knowing about climate change are important, but they are not enough. We therefore need to enlarge considerations of how knowledge influences actions, through experiential, emotional, visceral, tactile, tangible, affective and aesthetic ways of learning and knowing about climate change. This is a challenge confronted in the next chapter.

4

Ways of Learning, Ways of Knowing

In February 2007, the United States Congressional Committee on Science and Technology[1] held a hearing on the United Nations Intergovernmental Panel on Climate Change (IPCC) Fourth Assessment Report on the science of climate change (Solomon et al., 2007). This UN IPCC report had been published the week before and it immediately sparked discussion of its aggregation of findings that warming is "unequivocal" and that human activities have "very likely" been the main driver since 1950.

At the hearing, Republican Congressman Dana Rohrabacher questioned the report's authors about a period of dramatic climate change that took place 55 million years ago. In a rhetorical closing statement, Rohrabacher said, "We don't know what those other cycles were caused by in the past. Could be dinosaur flatulence, you know, or who knows? We do know the CO_2 in the past had its time when it was greater as well. And what happened when the CO_2 was greater since then and now? There have been many cycles of up and down warming." While some chuckled at what Rohrabacher later claimed was a joke that went awry, others crowed at the powerful misperception regarding what caused past climate change and how that is relevant to contemporary conditions. Part of the uncomfortable laughter may also have been attributed to Rohrabacher's influential climate contrarian stance on many climate science–policy issues addressed in the US Congress since 1989 (see Chapter 3 for more on climate contrarians).

Within these "multiplicities" of interpretation hanging in the balance (Emmerson, 2017), the tittering provided insights into the context-dependent and oft-fragile state of highly politicized climate discourse in the public sphere. Reflecting on the many responses to his comments years later, Rohrabacher

---

[1] This committee is now called the "The Committee on Science, Space and Technology," as a result of 112th Congress Chairman Ralph Hall's intervention in the 2011–2013 session.

conceded, "You have to be very careful when you're using humor" (Weigel, 2017).

Rohrabacher's foray into these arenas demonstrated that our pathways to knowing about climate change do not flow merely through scientific findings (see Chapter 3 on the information-deficit model). Everyday people (and elected officials too!) typically do not learn about dimensions of climate change by reading peer-reviewed literature, whether it is the latest IPCC report or new research from the journals *Science* or *Nature*. Further, most people do not learn about climate science and policy by attending lectures or attending college classes. Everyday engagement with climate change is rarely sparked by the latest scientific research or the persuasive oration from relevant-expert climate researchers. Instead, people make links between formal science and policy and their everyday lives by exposure (from relevant experts, influential actors, public intellectuals, talking heads, pundits, etc.) through a range of resonant and relevant communications and experiences. Such are our shared and complex subpolitical spaces in present times (Lemke, 2002).

There are many ways we may learn and know about climate change: experientially, viscerally, emotionally, affectively, tangibly and aesthetically. These pathways to knowing then elicit a range of responses. In fact, research by Emma Thomas, Craig McGarty and Kenneth Mavor (2009) has found that anger as an emotion can motivate particular segments of the public to confront social and political justice issues. They attribute this in part to social norming that then induces sustained commitments (Thomas et al., 2009).

There can sometimes be a fraught relationship between the concepts of knowing and belief. Public polling on climate change often asks people about whether they "believe" in climate change or not. But scholars such as Katharine Hayhoe have pointed out that this framing can imply a sort of religious faith that the climate is changing rather than a reliance on scientific facts that shape this knowledge. She has commented, "Science does not require belief for it to be true." Framing this as a question of "belief" can then promote evidence of climate change as dependent on whether you personally accept it or not, rather than dependence on evidence of climate change. Uncritical dependence on "belief" framing can be misleading in communications about climate change (Smith, 2015) (see Chapter 5 for more on framing).

These different ways of knowing feed our decision-making and choices from an everyday scale to long-term intergenerational considerations. Yes, these ways of knowing and learning may be informed by scientific ways of knowing about climate change. But while science is often privileged as the dominant way by which climate change is thought to be articulated, public understanding and engagement are embedded within a matrix of cultural, social, political and economic processes that make climate change meaningful in our everyday lives (O'Connor, 1999; Zaval, 2016). Commenting on the importance of experiential knowledge, David Harvey (1973) once noted, "I can find no other way of accomplishing what I set out to do or of understanding what has to be understood" (p. 17). Research by Dave Gosselin, Sara Cooper, Sydney Lawton, Ronald Bonnstetter and Bill Bonnstetter (2016) found evidence to support experiential learning as a pathway for knowledge retention. By tapping into these complementary ways of knowing, one can more effectively develop strategies of effective and creative climate communications.

In this chapter, I highlight some of these multiple pathways. In the context of contemporary climate change there is a need to better understand how to effectively harness the power of resonant communications and creativity to confront what works where, when and why in climate change discourse. Tapping into these complementary ways of knowing can more practically develop strategies of effective communications. I then focus on humor as a vehicle for creative climate communications, drawing on research conducted with University of Colorado Theatre professor Beth Osnes. Better understanding and accounting for these interacting factors can enhance our effectiveness in our ongoing communication endeavors.

Daniel Kahneman is a Princeton University psychology and public affairs professor. Among his many contributions over the past five decades, along with Amos Tversky he examined how biases shape judgment and decision-making (e.g., Tversky and Kahneman, 1973). In his 2011 book *Thinking Fast and Slow*, Kahneman (2011) wrote about the negotiation between intuitive/instinctive (fast thinking) and deliberative/contemplative (slow thinking) that shape our mental processing about climate change. He wrote that when confronted with issues such as climate change, "The machinery of intuitive thought does the best it can ... [but] the spontaneous search for an intuitive solution sometimes fails ... in such cases we often find ourselves switching to a slower, more deliberative and effortful form of thinking" (p. 13). In other words, we grapple with dimensions of climate change through different ways of mental processing.[2]

Recognizing these many ways of knowing about an issue such as climate change can help enrich strategies and tactics for effectively connecting with selected audiences in resonant ways. Lisa Zaval and James Cornwell (2017) have commented, "Time and again, rational deliberations about climate change are frequently overridden by faster, more intuitive processes based on associations, emotions, and rules of thumb" (p. 655). For example, local weather may

---

[2] These are also called System I (fast) and System II, as well as Automatic (fast) and Reflective (slow) (Sunstein and Thaler, 2008).

drive perceptions of climate change, as a form of "cognitive myopia" (Li et al., 2011). In fact, back in the late 1980s, a pattern of heat waves, fires and drought was found to have sensitized the general public to the threats of climate change more than scientific findings emerging at that time (Demeritt, 2001, p. 307) (see Chapter 3 for more).

Seeing (and Hearing) Is Believing (and Knowing)

Often times, climate communication pathways are charted through imagery. Imagery can include things like graphics, figures, charts, maps, logos, video games, cartoons, advertisements, photographs and films. And by extension, use of virtual reality is increasingly drawing on imagery to help understand climate change and risk management (e.g., Suarez, 2017). These endeavors can grab attention, promote comprehension, create awareness, change beliefs and reshape intentions, perspectives, reasoning and behavior (Hansen and Machin, 2008; Anderson, 2009; Moser, 2010; O'Neill, 2017).[3] They can also activate people's intuitive as well as rational and conscious senses.

Knowing, or "belief" about something through visual evidence, manifests in a variety of ways. For example, photographs capturing the size of the 2017 US presidential inauguration crowds – compared and contrasted with 2009 crowds – became a battleground of visual perceptions. At stake was the Donald J. Trump administration claims that the 2017 crowds were "the largest audience to ever witness an inauguration" (Hunt, 2017). However, the photos provided by the US National Parks Service (NPS) told a different story (see Figure 4.1). As follow-up investigations revealed, then-White House press secretary Sean Spicer and President Trump himself spoke independently with NPS director Michael Reynolds the morning after the event in 2017. From there, Reynolds then directed an NPS communications official that "President Trump wanted to see pictures that appeared to depict more spectators in the crowd" (Anapol, 2018a). Some photographs were then cropped to make the crowd sizes look bigger (Swaine, 2018).

In terms of visual representations of climate change, part of the challenge involves capturing not just an often invisible biophysical set of phenomena (Doyle, 2007), but also the political, human and cultural dimensions of climate change (Schneider and Nocke, 2017). In his book *What We Think About When We Try Not to Think About Global Warming*, Per Espen Stoknes (2015) noted, "The climate issue remains remote for the majority of us ... we can't see

---

[3] Among Jane Piirto's seven I's of creativity (see Chapter 1) is "imagery."

Figure 4.1 US National Parks Service photographs of the National Mall in Washington, DC during the 2009 inauguration of President Barack Obama (left) and the 2017 inauguration of President Donald Trump (right).

climate change. Melting glaciers are far away . . . and the heaviest impacts are far off in time . . . despite some people stating that global warming is here now, it still feels distant from everyday concerns" (p. 82).

However, the film *Chasing Ice* directed by Jeff Orlowski successfully represented climate change through the dramatic phase changes of melting ice and snow at the faraway polar regions of the planet. The 2012 documentary tracked renowned photographer James Balog and his Extreme Ice Survey team across Greenland, Alaska and Iceland as they captured time-lapse images and videos portraying real-time disappearing glaciers and impacts of global warming.

While texts are often privileged as primary means of climate communication, images (Doyle, 2007), visual arts and visual imagery have been found to enhance learning (O'Neill and Smith, 2015; Wozniak et al., 2015; Lewandowsky and Whitmarsh, 2018). Anthony Leiserowitz and Nicholas Smith (2017) have noted that affective imagery "plays an important role in shaping public risk perceptions, policy support and broader responses to climate change and other threats" (p. 14). Such considerations have motivated social science and humanities research into what works, how, why and under what conditions. For example, Alex Lockwood (2016) studied visual forms of affective imagery, focusing on expressions of "planetary boundaries" work by Johan Rockström et al. (2009) as well as the film *Cowspiracy* (Anderson and Kuhn, 2014). He sought for his work to better understand "the pivotal role of emotions within environmental communication" (p. 745). As another example, Cristina Ruiz, Rosario Marrero and

Bernardo Hernández (2018) examined how affect influenced public acceptance of an oil drilling project in the Canary Islands. Through a survey of residents on the seven islands, they found that negative emotions pervaded resistance to the project plan.

Julie Doyle (2007) has pointed out that images provide powerful ways to "bear witness'" to climate change (p. 131). But interpretation from this can certainly vary. Sander van der Linden, Anthony Leiserowitz, Geoffrey Feinberg and Ed Maibach (2014) tested how to depict scientific consensus that humans contribute to climate change (see Chapter 5 for more) through visual means. They concluded that the most effective communication approach was a "short, simple message that is easy to comprehend and remember" (p. 255). Meanwhile, Brendan Nyhan and Jason Reifler (2018) have found that charts and graphs "significantly decrease false and unsupported beliefs" (p. 1).

Saffron O'Neill and Sophie Nicholson-Cole (2009) empirically tested how images influence awareness and efficacy. Through their first study in the UK, they found that dramatic and fearful representations can successfully raise awareness but that these kinds of images were "also likely to distance or disengage individuals from climate change, tending to render them feeling helpless and overwhelmed when they try to comprehend their own relationship with the issue" (p. 375).

In follow-up research, Saffron O'Neill, Simon Niemeyer, Sophie Day, and I (2013) interrogated these trends in the Australian and US contexts. Results were remarkably consistent across these countries, indicating the potential "presence of a dominant, mainstream discourse around climate imagery" (p. 413). These results were also consistent with those of the previous study, in that "Imagery plays a role in either increasing the sense of importance of the issue of climate change (saliency), or in promotion feelings of being able to do something about it (efficacy) – but few, if any, images seem to do both" (p. 420).

Part of the considerations of climate imagery include data visualization, where climate science communication can find traction in climate policy and decision-making (Zhao, 2017; Boehnert, 2018). In one study, researchers Valentina Bosetti, Elke Weber, Loïc Berger, David Budescu, Ning Liu and Massimo Tavoni (2017) asked more than 200 policy actors involved in the 2015 Paris climate negotiations to predict how much global temperatures would increase over a finite time period, both before and after seeing a graphical representation that communicated modeling of this future. They found that even though these negotiators showed signs of hanging onto prior beliefs even in the face of incongruous visual evidence, "The gap between initial beliefs and scientific evidence can be partially reduced by using an adequate presentation format" (pp. 188–189).

One of the most well-known data visualizations is a depiction of temperatures in the Northern Hemisphere from AD 1000 to 2000 by Michael Mann, Raymond Bradley and Malcolm Hughes. The representation focused attention on departures from an "average" established by the mean temperature from 1961 to 1990 (Mann et al., 1998). As this figure was reproduced in the IPCC Third Assessment Report Summary for Policymakers (McCarthy et al., 2001), it became known as the "hockey stick" graph because the shaft of the hockey stick (temperatures from 1000 to 1950) give way to an upward blade (temperatures from 1950 to 2000) (see Figure 4.2).

In research on climate change visuals, Jordan Harold, Irene Lorenzoni, Thomas Shipley and Kenny Coventry (2016) have worked to improve accessibility of graphics for nonexperts. They examined connections among text and visuals, attention to visual representations, complexity of visuals and inferences made from visual portrayals and argued that "intuitive design does not equal improved accessibility" and "accessibility does not equal loss of scientific rigour" (p. 1082). Perhaps the most powerful example of this is the well-known Apollo 8 photo "Earth Rise," taken in December 1968. This was the first view that the general public had of planet Earth in its entirety (Cosgrove, 1994).

However, Julie Doyle (2007) has cautioned that "not all environmental problems can be seen" (p. 147). Doyle has argued that images depict "'what is' or 'what has been' rather than 'what may be'" (Doyle, 2007, p. 280). She has therefore argued that by the time climate change can be captured visually – a calving iceberg, a drowning polar bear – it might be too late to adequately address these problems. Considered in this way, images are consistently chasing time by forcing many invisible features of climate change to be expressed visibly. By relying on "picturing the clima(c)tic," we collectively capitulate into a more reactive stance vis-à-vis climate threats, as "photography cannot visualize the future as a present threat" (Doyle, 2007, p. 294). Related to these limitations, in a study of climate change imagery used in Spain's television news, Bienvenido León and Carmen Erviti (2015) found that a shortage of "attractive images is one possible explanation for why climate change is not covered more extensively" (p. 14).

Furthermore, the ways in which images are deployed can serve to stymy rather than promote engagement in the public sphere. Lynda Walsh (2015) has posited, "Habitual ways of visualizing climate change works against, not for, effective political action" (p. 361). Moreover, Birgit Schneider and Thomas Nocke (2014) have argued that climate change can be successfully visualized but this can still come up short in terms of imagining and fostering engagement that is needed. In fact, James Fleming (2014) has asserted that climate change is actually "unimaginable" (p. 345), in that the full extent of climate impacts and changes are perceived to be impossible to fully see and then fathom.

Anders Hansen and David Machin (2008) examined how the "Green Issues" Getty Images collection – part of a US$1 billion a year industry in which companies can buy images for their advertising, web pages and news reporting – effectively depoliticized and decontextualized climate change. They concluded that through a visual process of making climate change an abstracted and "generic" issue, the images effectively contributed to a "commercial appropriation ... [with] the effect of promoting greater consumption" (p. 792). In his *Society of the Spectacle*, Guy Debord interrogated how icons and images have been mobilized in representations of contemporary challenges. Debord warned that in the process of reconciling imagery and capitalism in the way that Hansen and Machin illustrated, "the real consumer becomes a consumer of illusions. The commodity is the factually real illusion, and the spectacle is its general manifestation" (Debord, 1983, p. 47) (see Chapter 5 for more).

Climate imagery can provide powerful ways of knowing about climate change, but not in any particularly prescriptive set of ways. Instead, images are "co-constructors of environmental narratives that, in combination, convey complex and multi-dimensional messages" (DiFrancesco and Young, 2011, p. 520). In his seminal 2012 book *Visualizing Climate Change*, through a focus on landscapes and community mapping Stephen Sheppard explored how to make carbon and climate change visible in our everyday, and in our immediate surroundings in order to help build awareness and detonate engagement and action. He concluded the book with a rhetorical question and answer: "Can imagining such future realities [of climate change] with the help of visual imagery and visioning techniques really get people to enact climate change solutions? Perhaps not by itself, but it can certainly help. If people can learn to see things differently, if they have more compelling and informative ways to envision and plan their future and pressure their politicians, it can only be a good thing" (Sheppard, 2012, p. 468).

A less studied learning and communication pathway on climate change is through sound, from sonification of data sets to music interpretation. Auditory ways of knowing are most often considered through music, from the resource intensity of music production (e.g., Bottrill, 2010) to the communication of environmental topics through sonification and music (e.g., Pedelty, 2011). For example, Radiohead frontman Thom Yorke's song "Hands Off the Antarctic" gave voice and raised attention to ongoing Antarctic protection efforts through a twenty-five-government Commission for the Conservation of Antarctic

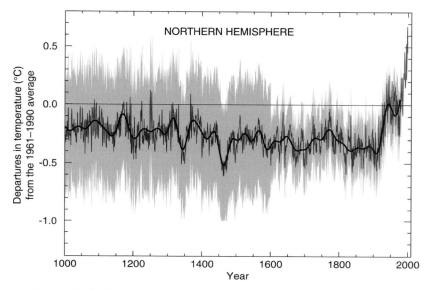

Figure 4.2 The "hockey stick" graph by Michael Mann, Raymond Bradley and Malcolm Hughes (1998).

Marine Living Resources (Breyer, 2018). A great deal of analyses that have been done on music's influence on social, political, cultural and environmental issues have been conducted through the disciplines of cultural studies and geography (e.g., Goodman, 2010; Kanngieser, 2015; Comstock and Hocks, 2016).

To date, sonification is, however, an underexplored research endeavor. As an exception to this general "climate silence" (to borrow a term) to date within the environmental communication literature, Harriet Hawkins and Anja Kanngieser (2017) considered how "sonic art" along with visual representations can "help in understanding, imagining, and even bringing about alternative climate relations and futures" (p. 3). They found that these approaches activated emotional and experiential responses in productive ways, overcoming key "issues of insensibility, abstractions and distancings" (Hawkins and Kanngieser, 2018, p. 8). As a second example, Josh Wodak (2018) has explored how music provides "a soundtrack that narrates the rapidity of contemporary biophysical change ... where music affords otherwise unobtainable engagement with environmental themes" (p. 2). He commented, "Perhaps the youth generations coming of age in this epoch ... will write a new soundtrack of popular music as novel as the geological era they were born into" (pp. 11–12). In terms of experimentation and practice, long-time journalist Andrew

Revkin's pursuits of climate communication became intertwined with his musical interests through his album entitled "A Very Fine Line."[4]

It is important to remember that when deployed effectively, these audio and visual pathways – from graphs and figures to music and films –can reach low-literate or illiterate audiences (often members of vulnerable communities) in ways that texts clearly cannot (Maes, 2017). So continued investment in better understanding what is simple, robust and accessible is wise and worthwhile. As such, there is no silver bullet image for communicating about climate change; rather there are many ways to engage through imagery, with various effects. Much experimentation and reworking (as the section heading here suggests) is warranted.

Laughing in the Face of Climate Change

In recent years of apparent saturation of somber and science-led climate change discussions, comedy and humor – as a pathway to emotional, affective visceral and experiential ways of knowing – are increasingly looked to as potentially useful vehicles to meet people where they are on climate change (e.g., Anderson and Becker, 2018). While much public discourse on the topic of climate change has relied primarily on scientific ways of knowing over the past decades, prominent culturally resonant framings have focused on climate change as a "threat" with associated doom and gloom, psychological duress (O'Neill et al., 2013; Clayton et al., 2015). For example, in October 2018 journalists Chris Mooney and Brady Dennis of the *Washington Post* wrote about the work by the IPCC in October 2018 to publish a special report on the potential for keeping warming to an aspirational 1.5°C temperature target noted in the Paris Climate Agreement. The headline read "Climate scientists are struggling to find the right words for very bad news" (Mooney and Dennis, 2018). Similarly, Frédéric Simon (2018) penned an article in the days that followed entitled "'Bad news' and 'despair': Global carbon emissions to hit new record in 2018, IEA says." This may feel sobering and overwhelming, leading many to increasingly ambivalent states by these sad events, developments and activities.

But among other possible pathways of knowing about climate change, humor is generally underutilized; yet comedy has power to connect people, information, ideas and new ways of thinking/acting. Comedy brings to the fore multiple truths and ways of knowing, in its oft-deployed delight in the

---

[4] See https://dotearth.blogs.nytimes.com/2013/03/03/lifes-very-fine-lines/ for more.

multiple meanings of single ideas, statements or even words. Comedy can then exploit cracks in arguments, wiggle in, poke, prod and make nuisance to draw attention to that which is incongruous, hypocritical, false, or pretentious (Berlant and Ngai, 2017). Comedy wields power to destabilize and threaten fundamentalist thought and practice through more nuanced and conditional interpretations of truth (e.g., Osnes, 2008) (see Chapter 1 for more about "truth").

Power flows through comedy to enable movement between "authorized" and revealed alternatives and can seed fertile locations for subversion, resistance and liberation as it opens up additional dimensions of understanding of the world (Foucault, 1984). In contrast to brash imposition of law or mandate, comedy and humor – and by extension, laughter – have the power to reconstitute subtle power-knowledge regimes that permeate and create what becomes "permissible" and "normal" as well as "desired" in everyday discourses, practices and institutional processes (Foucault, 1975).

For example, the subtle destabilizing power of laughter was on display during US president Donald J. Trump's speech at the United Nations General Assembly in New York in September 2018. In his speech, the president boasted of his achievements to date, declaring that his administration "has accomplished more than almost any administration in the history of any country." However, on the floor of the full UN hall, his comments were not met with admiration from the assembled audience. Instead, they prompted chuckling and laughter. This reaction clearly surprised and destabilized the president. After the laughing subsided, he somewhat sheepishly responded "I didn't expect that reaction, but that's okay."[5]

In this landscape, a number of scholars and practitioners have argued that humor has the potential to productively spark awareness and engagement for people across political, cultural and social arenas on important social issues such as climate change (e.g., Chattoo, 2017). In a study of political humor, Jenny Davis, Tony Love and Gemma Killen (2018) found that humorous content achieved "disproportionate attention" in the public sphere (p. 3193). Comedian Shane Mauss has connected these worlds through humorous approaches to asking questions about science (Young, 2016).

The Laughter Report: The [Serious] Role of Comedy in Social Change, led by Caty Borum Chattoo and published by the Center for Media and Social Impact at American University, reviewed research and findings across multiple disciplines in an effort to understand the potential impact of humor on social

---

[5] See https://video.foxnews.com/v/video-embed.html?video_id=5839826114001 for the footage of this portion of the president's speech.

change. The report highlighted multiple examples of positive social change resulting from humor initiatives. For example, a John Oliver episode of *Last Week Tonight* in 2015 is credited by many as "having influenced New York City officials to change their city's bail protocol," which Oliver asserted was previously used to lock up the poor, even when their guilt was not proven (Chatoo, 2017, pp. 4–5).

Michel Foucault (1975) has written, "It is not the activity of the subject of knowledge that produces a corpus of knowledge, useful or resistant to power, but power-knowledge, the processes and struggles that transverse it and of which it is made up, that determines the forms of possible domains of knowledge" (pp. 27–28). In other words, the dynamism and nonlinearity that comedy can bring to discourse and awareness thereby provide potential sites of powerful resistance and legibility amid adversity.

Smuggled in through the oft-unsettling and uneasy contours of comedy and humor, one can analyze the extent to which new discourses and framings could take shape through relationships of power, and where knowledge and meaning arise through discursive struggle (Hall, 1988). Insights from Michel Foucault can help shed light on the interactions of power and knowledge at the human–environment interface, through conceptions of biopower (Van Assche et al., 2017). In addition, Ben Anderson's (2017) treatment of affect, along with additional literatures in environmental communication and emotional and affective geographies (e.g., Negri, 1999; Davidson et al., 2012), provides further insights. With these tools of analysis, comedy and humor can be interrogated regarding how power with/through people is exerted to both connect and distract and to shape new ways of thinking/acting about anthropogenic climate change. More widely, these approaches help consider how humor influences experiential, visceral, emotional and aesthetic ways of knowing about climate change. As such, power *within* rather than power *over* the process of and effects from comedic communications of climate change can be examined. The dimensions of Foucaultian power course through the veins of a shared social body at particular times and places (Thrift, 2004) to comprise "a politics of life" (Anderson, 2012).

Comedy has had the power to influence public thought, understanding and behavior over time. By examining the addition of humor to the climate change communication mix, one can understand how power has been and can be harnessed to positively contribute to effective climate communication. It is useful to explore how the historic roots of comedy have shaped contemporary uses, and analyze how comedic approaches can uniquely contribute to ways of knowing and understanding through both theory and practice of climate change communication.

The term "comedy" is derived from an amalgamation of the Greek words *komos* or *komai* and *oda*. *Komos* translates as "revel," while *komai* comes from the word for "village" (Stott, 2005). As such, the essence of comedy involves shared experiences of levity and fun. In classical Greek society from 400 to 300 BC, Plato (through the *Republic*) and Aristotle (through *Poetics*) both treated comedy as the profane, as a lower form of expression. They did not dismiss comedy as art but they treated it an approach from a different vantage point than from Greek tragedy (Weitz, 2009). Aristotle (1996) argued, "The laughable is a species of what is disgraceful" (p. 9).

Moreover, in 411 BC the Greek playwright Aristophanes first presented what is called "Old Comedy" through *Lysistrata*: a play about the far-fetched idea that women could end the Peloponnesian War by refusing sex to their men. In the comedy, this premise leads to all sorts of antics. Yet the social commentary throughout is of heavy consequence while Aristophanes puts forth a seemingly absurd solution to effectively expose the far greater absurdity of war.

Old Comedy was seen as distinct from "New Comedy," taking on themes and issues less consequential to civic life. "New Comedy" became known primarily from the substantial papyrus fragments of Menander around 300 BC. Menander's comedies tended to be more about the fears and foibles in the domestic scene, including personal relationships, family life and social mishaps rather than politics, philosophy and public life. Taken together, comedy from the classical Greek and Athenian period significantly informed humorous expression from Rome, Europe, North America and beyond. Though there are historically many forms of comedy from which comedic traditions in the United States are influenced – such as the West African griot who served traditionally as storyteller, social commentator, satirist and historian – the formal divisions of comedy inherited from ancient classical Greece tend to dominate the thinking and practice of comedy (Appiah, 2016).

Perhaps it is through these early bimodal roots that comedy uncomfortably maps onto climate change discourse in the twenty-first century. Seen as a collective action problem, climate change is an issue filled with many deep-seated paradoxes, making it a topic that exposes both the frivolity and failings of the human race. Examples include the paradox that as greater (scientific) understanding improves, rather than settling all associated questions, the process unearths new and more questions to be answered. Common and shared adversity alongside innovations in climate communication research has led some to turn to comedy to creatively confront the multifarious threat of contemporary climate change.

In the name of lost *ethos*, Philip Smith and Nicolas Howe (2015) have interpreted Aristotle to posit that comedic irony is not a good communication strategy for climate change. They have consequently argued, "We must speak seriously of serious things" (p. 203) and therefore pegged deficiencies of the accomplishments of comedic characters like the Yes Men, Reverend Billy and others as failing in their humorous messaging. Smith and Howe have postulated, "Words like 'prank', 'hoax', 'imposter' and 'fake' invariably accompany accounts of their actions, and . . . such terms do not line up easily with those that Aristotelian theory predicts are more helpful, such as 'trust', 'transparency' and 'goodwill'" (p. 204).

However, in line with many other scholars and practitioners, many still feel that the risks are worth the potential analytical rewards. From this perspective, climate change communications through the vehicle of comedy have the potential to increase salience of climate change, potentially exposing audiences to new ways of learning about associated threats, challenges and opportunities. Those who may not otherwise pay attention to climate change find pathways to engage with it through the comedy of Larry Wilmore (former host of *The Nightly Show* featuring a panel including Bill Nye the Science Guy to discuss presidential candidates' stances on climate), Seth Myers (*Late Night* host who occasionally has targeted US federal administration stances on climate change), Samantha Bee (who reports on *Full Frontal* about news regarding climate denial), and former host Jon Stewart and current host Trevor Noah of *The Daily Show* and former *Colbert Report* host Stephen Colbert (now host of *Late Night with Stephen Colbert*), who have all engaged with occasional climate change-related segments. And, they may remember it better while sharing it with others.

For example, Sarah Silverman took time during her 2018 Hulu show *I Love You America* to address the need for climate action. In her monologue, she focused on how climate change is mainly fuelled "by the interests of a very small group and absurdly rich and powerful people." Silverman then observed, "We cannot give up," adding "the disgusting irony of all of it is that the billionaires who have created this global atrocity are going to be the ones to survive it. They are going to be fine while we all cook to death in a planet-sized hot car." As another example, in his 2008 mockumentary film *Sizzle: A Global Warming Comedy*, Randy Olson sought to make people laugh while addressing pertinent and vital themes of climate skepticism and alarmism. Casting himself as the lead protagonist, *Sizzle* was a film about Randy's fictitious quest to make a documentary about global warming.

Moreover, like the auditory and visual pathways I mentioned earlier, comedic approaches offer new routes to "knowing" about climate change, overcoming often sober or gloomy scientific assessments through experiential, narrative, emotive and relatable storytelling (see Chapter 2 for more). In addition, humorous treatments can help increase one's accessibility to the complex and often-distant dimensions of climate change while bringing a long-term set of issues into the immediate social context. While comedy can provide relief amid anxiety-producing evidence as an emotional salve and tool for coping,[6] it also serves to bridge difficult topics and overcome polarized discussions through entertaining and nonthreatening ways in order to recapture a missing middle ground.

Through multiple comedic pathways – from the satirist mocking subjects from a superior position to the humorist who stands arm and arm with its subject affectionately to the comedian who stands below as a victim of the system who can self-deprecatingly punch up – comedy and humor ask questions and therefore exert power to create new ways of considering climate change. As a high-stakes, high-profile and highly politicized challenge in the new millennium, climate change has the potential to overwhelm everyday people. Comedic approaches help to alleviate these feelings and can make these issues more approachable and manageable. Comedy can make chunks of climate information more digestible through compelling stories, making communications more palatable and easier to swallow.

Comedic approaches offer potential to shepherd in new pathways of knowing through experiential, emotional and aesthetic learning. Climate change comedy and humor can:

- increase salience of climate change issues (exposing audiences to new insights);
- offer new routes to "knowing" about climate change (through experience/emotion);
- help increase accessibility of a complex, often-distant, long-term set of issues;
- engage new audiences;
- increase retention of climate change information through effective storytelling;
- provide relief amid anxiety-producing evidence of causes and consequences of climate change; and
- bridge difficult topics, overcoming polarized discussions through often entertaining/nonthreatening ways.

Past studies of comedy and climate change have examined these potentialities in varying contexts. These have focused largely on television programming.

---

[6] Sally Weintrobe has pointed to anxiety as a leading impediment to climate change engagement (2013). See also work by Angela Mauss-Hanke, Johannes Lehtonen and Jukka Välimäki for more (2013).

Lauren Feldman (2013) studied comedic reports on climate contrarianism on the popular US programs *The Daily Show* and *The Colbert Report* and found the programs were able to more effectively question dissenters (also commonly called "contrarians" or "deniers") in ways that an 'objective' journalist could not. In follow-up work, Feldman (2017)considered these findings in the context of larger considerations of potential misreading of satirical intent, of the capacity of satire to communicate the often-serious issues within science.

Ailise Bulfin (2017) explored the role of ecological catastrophe narratives in current popular US culture through US programs such as the comedic sit-com *Last Man on Earth*. In a separate study of *The Daily Show* and *The Colbert Report*, Paul Brewer and Jessica McKnight (2015) argued, "Satirical television news may provide an alternative route for influencing public perceptions of climate change by presenting information in an entertaining format that draws otherwise unengaged viewers" (p. 17). Brewer and McKnight (2017) have also analyzed the US show *Last Week Tonight* with John Oliver and found that watching particular segments about climate change increased viewers' beliefs about global warming.

Through a study of BBC programming, Joe Smith (2017) argued that comedic approaches provided exceptional openings to consider links between sustainability, material consumption and climate change . Robin Nabi, Abel Gustafson and Risa Jensen (2018) found that emotional experienced enhanced climate change policy advocacy among the experimental study group partici-pants of US undergraduate students. Furthermore, through an experimental study of sarcastic humor Ashley Anderson and Amy Becker (2018) found that the levity of video communications in *The Onion* served to raise beliefs that the climate is changing and heightened perceived risk of climate change among respondents who did not previously believe climate change was a serious issue.

Adeline Johns-Putra (2016) found that these new communication avenues – evident particularly through UK dramatist John Godber's 2007 *The Crown Prince* – have widened audience engagement on climate change. Similarly, Gravey et al. (2017) found that comedy fostered enhanced learning on climate change and sustainability topics. In analogous work to address poverty and environmental degradation through climate adaptation, Sreeja Nair (2016) found that participatory comedy skits exhibited potential to enhance social learning.

However, not all studies to date have pointed clearly to comedy and climate change as an effective pairing that shapes attitudes about climate change. In one study, Christofer Skurka et al. (2018) found that humorous video communica-tions produced greater intentions to take action on climate change, but did not heighten risk perceptions. In a study of knowledge and attitudes on climate change among university students, B. Elijah Carter and Jason Wiles (2016) found that comedic approaches did not enhance learning and engagement. They surveyed undergraduate students about their attitudes and opinions regarding climate change after watching one of two videos: one, an authoritative educational film from the IPCC; the other, a comedic video about climate change by Jon Stewart, Stephen Colbert or John Oliver. They wrote, "Despite our expectations about differential effects between IPCC and comedy videos, little difference was observed. No group was significantly more or less likely to change their opinion about climate change" (p. 17).

In social science and humanities studies of the efficacy of comedy for effective communication beyond climate change, some research has shown that perceptions of distant threats may impede expressions of concern and engagement on various social topics (e.g. McGraw et al., 2012, 2014). Moreover, research by Peter McGraw and colleagues has shown that positive emotions can actually serve to inhibit the sense that something is wrong and needs to be addressed through problem-solving behaviors (McGraw et al., 2015). These findings are tethered to links among distancing effects, affect and risk perception (Johnson and Tversky, 1983). A theory of "benign violations" – that something is wrong yet nonthreatening – developed by Peter McGraw and Caleb Warren (2010) has helped to illuminate how the vehicles of humor and comedy possess power to find traction in public discourse.

For example, soon after the release of the IPCC Special Report on 1.5°C (see Chapter 3 for more), in a segment on *The Daily Show*, Trevor Noah observed, "You know the crazy people you see in the streets shouting that the world is ending? Turns out, they're all actually climate scientists." As another example, on his ABC show *Jimmy Kimmel Live!* comedian Jimmy Kimmel commented, "There's always a silver lining. One planet's calamity is another planet's shopportunity." He then cut to a spoof going-out-of-business advertisement for planet Earth that included statements like "Everything must go! 50% of all nocturnal animals, insects, reptiles and amphibians . . . priced to sell before we live in hell. But you must act fast because planet Earth is over soon. And when it's gone, it's gone." Cutting back to Jimmy Kimmel, he embodies the theory of benign violations as he then comments, "It's funny because it's true." However, a deeper consideration is that this violation may not be too benign after all.

Through studies of humor and comedy over time, it is clear that context and content both matter to the efficacy of humorists-as-messengers (or "claims makers") and to their messages (or "claims"). And these elements then connect to issues of trust (see Chapter 1 for more). For example, when asked why so many people question whether the climate is changing, comedian Lewis Black responded, "I don't know if it is that many people. But it's enough of them that it is irritating . . . I don't believe it is up for argument anymore, it is real."[7]

Jokes emerge from complex and dynamic subpolitical spaces and from historically contingent social frameworks. Anthropologist Mary Douglas (1975) – in her work on joking relationships in traditional cultures – has pointed out, "All jokes are expressive of the social situations in which they occur. The one social condition necessary for a joke to be enjoyed is that the social group in which it is received should develop the formal characteristics of a 'told' joke: this dominant pattern of relation is challenged by another" (p. 98).

To illustrate, on *The Late Show with Stephen Colbert* on July 28, 2017 host Stephen Colbert assembled a short skit called "Al Gore's climate change pick up lines."[8] This was inspired by Gore's visit to the show earlier that week promoting his film *An Inconvenient Sequel* when Gore called it a "hot date movie." With the funny fueled in large part by the efficacy of the claims makers (over the claims themselves), Colbert and Gore take turns offering pick-up lines in front of a pastel backdrop and sultry slow-jam background music. As examples, Gore offered, "Are you climate change? Because when I look at you, the world disappears." Colbert chimed in, "I'm like 97% of scientists, and I can't deny . . . it's getting hot in here." Then Colbert joked, "Is that an iceberg the size of Delaware breaking off the Antarctic ice shelf, or are you just happy to see me?" Gore followed, "I hope you're not powered by fossil fuels, because you've been running through my mind all day."

In William Shakespeare's 1605–1606 play *King Lear*, the fool is one who mocks the king of Britain's misfortune. He is one who is an early comic entertainer, someone who had a license to speak penetrating truths to the king about his choices and behavior. The fool was a valuable character in that it served kings well to have someone nearby who dared to provide honest feed-back, and he did it wisely through folly and effectively through vehicles of humor and wit. The name "fool" – while seemingly unfortunate in the context of contemporary parlance – could be also seen as a manner of protection for the comedian, given the age-old excuse when getting dangerously too close to a painful truth that "it was only a joke." As such, playing "the fool" can be a clever devise for diffusing the situation or denying intentionality or serious intent. Even the comedian Jon Stewart has been referred to by himself and others as a contemporary court jester (Leopold, 2015).

Jokers themselves draw on power from the immunity of wit, as these comedic acts carry insight into a deeper consciousness. Humor and comedy

---

[7] View the clip here: https://climatecrocks.com/2018/09/17/lewis-black-on-climate-change/amp/
[8] The segment can be viewed here: www.youtube.com/watch?v=FCXxT94NJmA

then have the potential to generate emotion and, in turn, leverage power to enable movement between "authorized" and revealed alternatives. Thus, their expressions cultivate fertile locations for subversion, resistance and liberation as they open up additional dimensions of understanding of the world. Dan Chapman, Brian Lickel and Ezra Markowitz (2017) have emphasized that context matters in determining how emotional ways of knowing shape awareness and engagement. As such, they have noted, "At a practical level, targeting specific emotional reactions in an effort to promote productive engagement with climate change is unlikely to produce consistent and predictable effects because few if any messages can be designed to produce the same emotional response in all people" (p. 852). Ultimately, responses are driven by perceived and targeted audiences (see Chapter 1 for more) as well as what is relevant to them (Graves, 2018).

Through capillaries of power that manage the conditions of our shared everyday lives, it is worth examining how comedic communications of climate change mobilize power and produce as well as contest discursive trends. From that one can then analyze how these approaches can effectively facilitate radical reimagining and (re)engagement of attitudes, beliefs, perspectives and behaviors as they relate to climate change. Power mobilized through comedy is potentially power to question carbon-based industry interests and their mouthpieces, and power to destabilize status quo cultural perceptions of consumption and environmental impact.

Stand Up for Climate Change

Along with colleagues Beth Osnes and Rebecca Safran, we have pursued experiential, visceral, affective and emotional pathways to knowing as they map onto potentially effective climate change communications. We took up these considerations through what we called the "Stand Up for Climate Change" comedy and climate change project (Boykoff and Osnes, 2019). In this project, we focused mainly on what many might deem an absurd solution: that comedy may serve as one effective way to meet the seemingly impossible task of avoiding the worst impacts of climate change. Yet remembering the power that political comedy (Greek Old Comedy) once held, it was a way we substantiated the contemporary experiment.

Part live performance, part international video competition, this became a case study of how comedy and humor could influence perspectives, attitudes, intentions, beliefs and behaviors on climate change. To better understand these dynamic interactions, we interrogated how productive this comedic experiment was for both the participants and the audiences in terms of climate communication. We were then able to gain some insights regarding how comedy – as a multimodal communication pathway – is able to meet people where they are on climate change.

The "Stand Up for Climate Change" initiative took place over from 2016 to 2018, as part of the Inside the Greenhouse (ITG) project at the University of Colorado (USA) (see the Preface for more about ITG) (see Figure 4.3). Live performances of sketch comedy, improvisation and stand up featured primarily students creating and performing stand up and sketch comedy for an audience composed largely of their peers.[9] Video entries were submitted from around the world.[10] We embarked on this endeavor through motivations to better understand comedic and humorous sites as "ways of knowing" to then analyze how power flows through discourses and actions the public arena. The initiative grew from ongoing efforts to find ways to connect with different audiences to make climate change more relevant and meaningful through humor and for the students to experience comedy as a viable mode of climate communication.

To analyze the efficacy of these efforts, we conducted surveys of participants and audience members and supplemented these findings through content analysis of the performances as well as participant observation throughout the process.[11] These approaches are multimodal, where modes are open-ended and multifaceted systems through which meaning is communicated in the spirit of finding ways to connect with different audiences to make climate change more relevant and meaningful (see Chapter 1 for more).

The Stand Up for Climate video competitions were held in 2016, 2017 and 2018. Across the three competitions, thirty-three entries were received from nine different countries. Calls for entries were circulated through numerous email listservs, message boards and social media outlets. The announcement noted that successful entries will be those that can "find the funny" while relating to climate change issues in less than three minutes. Entries that were produced within the previous calendar year to the deadline were considered. Panels of faculty, graduate and undergraduate students at the University of

[9] Navigating an audience that varied from peers to minors and older adults was a challenge for content creators.
[10] More information on the video competition can be found here: http://insidethegreenhouse .colorado.edu/news/winners-announced-2016-comedy-climate-change-video-competition and here: http://insidethegreenhouse.colorado.edu/node/2017 while more information about the live performances can be found here: http://insidethegreenhouse.org/project/comedy-climate-change
[11] Further details about the methods in this study can be found in Boykoff, M. and Osnes, B. (2018). A Laughing Matter? Confronting Climate Change through Humor. *Political Geography*, https://doi.org/10.1016/j.polgeo.2018.09.006

Colorado then ranked the submissions each year and winners were determined from an equally weighted ranked pool (see Figure 4.4).

As examples, Jeremy Hoffman from Oregon State University in the United States won third prize in the 2016 contest with a piece called "The Sound of Skeptics." This was a satirical parody to the tune of Simon and Garfunkel's 1964 classic "The Sound of Silence." Hoffman described his motivation behind the composition as a creative way to approach "the struggle of the climate science community in dealing with the increasingly loud but remarkably small population of 'climate change skeptics' that willingly deny the impacts of anthropogenic CO_2 emissions on global climate change." In 2017, the composition "The Summit" by Giovanni Fusetti and Tejopala Rawls from Australia won first prize. This was a piece in which nine performers dressed in formal suits acted as delegates to "this country" and "that country" in ongoing international climate negotiations. They satirically debate about terminology and action at the ocean's edge while the waters rise around them. While arguing about "multilateral" and "bilateral," and "committee" and "sub-committee," they eventually are silenced by the enveloping waters just as "the chairman" calls for a vote. The final text reads, "It doesn't have to go like this . . . it is time for action."[12] In 2018, Madeleine Finlay and Sarah Barfield Marks from the United Kingdom won first prize with their creative take on "Peer Review" in which they pointed out the contrasts and confusion between relevant expert reviews and the court of public opinion.

The comedy shows were held in 2016 and 2017. These were a mix of live performances with some preproduced video compositions.[13] Both performances occurred in spaces that each accommodated approximately 150 people. The majority of the student participants across both events were Environmental Studies majors with little to no experience performing comedy. This lack of experience posed a particular challenge, as many theatre professionals deem comedy to be the hardest to master (and the most obviously exposed when done poorly). Other major areas of study included Astronomy, Atmospheric and Ocean Sciences, Communication, Ecology and Evolutionary Biology, Geography, International Affairs, Journalism, Political Science, Sociology, Theater and Dance. In these two performances, participants chose either to perform individually (in each case performing stand up) (see Figure 4.5) or in small groups of two to five people (performing stand up, improv or sketch

[12] www.insidethegreenhouse.org/short-films?field_sub_category_tid=Allandcombine=summitand field_age_range_value=Allandfield_length_value=
[13] These preproduced videos included the contest winning entries along with some previous Inside the Greenhouse works, and some additional outside contributions.

Figure 4.3 Scenes from "Stand Up for Climate Change" performances at the ATLAS Black Box Experimental Studio in the Center for Media, Arts and Performance (left), and at the Old Main Chapel (right) at the University of Colorado.

comedy). In year one, participants devised compositions on any theme or topic they chose. Each composition just needed to use comedy as a communication vehicle about a particular dimension or set of dimensions related to climate change. In year two, participants created works that related to the theme of communicating humanity's relationship with energy and climate change. In both years, all participants were required to carefully consider and articulate who was their imagined/intended primary audience and what was the principle message they sought to communicate through each composition.

As an example of a multimodal performance piece, two students, Trevor Bishop and Tanner Biglione, created an Academy Awards skit, cutting to the portion of the event when the award for Best Picture is announced. The skit began as the ceremony played a video segment of a fictitious film called *Wild Pollution* in which a protagonist is out in the wild catching and corralling various forms of garbage and waste (e.g., cans, paper, glass). The lights were then raised and the live emcee announced the winning film. Drawing on the February 2017 Academy Awards mishap where *La La Land* was mistakenly awarded Best Picture before the organizers corrected the announcement, the live emcee performed a similar mishap before then naming the fictitious film *Wild Pollution* the winner of Best Picture. Trevor and Tanner – the fictitious director and star of the film – then went onstage to accept the award. Trevor exclaimed, "Thank you, thank you, we understand we deserve this significantly more than any of the other films submitted" and Tanner followed with "I mean what can I say, this is an incredible honor and we really just gave the people what they wanted . . . TRASH." Trevor then continued, "We would like to thank our President for removing any kind of environmental regulations, making this documentary possible" and Tanner finished by saying, "We would also like to thank the people. Without your gross negligence, we would not have been able

Figure 4.4 Selected video competition entries: 2016 winner "Weathergirl Goes Rogue" by Heather Libby [Canada] (left); 2017 winner "The Summit" by Giovanni Fussetti and Tejopala Rawls [Australia] (center); 2018 winner "Peer Review" by Madeleine Finlay and Sarah Barfield Marks [United Kingdom] (right).

to observe these pollutants in their unnatural habitats." The two then exited the stage to crowd applause. Through this exaggerated composition, the multi-modal piece sought to communicate a serious message about the consequence

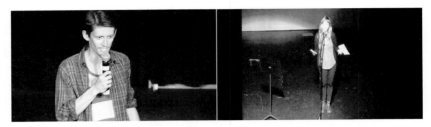

Figure 4.5 Nick Spencer performing stand up at the 2017 event (left); Lauren Gifford performing stand up at the 2016 event (right).

Figure 4.6 Jeremy Stein, Curtis Beulter, Caitlin Lizza and Garrett Hernandez-Rimer performing "The Presidential D(eb)ating Games" in 2016 (left); a scene from "Wild Earth" by Trevor Bishop and Tanner Biglione in 2017 (right).

of low regulatory environments matched with cultural consent through a humorous case of wild trash (see Figure 4.6).

Participants prepared for these performances by discussing contemporary peer-reviewed materials about climate change communication, by completing performance-based exercises in the months preceding the event and through conversations with visitors who shared varied expertise in communication.[14] For example, Carrie Howard, Jane Saltzman and Ian Gibbs from the Front DeRanged Improv Comedy Troup visited the group to share skills and approaches for improvisation. The performance-based exercises consisted of vocal and physical exercises and activities to release tension, increase their artistic commitment, expand their expressive range, exercise imagination and loosen inhibitions. To study improv (short for improvisation) as a part of this

---

[14] Visitors included Professor Max Liboiron (Department of Geography, Memorial University of Newfoundland), Professor Peter McGraw (Psychology and Marketing, University of Colorado), Professor David Poulson (Knight Center for Environmental Journalism, Michigan State University), Professor Zoe Donaldson (Department of Molecular, Cellular and Developmental Biology, University of Colorado) and Lauren Gifford (Department of Geography, University of Colorado).

experience is to work to overcome conditioned responses and behaviors. The difficulty in learning improv is largely the difficulty in letting go, having an open mind and embodying a willingness to try out new realities among other people. It takes training and rehearsal to develop skills, control, confidence, mental agility, trust and spontaneity that help with improvisation (Atkins, 1994). As an example, one exercise began with all would-be performers standing in one large circle. Someone began the activity by holding and then throwing an imaginary ball to another participant. That second participant then would catch, meld and shape the ball, providing information about the ball's size, weight, bounciness, even texture and smell. That participant would then throw the ball to another in the large circle. Embellishment of catching, feeling and throwing the imaginary ball was encouraged throughout. This activity was designed to help participants move from cerebral to performative spaces and to encourage and enhance commitment and creativity through mindful performance.

Finding the Funny

From these performance activities, participants then assembled a collaborative list of "what was funny" from their actions. (In exploring these areas of communication, it is notable that analyses of the funny are not inherently funny in and of themselves.) Prominent were an acknowledgment that exaggeration, full commitment, the introduction of ridiculous ideas into an otherwise logical world, suspense, surprise, clever recognition of truths, imitation (e.g., the human as mechanical), honesty, timing, incongruity, absurdity and specificity all played parts in constructing effective comedy.

Participants in each event completed postevent surveys. Questions involved asking participants to describe in detail: (1) risks and benefits they encountered in utilizing comedy for climate change (from both pre- and postevent perspectives), (2) a moment that stood out to them from the performance, (3) how the use of comedy for climate communication can be most effective in these settings or in others and (4) what techniques proved useful for climate communications through comedy. Each of these questions solicited open-ended responses. In addition, audience surveys were administered immediately after each performance. Attendees were asked how much comedy might have succeeded in making them think, feel and engage with climate change, whether the use of comedy seemed to trivialize issues associated with climate change and if they felt that comedy could make a serious contribution to ongoing climate change conversations. Together, these surveys provided insights into the practice-based research involved in these participatory activities, by identifying and describing successful and unsuccessful techniques[15] deployed to achieve stated objectives of climate communication through comedy.

Present-day comedic approaches to sensitive issues in politics, environment, culture and society have consistently found traction through creative comedy communications. These approaches have shown themselves to productively question, challenge and point out things that audiences may not have noticed, or may have overlooked. In context, these approaches have also drawn from experiential and emotional ways of knowing to point out hypocrisy and hubris.

For example, at the 2017 Academy Awards host (and comedian) Jimmy Kimmel articulated some angst about the 2016 presidential election as it related to continuing challenges of race in the United States. In a concise yet piercing observation in his introductory comments on stage, he commented, "I want to say thank you to President Trump. I mean remember last year when it seemed like the Oscars were racist? It has been an amazing year for movies. Black people saved NASA and white people saved jazz. That's what you call pro-gress." In these jokes, Kimmel effectively made explicit some racial under-currents that have been flowing through cultural practices in the entertainment industry and in society. He also drew attention to absurdities associated with surface-level interpretations of progress on race, *not* necessarily demonstrated by what the academy had chosen to congratulate as a top film. Furthermore, the consistent selection of a comedian to emcee these events points to their dual value in providing entertainment but also to holding "stars" accountable through comedic truth-telling. This point was not lost on our students, espe-cially the two who created the skit entitled "Wild Pollution" which utilized the fictional setting of the Academy Awards to humorously expose the conse-quence of the weakening of environmental regulations.

These threads of context-dependent honesty, timing, exaggeration, surprise, clever recognition of truths and full commitment all compelled the efforts in the "Stand Up for Climate Change" initiative forward. We wanted to design an experience in which students could experiment with ways of finding traction for the communication of climate issues through comedic approaches. Moreover, the capacity- and confidence-building dimensions of the enterprise added further texture to the endeavor. Early in the process, the challenges of combining comedy and climate change were abundantly clear. A number of students voiced concern that the remit of comedic delivery on climate-related content was as easy as mixing oil and water, where it was very difficult to make

---

[15] Technique here is defined as a way of carrying out a particular task, especially the execution of performance of an artistic work or a scientific procedure, a skillful or efficient way of doing or achieving something. In the field of performance, identifying techniques is one manner in which new knowledge contributes to the fields of interest and involvement/investment.

something so serious also funny. They also expressed anxiety about the challenge to effectively connect on climate change through humor. These expressions drew on the complex affective/emotional costs and benefits that can be associated with acting or thinking outside one's comfort zone.

Adrian Chappell (2006) has found productive levels of anxiety to be "a highly desirable and necessary hallmark of independent and self-motivated individuals" (p. 26); however, David Bissell (2008) has pointed to nuances of (dis)comfort as they relate to creativity and capabilities. With these complexities in mind, a participant reported, "This project took me out of my comfort zone" while another recounted the "anxiety of performing live." However, that same participant then related a sense of satisfaction and "boost of confidence speaking in front of large groups . . . it honestly doesn't get much scarier than doing stand-up in terms of anxiety around social performances." These comments are consistent with findings from Ben Spatz (2015) regarding successful knowledge generation through techniques and practices of embodiment. Furthermore, participant and audience responses, along with observations of the process, pointed to the capabilities of humor to provide relief as an embodied and effective coping practice amid an otherwise distressing set of considerations (McCormack, 2003).

While six participants performed stand-up comedy, the other seventy-six live performers performed elements of improv, situational comedy and sketch comedy.

A subset of these performances found familiarity and resonance through popular televisions shows. For example, in 2016, Jeremy Stein, Curtis Beulter, Garrett Hernandez-Rimer and Caitlin Lizza created and performed "The Presidential D(eb)ating Games," playing off the popular US 1970s/1980s game show *The Dating Game* . In this sketch, performers played the roles of Hillary Clinton, Bernie Sanders and Donald Trump arguing about climate and environmental topics in an Oxford-style debating format while they vied for a date with Mother Nature. The student performing as Bernie Sanders cited his reason for being chosen as the high albedo of his gray hair that reflected incoming short radiation back out to space to prevent the Earth from warming. Here the tension surrounding the 2016 US presidential election provided a fertile context for humor. Also in 2016, "Always Sunny in Boulder, Colorado" by Sean Christie, Emily Buzek, Clarissa Coburn and Alex Posen played on the *Always Sunny in Philadelphia* sitcom but drawing on climate change themes. In 2017, *The Bachelorette* by Andrew Taylor-Shaut, Yue Li, Gustaf Brorsson, Maggie Patton, Hannah Higgins and Enric Sabadell anthropomorphized coal, oil, natural gas and the sun vying for a long-term relationship with Mother Earth.

Subversion and Sedation

Participant survey feedback consistently recognized the risks of trivializing a critically important set of issues. These survey responses also consistently pointed to the central importance of audience and context when considering whether compositions were successful. One participant reported, "It showed me how fun climate communication can be, which helps to reframe the whole conversation in a way that feels more manageable," while another shared that comedy "made it easier to bring up the subject of climate change without being depressing." Another participant wrote that this "different form of climate change communication allows communicators to reach broader, otherwise disinterested audiences."

As such, many of the participants chose to focus on social cultures of partying and celebration. For example, "Teach It 'n Preach It" in 2016 by Elana Selinger, Alaire Davis, Greg Chancellor and Blake Ahnell portrayed a conversation about climate impacts while socializing one evening in a friend's apartment. Based on what the audience laughed at during the scene, the humor appeared to be context dependent and stemmed from the honest portrayal of their daily lives in which one person lamented about the impact on global warming of melting snows as two others entered wearing full snowboard gear carrying their boards. They were told by the climate-worried roommate that their beloved snow had melted due to increased temperatures on the Earth's surface. In another performance that same year called "Party on the Hill" [referring to a neighborhood called University Hill in Boulder Colorado] creators Blaine Hartman, Tommy Casey, Meagan Webber and Ashley Seaward performed different archetypal characters – a "frat guy," a "hippy," a "Starbucks girl" and an environmentally conscious student. They discussed environmental awareness and (dis)engagement, along with their different priorities for environmental conservation and stewardship in clever and context-sensitive ways. The conservative student chastised the hippy, saying, "Dude, you're higher than sea-level rises," cleverly chiding his friend while alluding to the problem of global warming contributing to rising sea-levels. In 2017, "A Greenie in a Greensuit" by Clinton Taylor, Heddie Hall, Tori Gray and Forrest Dickinson portrayed a situation in which two University of Colorado friends are visiting two other friends at another university on St. Patrick's Day. They are at a college party playing drinking games and debating varying campus cultures of environmental (un)consciousness. The humor derived from the exaggerated costuming of the two visiting environmentally minded CU students (where that is a stereotype) who were dressed in full, skin-tight, green outfits. Another source of the humor was their full commitment in wearing these ridiculous costumes while engaging in otherwise perfunctory conversation.

Through the process – and products in the show – audience feedback through survey data, the general atmosphere on the night and further feedback from the students after the event all indicated that the experiment was seen to largely prove uncomfortable at times, but ultimately productive. One participant commented, "It helped me not only become a better climate communicator, but also built my confidence in the academic and social realms." Another recounted, "Never in my whole life have I been so nervous and stressed out . . . and yet, I've never walked away from a presentation or in a class event feeling as proud of myself as I did [when completing this work] . . . " Another participant shared, "The project has made me think more openly about the many ways people can establish common ground between one another surrounding important issues such as climate change." An audience survey respondent commented, "In the realm of comedy and satire, it seems that regardless of your personal biases or political affiliations, anyone who doesn't take themselves too seriously can participate and appreciate the art form. In other words, the approachability of humor transcends . . . barriers because of the humble pretence that funny is *funny.*" Another audience survey reported, "It is a unique and seemingly unconventional way to reach out to people about this issue."

Many performances found traction through familiarity of popular US television game shows. For example, Tiana Wilson, Alec Nimkoff, Joseph Meyer and Reghan Gillman played off the popular show *Jeopardy!* to perform 'Climate Jeopardy' with climate change answers and questions. When asked what was a change that starts with the letter "C," contestant Hillary Clinton responded, "Campaign. My campaign." This content cleverly alluded to the accusation that she changed her mind on key issues within her campaign and that she was so obsessed with her own campaign. Another example was the futuristic "2050 Price Is Right!" by Edwin Chambers Zachary Lautmann, Katelynne Knight and Jennifer Stodgell, where contestants "Jen," "Dom" and "Billy" competed for prizes. Part of the humor was derived from the timing, having the year be 2050 and the surprise of learning that their "beach" holiday they had won would be in the now warm and sunny shores of Alaska.

The "Stand Up for Climate Change" initiative wrestled with Peter McGraw's humor code navigating between the systematic science and the elusive art of comedy (McGraw and Warner, 2014). Particularly with climate change – one of the most polarized issues in contemporary US culture and politics – this can be seen to open up spaces of engagement and "middle grounds" that are otherwise not accessible. Participant and audience responses, along with observations of the process, all pointed to the capabilities of humor to provide relief as an embodied and affective coping practice amid an otherwise distressing set of considerations. As such, the both disarming and subversive power of comedy served to open up everyday spaces for reflections and expressions of opposition and resistance to contemporary climate change causes and consequences. From this, while the potential for distraction and trivialization lurked, the realities were that the power of comedy as a vehicle for social, political, economic and cultural change was revealed. One participant reflected, "Laughter may cause people to drop their defences and be open to listening to other ideas and points of view" while an audience survey respondent commented, "Humor can be used to motivate problem solving, which is exactly what we need more of now."

The process and products therefore made evident the power of comedy to lubricate sites of subversion as well as sites of distraction. Robin Nabi, Abel Gustafson and Risa Jensen found that emotions (in particular hopeful appeals) are persuasive in shaping climate change communication and engagement. Those involved observed that oft-involuntary or subconscious laughter – an immediate meter of comedy's success or failure – sometimes moves considerations only part of the way toward needed social change (Chattoo, 2017). The social function or philosophical value of humor and comedy (apart from giving pleasure and entertaining) remained an open consideration after these "Stand Up for Climate Change" experiences. There is evidence that humor and laughter can help to elicit action (e.g., Berlant and Ngai, 2017; Elias and Parvulescu, 2017), reduce stress in adverse environments (Newman and Stone, 1996; Martin, 2002), alleviate suffering (Osnes, 2008) and effectively attend to grief and pain (Zillmann et al., 1993; Keltner and Bonanno, 1997). There are also indications that humor and comedy may prove to lessen the importance and seriousness of issues (Valdesolo and DeSteno, 2006; McGraw et al., 2015).

Mikhail Bakhtin developed a theory of carnival to draw out the argument that humor and comedy, more broadly entertainment, can possess great power to distract (Bakhtin, 1984). In this approach, carnival is viewed as a vehicle of an authentic proletarian voice contending with oppressions of the ruling classes, where carnival represents a temporary suspension of social rules, codes of conduct and deference. Gill Ereaut and Nat Segnit (2006) pointed out dangers of mere "small action" engagements through this focus on the individual (p. 23) (see Chapter 5 for more). In this regard, Stephen Greenblatt (1988) has pointed out, "this apparent production of subversion is . . . the very condition of power" (pp. 44–45) while Andrew Stott (2015) has posited that inversion and misrule then exist within a confined space of "licensed transgression" (p. 35).

The ability of comedy in recent decades to cause discomfort to those most powerful stands provides a subversive riposte. Beth Osnes has documented how members from a famous traditional comedic troupe in Burma known as "the Moustache Brothers" were imprisoned for performing various jokes at an auspicious public gathering at the compound for Aung San Suu Kyi on the Burmese Independence Day celebration in 1996. Likely proof that these comedians acted very much outside the overall design of those in power resides in the harshness of their sentence, seven years of hard labor in a stone quarry at a prison for hardened criminals. All other prisoners wore chains between their legs, but solid iron bars were placed between the legs of the comedians, making sleep and work extremely difficult (Osnes, 2008).

This negotiation of power and the temporary suspension of social rules governing that power were brought into play figuratively and literally in the second year of experimentation under study. One student doing stand-up took to task a prominent US politician for his dismissal of climate change. In the weeks before, he shared his script for feedback and approval. However, during the performance, the encouraging cheers from the crowd emboldened him, and in the absence of traditional class performance pressures he jumped the script and insinuated off-color accusations about this politician. From back stage the professors intervened and gestured firmly for him to halt the performance and exit the stage, pulling him back into the social rules surrounding these particular events. In the following days, the student earned a lower grade because of the perception that his comments were distasteful. He replied that he understood, even though he pointed out that he had still worked diligently and rehearsed extensively for the performance. As a performer onstage, the intoxicating freedom he felt in that moment may have released inhibitions that led him to speak against his better judgment. In this example, Foucault's approach through biopower helps us to understand this interaction with humor as a politically saturated and power-laden micro-event. The both liberatory and potentially damaging process revealed here contributed to a more nuanced understanding of how the use of comedy in a live setting provides for the maintenance, contestation, construction and challenging of wider discourses.

Despite the general feelings of advances made through these activities associated with the "Stand Up for Climate Change" initiative, it was apparent that performances and artefacts walked this tightrope between sites of subversion and sites of sedation. While seeking to offer counterweights to common vernacular and thought regarding contemporary action on climate change, many participants ran risks of appearing to engage in radical innovation but effectively reinscribing norms of climate injustices and inequalities perpetrated across gender, class and culture. These efforts were then seen to potentially foreclose on the possible reimaginings that participants sought to articulate and create (Foucault, 1996). Yet in taking these latent prejudices and assumptions out of hiding and putting them on the stage, the participants thereby made them visible, uncomfortable manifestations of what lurked beneath daily behaviors and societal structures, which then had the potential to spur conversation that could lead to possible resolution, healing and progress.

Some survey responses from this University of Colorado class also reflected the realities that this tailored approach through comedy is not always success-ful, nor should it be expected to be. Those responses cohere with experimental work by Ya Hui Michelle See, Greta Valenti, Angeline Ho and Michelle Tan. Their research revealed many ways in which framing can (at times unexpect-edly) backfire (See et al., 2013). In the context of climate change communica-tions, Dan Kahan (2013) has found that subtleties in messaging can activate strong ego-defensive attitudes as well as produce ineffective or even counter-productive results.

While one participant observed that the process "showed me how hard it is to think of good ways to discuss this issue without being too 'in your face,'" an unspoken and often underconsidered challenge was one of being "in the face" of power too little.

John Morreall (1983) has written that resilient humor can effectively provide defense against tyranny, but the seductiveness of comedy and laughter may remain effectively encased in the confines and logics of instrumentalized, commodified and reified structures and processes of late capitalism, and there-fore innocent of relations to power (Adorno, 1991). As such, the "Stand Up for Climate Change" project and wider efforts to address climate challenges through comedic communication tools continue to brace against and lean into the forces of both subversion and sedation when assembling and carrying out their communications work. It is important to remember, however, that appear-ances of relief that can be perceived as moments of sedation can also be useful as moments of respite from which to draw strength to confront tyrannical behaviors through comedy in moments to follow. Going forward, a resolute mindfulness of the longer struggles involved – and how comedy can serve as both salve and seed – help to attend to power-laden processes that foster business-as-usual carbon-based capitalist practices that contribute to current changes in the climate.

Complaining Done with Charm

Lorne Michaels – creator and producer of the US-based live sketch comedy and variety show *Saturday Night Live* – has commented that comedy is "complaining done with charm" (Bevis, 2013). In the context of endeavors to communicate about climate change through comedy, this may provide a window into coping strategies-as-response to shared dread and uneasiness about the challenges that anthropogenic climate change carries. While anthropogenic climate change is one of the most prominent and existential challenges of the twenty-first century, this can be a frightening notion where humor and laughter are inappropriate and incoherent. However, social science and humanities research into affective and emotional dimensions of learning have provided insights into the ability of laughter to enact, disrupt and reconfigure relationships at the human–environment interface (e.g., Emmerson, 2017).

Through a dialectic of connectivity and difference (Castree, 2010), these collective ways of knowing through humor can unleash productive and creative forces that laughter from humor, comedy can bring to light power configurations and relations (Williams, 2016), and it can further entrench these constellations of power. As an example of the latter, after a heavy Washington, DC snowstorm in February 2015, US Senator James Inhofe (R–OK) carried a snowball to the podium on the floor of the US Senate in an apparent effort to call the warming planet into question. As he pulled his snowball from a clear plastic bag, he commented, "In case we have for-gotten – because we keep hearing that 2014 has been the warmest year on record – I ask the Chair: 'do you know what this is?'" Inhofe then waved the snowball and said, "It's a snowball, from outside here. It is very very cold out, very unseasonable." He then threw the snowball underhand to the Senate president Bill Cassidy (R–LA), saying, "Here Mr. President, catch this," while chuckling in delight. Originally airing on C-SPAN, this went viral through news and social media.[16] This stunt was prompted by Inhofe's motivation to disprove anthropogenic climate change by way of a cold front passing through the US capital in the winter of 2015. Though some saw this as clever questioning, most who viewed it saw it as hilariously flawed logic (see Figure 4.7).

Noam Gal (2018) has actually documented how humor and irony commu-nicated through social/digital media have great potential to be misinterpreted, effectively "deepening existing social gaps" rather than bridging them. While this event also generated public discourse of Senator Inhofe's many ties to fossil fuel industry power brokers, overall, the escapade demonstrated yet again that comedy powerfully cuts many ways as it flows through the shared social body.

---

[16] The segment can be seen here: www.washingtonpost.com/video/national/sen-james-inhofe-brings-snowball-to-senate-floor/2016/06/02/3ca067d0-28da-11e6-8329-6104954928d2_video .html?utm_term=.262166b9b288

Figure 4.7 US senator James Inhofe (R–OK) on the floor of the US Senate with a snowball in February 2015.

Comedy helps to bridge between levels of social systems: micro, meso and macro (see Chapter 4 for more). These multiscale comedic approaches then help to explore how agency, social structure, culture, institutions, inequality, power and spatial dimensions of these issues shape how we address present-day climate change (Ehrhardt-Martinez et al., 2015). In so doing, power saturates social, political, economic and institutional conditions that shape these relations and interactions (Wynne, 2008). Through the wider and context-sensitive lens taken up here, we have begun to interrogate how these interactions shape and threaten/manage the conditions and tactics of our social lives (de Certeau, 1984) and how knowledge, norms, conventions and (un)truths can be maintained and/or challenged (Foucault, 1980) (see Chapter 1 for more). However, additional considerations of the affect help to understand different entry points into awareness and efficacy and into diverse ways of knowing. Affective politics then "affirm Foucault's important caveat – 'it is not that life has been totally integrated into techniques that govern and administer it; it constantly escapes them'" (Anderson, 2012, p. 41).

For example, in work in Senegal to confront extractive forestry practices, Jesse Ribot and colleagues drew on creative communications to show comic films to the people they depict in order to open up conversations. He observed, "Scholarly work does not seem to make headway. So, we turned to humor and theatre ... Humor, irony and the absurdity of everyday reality are powerful. They are real – more palpably real than cold scholarly analysis ... Humor works where it states the obvious but unspeakable" (Ribot, 2014, p. 6).

In another instance, on *The Daily Show*, Trevor Noah drew on resonances across scale when challenging US president Donald Trump to care about climate change, on the heels of the April 29, 2017 "People's Climate March." He commented, "If people really want to get Trump's attention [on climate change], go protest on the golf course. Or even better, get a tee time ahead of him and go really slow . . . and he will be like 'hey assholes, you're not doing anything' and be like 'yeah, it's a metaphor for your Presidency!'" Later in the skit, he commented, "There is one thing that just might get Trump to care about climate change . . . his properties." Noah then commented, "President Trump, you may not care about climate change, but I know you care about winning, which is why you're not going to let climate change kick your ass by flooding your winter wonderland [Mar-a-Lago]. Come on President Trump, it's time for you to stand up and tell the world: 'Nobody sinks your properties but you.'"[17] Through humor, Noah tugged on connected strands of political, economic and institutional relations woven through our shared, lived experiences and finds humor as a vehicle for serious social and environmental commentary.

Comedy and humor – as pathways to emotional, affective visceral and experiential ways of knowing – therefore offer capacities and potential to provide useful complementarity and counterweights to often-gloomy scientific ways of knowing about climate change. As such, these approaches can enable resistance and reimagining of collective futures in the face of climate change in the twenty-first century. Friedrich Nietzsche (1887) has said, "Perhaps if nothing else today has any future, our *laughter* may yet have a future." There is seductiveness here, as humor and comedy provide a refreshing change from doom and gloom. In an opinion piece in the *Washington Post*, Michael Mann, Susan Joy Hassol and Tom Toles (2017) argued that "climate doomism" (commentary that portrays climate change not just as a threat that requires an urgent response but also as an essentially lost cause, a hopeless fight) regrettably "leads us down the same path of inaction" as climate denial also does. However, while the saying goes that "laughter as best medicine" there are dangers raised by Max Horkheimer and Theodor Adorno (1947, touched on earlier) that this medicine could be an opiate that manipulates a public citizenry into passivity. Moreover, bemoaning jokes as narcotics they have lamented how "it makes laughter the instrument of fraud practiced on happiness" (Horkheimer and Adorno, 1947).

Conclusions

Going forward, more examinations of creative and effective projects and practices through emotional, affective, experiential, visceral, tactile and

---

[17] The segment can be viewed here: www.youtube.com/watch?v=Z7msRhXgKXk

aesthetic ways of knowing about climate change are warranted; more efforts also must be made to build capacities and provide feedback to practitioners and everyday citizen communities to enhance climate awareness and engagement. John Schwartz (2018) has commented, "We don't usually get a lot of laughs out of climate change." However, these humorous approaches can tap into complementary ways of knowing in order to more effectively develop strategies for effective and creative communications about climate change, specifically by experimenting in key areas of awareness, efficacy, feeling/emotion/affect, engagement/problem solving, learning and new knowledge formation. These enlarged approaches can then feed into expanded pathways to science–policy engagements (Oreskes and Conway, 2018).

Moreover, participatory activities are potentially powerful avenues through which people consider resonant climate challenges (Osnes, 2014). Together, these are powerful forces contribute to "the visceral climate experience" (van Rennsen, 2017) that shape our perceptions on climate change as well as a range of other issues. In some cases, these help move perceived engagement from third-person perspectives to the first-person (Rosenthal and Dahlstrom, 2017).

This movement has been seen to engender further substantive engagement in climate and other environmental issues (Hoewe and Ahern, 2017). This has been evident in social science and humanities studies of public service announcements (Sun et al., 2008), emotional advertisements (Gunther and Thorson, 1992) and environmental documentaries (Lin, 2013; Cooper and Nisbet, 2017) where emotions generated from placing oneself in the storyline have served as a pathway to advocacy and engagement (Hammond, 2018). Jeanne Ellis Ormrod (2017) has referred to emotionally infused cognitive processing, stating, "We're more likely to pay attention to, think about, and remember objects and events that evoke strong emotions" such as excitement, laughter, joy or sadness (p. 146; see also Pintrich, 2003) (see Chapter 5 for more).

In this spirit, Amy Luers (2013) has called for the development of a learning culture to prioritize effective engagement methods. Furthermore, Mike Hulme (2009) has commented that research "must stop viewing global change as yet another opportunity to apply our existing tool kit. We must view the problems of global change as an opportunity to better recognize the limitations of current tools, and as a test bed in which to develop new formulations and analysis methods" (p. 279) (see Chapter 7 for more).

Scientific groups are increasingly "smartening up" their communications. For example, the National Academies of Sciences, Engineering and Medicine produced a 2017 report called "Communicating Science Effectively: A Research Agenda." Among their recommendations, the experts involved stated, "use a systems approach" to guide science communication (National Academies of Sciences, Engineering and Medicine, 2017, p. 8), acknowledging that scientific ways of knowing blend with other ways of knowing in audiences, and these are dynamic spaces of interaction.

Going forward, social science and humanities research in these areas – from short- to long-term impacts of experiential, affective, visceral, tactile, aesthetic and emotional influences – as they relate to climate change engagement can and should continue to expand. Through creative communications that account for these multiple pathways, we can more capably identify and access potential sites of engagement. As such, our struggles to continue to make sense of these dynamic interactions are struggles worth continuing to interrogate and pursue going forward.

5

It's Not You, It's Me ... Well It's Actually Us

In his 2012 speech at the Democratic National Convention, then-US senator John Kerry remarked, " ... an exceptional country does care about the rise of the oceans and the future of the planet" and that it is "a responsibility of the leader of the free world." Four years later, then-CEO of ExxonMobil Rex Tillerson commented, "just saying 'turn the taps off' is not acceptable to humanity ... The world is going to have to continue using fossil fuels, whether they like it or not" (Neate, 2016). Each made these comments just months before becoming US Secretary of State in the Obama and Trump administrations, respectively. While taking up the same position consecutively in the US government, their comments represented exceptionally divergent views and framings of the climate and associated decarbonization challenges.

In *Why We Disagree about Climate Change*, Mike Hulme (2009) contended, "Disagreements about climate change are as likely to reveal conflicts within and between societies about the ideologies that we carry and promote, as they are to be rooted in contrary readings of the scientific evidence that humans are implicated in physical climate change" (p. 33). This seminal book effectively challenged readers to critically (re)consider the dynamically changing physical and cultural dimensions of the idea of climate change over time, and to (re) examine the notion that the idea of climate change has been harnessed to promote various ideological projects. In so doing, he effectively has helped readers ponder how we think about, discuss and formulate actions about climate change. He concluded, "The sources of our disagreement about climate change lie deep within us, in our values and in our sense of identity and purpose" (p. 364). He argued, "Our disagreements should, at best, always lead us to learn more about ourselves ... " (p. 364). Thus, improving our considerations and understanding of these elements can help us collectively get to the root of our climate quarrels.

In this chapter, I build on these insights to consider the importance of how we choose to discuss climate change in various contexts and scales. To do so, I explore processes of framing at different levels of engagement, with a perspective that there is no "correct" framing that solves all climate communications challenges, nor eliminates any resistances to engagement. However, social science and humanities research to date has provided key insights into what works, how, when, why and in what contexts. Lorraine Whitmarsh and Adam Corner have pointed out that "while there are no 'magic words' that can overcome deep-rooted cynicism or disinterest, people have clear preferences for different narratives" (p. 132). Rather, disagreements or varied perspectives can be harnessed as productive and revelatory rather than bothersome issues that need to be eradicated. I work through case study considerations of "More than Scientists" and a "97% consensus" meme as well as a Gateway Belief Model to explore these dimensions. Doing so illustrates some key pathways where there are creative and hence effective ways to reframe these issues in audience- and context-sensitive manners, then maximizing opportunities to find common ground on climate change. These effective ways constitute a silver buckshot approach to climate communications.

Framing Climate Change

Framing takes place both through one-way and dialogical communications. It is a way to mobilize words, images, sounds and aesthetics to shape other's attitudes, intentions, beliefs and behaviors (Bolsen and Shapiro, 2017). Framing is a mechanism that both consciously and unconsciously privileges certain interpretations and "ways of knowing" over others, within a larger current of dynamic activities. Framing is a mechanism for allowing certain interpretations and "ways of knowing" over others, within a larger current of dynamic activities. Robert Entman (1993) has commented, "To frame is to select some aspects of a perceived reality and make them more salient in a communicating text, in such a way as to promote a particular problem definition" (p. 52). Framing can help build citizen and decision-maker competence (Druckman, 2001). While some may dismiss framing as manipulation, an expanded perspective views it as a way to more mindfully meet people where they are on climate change. Essentially, framing involves a process of selecting what elements you choose to present and emphasize through the fire hose of information in everyday life.

Mike Schäfer and Saffron O'Neill (2017) have traced the concept of framing through Gestalt psychology, interpretive sociology, political science understanding of new social movements and communication studies. They pointed out that from the outset this concept has been interdisciplinary. Dietram Scheufele (1999) has advanced research in these areas by interrogating frame building and frame setting as they relate to communicators, content, context and audiences. In the case of exposure to public health and climate change framing, Teresa Myers and colleagues have found that different frames can prompt widely varying responses, from hope to sadness as well as apathy and indifference (Myers et al., 2012). Stephan Faris, Steven Lipscombe, Sonia Whitehead and Damian Wilson (2014) have urged communicators to use language that resonates with everyday people, to listen to what stakeholders need and pay attention to sociopolitical realities on the ground when assembling discursive and material interventions (pp. 21–22).

Over time, successful climate communications have involved careful framing strategies. One key has been found to be the mindful treatment of gains and losses/sacrifice frames. Cheryl Hall (2013) has argued for a framing of climate change engagement as one of (realistic) sacrifice to achieve sustainability. There are subtleties here. For instance, Catriona McKinnon (2014) has pointed out that sacrifice is distinct from deprivation. Meanwhile, Mark Hurlston, Stephan Lewandowsky, Ben Newell and Brittany Sewell (2014) found through experimental research in Australia that framing sacrifice as small reductions in future income rather than present losses is more effective for public support for climate policy action. In other research, Matt Nisbet has found two frames that most promisingly compel engagement with the threat of climate change: economic development and moral/ethical concerns. Monica Aufrecht (2017) has explained their efficacy by noting that these frames inspire and "reorient the discussion to positive human activity" (p. 4). She continues, "The morality and ethics frame draws on people's sense of duty, and evokes images of human courage rising to the challenge. The economic development frame focuses on human ingenuity and triumph over the intellectual and technological struggle. These frames place humans as victors, not as victims; as saviors, not as criminals. They draw on the best aspects of humanity" (p. 4). Aufrecht calls it the frame of "positive human action" (p. 15).

However, through experimental framing research Stephen Flusberg, Teenie Matlock and Paul Thibodeau (2017) found that the metaphors "war" and a "race" against time effectively induced feelings of urgency and willingness to modify behaviors. Furthermore, Kayla Gabehart (2015) has argued that climate change communications should draw principally on "national security" and "energy security" frames to successfully combine pro-environmental behavior with economic and patriotic values. And Carena van Riper and colleagues have examined how place-based motivations associated with the

Leave No Trace program in the USA have helped to close a "value-action gap" on climate change awareness and pro-environmental engagement (van Riper et al., 2018).

Framing has been executed in various ways. For example, members of the Trump administration in the Environmental Protection Agency have framed a number of studies that document threats such as pesticides on people and greenhouse gas emissions on ecosystems as "secret science" because the identities of the test subjects cannot be revealed because of data protections (Hakim and Lipton, 2018). At the surface level, a plea for transparency is actually itself revealed as nefarious by eliminating key evidence on threats to public health and environments once one digs deeper. But in the meantime, while everyday citizens do not have time or inclination to dig deeply into these studies, the damage to perceptions through suggestions of "secrecy" is done. Materially, communities – both humans and ecosystems – suffer. In 2018 testimony David Michaels, epidemiologist and Professor of Environmental and Occupational Health at George Washington University School of Public Health, described this as a "cynical approach . . . best described as 'weaponized transparency'" (Hakim and Lipton, 2018). Michaels – who was also the former Assistant Secretary of Labor in the Occupational Safety and Health Administration (OSHA) in the US government – drew parallels to similar tobacco industry tactics in the 1990s to demand access to raw data of studies linking smoking and cancer in order to delay regulations. Such tactics have been documented by Naomi Oreskes and Eric Conway in their seminal 2010 book *Merchants of Doubt*.

Disembodied analyses of framing and rhetoric that do not take contextual features that give rise to those articulations into account then provide only a partial accounting of what works, how, when and why (Brulle, 2010; Lakoff, 2014). John Dryzek and Alex Lo (2015) have pointed out that effective "rhetoric must always be context-specific: it works because the rhetorician knows something . . . about the particular dispositions of the audience" (p. 14). Still, over time many professional think tanks and research groups have sought to decipher the "right" framing of messages on climate change in order to elicit proper engagement in the public or among decision-makers. These endeavors have then extended into queries such as "What image is best?" and "How many polar bears will compel action?" Overall, Kristin Olofsson, Christopher Weible, Tany Heikkila and J. C. Martel (2017) have noted that "failing to establish how an issue is being framed or narrated limits the potential for action by indiscriminately accepting the existing communication strategies" (p. 15). Furthermore, Thomas Bernauer and Liam McGrath (2016) have cautioned that "simple reframing" of communications is unlikely to overcome critical pockets of resistance to various forms of climate policy action. There is no recipe book, set of directions or tool kit for success (see Chapter 7 for more). More deep consideration through ongoing research across the social sciences and humanities is needed and warranted. In this process of expanded considerations, both spatial and temporal contexts matter (see Chapter 7 for more).

Scale of the Challenge, Scale of Responses

On the campaign trail in 1988, George H. W. Bush declared, "Some say these problems are too big, that it's impossible for an individual, or even a nation as great as ours, to solve the problem of climate warming ... it can be done, and we must do it."[1] While that statement shows how far the politics of climate change has shifted since that time,[2] it also demonstrates an awareness of the need for both collective engagement and deep determination. And it communicates that this is a global-scale problem, and it necessitates a multiscalar set of responses.

Through examinations of climate change and society, Karen Ehrhardt-Martinez, Thomas K. Rudel, Kari Marie Norgaard and Jeffrey Broadbent (2015) mapped out needed engagements at micro- (e.g., individual), meso- (e.g., community) and macro- (e.g., global) scales. Their arguments that solutions to global environment challenges need not necessarily manifest at any particular scale align with Eleanor Ostrom's (2010) view of polycentric systems for coping with collective action and global environmental change. This has been evidenced in many social scientific studies of action in the face of current climate challenges. For example, through examinations of policy entrepreneurship in the C40 Cities Climate Leadership Group and the Carbon Disclosure Project, Michael Mintrom and Joannah Luetjens (2017) found that multiscale engagements can work to effectively promulgate solutions for climate challenges beyond the cityscape to other places and registers. Similarly, Angel Hsu et al. (2017) evaluated climate change pledges by cities, states and regions, representing about half a billion people on planet Earth, along with 2,000 private companies. Even though they found significant projected reductions in greenhouse gas emissions, they observed, "When we look at the individual pledges [by cities, regions and businesses] the impact

---

[1] www.c-span.org/video/?4248–1/bush-campaign-speech
[2] In an October 2018 interview with Lesley Stahl on the US program *60 Minutes*, while walking back a previous statement via Twitter that climate change was a hoax, the US president said, "I don't think it's a hoax ... but it – I don't know that it's manmade." www.cbsnews.com/news/donald-trump-interview-60-minutes-full-transcript-lesley-stahl-jamal-khashoggi-james-mattis-brett-kavanaugh-vladimir-putin-2018–10-14/

isn't that large so we absolutely need national governments to pull through and do a lot of the heavy lifting" (Milman, 2018). Moreover, researcher Klaus Lackner has commented, "Cities getting involved is good and important but we haven't really acknowledged how big and serious the challenge is. We are whistling in the dark" (Milman, 2018).

Research in the social sciences and humanities has helped us to better understand how many of these issues in fact are multiscale problems that require a collective and multipronged set of responses. For example, Robert Gifford and Louise Comeau (2011) found that positive actions such as taking public transport, composting and buying green products are more effective than demands to drive less and cut your energy consumption.

However, research has also provided insights into a realization that effective engagement is not simply a matter of "meat free Mondays," installation of compact fluorescent light bulbs, buying local vegetables and putting a solar panel on your rooftop (if you own your home). While these can potentially be seen as helpful on environmental grounds, they alone are insufficient responses in the face of the scale of the problem. However, the locus of agency often nonetheless remains stuck in focus at the individual level. In some ways, this is very logical: when considering how we are situated in this global issue, a first step may be to ponder how our individual actions may contribute to the problem. Moreover, the natural and most easily controllable way to influence change is often through one's self (Tjernström and Tietenberg, 2008). This is not to say that individual engagement is a bad thing. Everything counts and this is often the location where most people feel they can make change. However, I do argue that individual-scale engagement is problematic when it is considered as the *only* site of engagement.

Challenges arise when attention paid to individual actions and claims subsumes deeper institutional questions as well as larger-scale contemplations. Aysha Fleming (2014) and colleagues have noted that individual "culture of consumption discourses" can be traps that effectively constrain social action on climate change. Through attention paid to "food miles, waste, excessive packaging and our insatiable appetite for cheap products ... travel behaviour, especially their personal vehicle use and air travel" people can effectively displace responsibility onto others who may consume more than they do (p. 414). In her book *Radical Consumption: Shopping for Change in Contemporary Culture*, Jo Littler (2009) has commented that political economic and societal dimensions of environmental challenges are often overlooked when coverage is overly focused on atomized alternatives for action.

When climate change action becomes a consideration of responsibility of the individual instead of the decision-makers or regulators who might affect significant policy changes through altering production and distribution, responses pale in the face of the scale of the climate challenge. While such trends may appear to open up space for the development of more pro-environmental and (pro)activist stances, this individualization tends to atomize social, economic and environmental movements for change.

In the early 1900s, philosopher Thorstein Veblen pondered the complex processes of social relations embedded in market economics, particularly focusing on tensions of efficiency and safety in consumer society. In contrast with Karl Marx's argument that working classes should revolt against capitalism and capitalists, Thorstein Veblen posited that working classes were instead inextricably bound up with capitalist classes through common indoctrinations. Therefore, Veblen argued that instead of endeavoring to topple the bourgeoisie, members of the working class instead seek to emulate them. This inclination to imitate rather than oust was the centerpiece of Veblen's theory of social stability and "conspicuous consumption" (Heilbroner, 2011).

Through mediatized interactions and "normalizing discourses" in these contemporary times (Benedetti and Lewis, 2019), our consumer lifestyles have been found to be largely maintained (Kurz et al., 2010). Kristina Diprose (2018) and colleagues found that British media coverage of sustainability was largely encased in neoliberal technocratic solutions such as corporate social responsibility and "buying local." But Tim Luke has cautioned that trends toward individualized consumerism have constricted substantive engagement with climate change. This is the conspiracy of the personalized and individualized framing of climate change. He has argued that green consumption at this individual level has effectively become a form of lazy or impotent environmentalism that displaces considerations of one's role in collective action (Luke, 1997, 1999). In these ways, Luke (2008) has warned that when we view our engagement with environmental challenges as individual consumer decisions, these remain "too entwined in the reproduction of most existing power relations and global market exchange" (p. 1811). Moreover, Robert Brulle (2010) has argued that an excessive mass media focus on individuals "works against the large-scale public engagement necessary to enact the far-reaching changes needed to meaningfully address global warming" (p. 94).

This spotlight on individuals-as-consumers also has had the potential to absolve political leaders and policy actors from responsibility to work on larger scale changes (e.g., public deliberation of climate-related policy proposals), as the suggested solutions emanate from (green) consumer behavior. There is the potential for free ridership at many scales, which relates to a bystander effect, defined by George Marshall as "the more people we assume know about a problem, the more likely we are to ignore our own judgment and watch the behaviour of others to identify and appropriate response" (Marshall, 2014, p. 26). This can become a problem where there is a normalizing of inaction as we await others to act on climate change. As climate scientist Roger Pulwarty has characterized it, the mindset has been, "Let's be proactive, you first."[3] So buying electric cars, taking shorter showers and committing to meat-free diets may be helpful in the mindset that "everything counts." Moreover, these behaviors are useful insofar as modeling such climate-friendly behaviors catalyzes pro-environmental behaviors in others people too (Karp, 1996). However, when these individual actions remain a collection of individual actions, they still are feeble in response to the scale of the collective challenge. Therefore, these individual actions must also connect with collective and larger-scale actions.

Furthermore, mere engagement in these individualized and "green" consumer behaviors can also spark a "greener than thou" perspective, where an individual not only feels great about his or her (limited) consumer actions but also judges and potentially shames others who are not engaging similarly through the marketplace. This can be toxic as well as counterproductive. A central concern here is that when the focus remains solely on the individual, atomized and "virtuous" movements can effectively distract and divide. While green consumerism may serve as a balm soothing one's consciousness, without awareness of engagement at larger scales it remains a paltry attempt to confront the climate challenge effectively. Martin Lukacs (2017) has argued that we have been conned into fighting issues such as climate change as individuals, thereby stymying our abilities to sufficiently address the challenge.

For example, from March 2010 through September 2011 Monterey Bay Aquarium in California staged the *Hot Pink Flamingos: Stories of Hope in a Changing Sea* science exhibit. This exhibit was a multimodal communication feast, from live animals in aquaria to videos and interactive stations. The exhibit invited visitors to consider how energy derived from fossil fuels contributes to carbon pollution and ocean acidification. Merav Katz-Kimchi and Lucy Atkinson (2014) examined these themes in the exhibit, and found that "individual, marketplace-based action on climate change [was prioritized] over solutions requiring large-scale social change or collective action" (p. 1). While engagements about how to change individual daily habits was seen as positive,

---

[3] This was a comment from the National Oceanic and Atmospheric Administration researcher at the October 10, 2018 "Leadership Roundtable: Science and the Role of Universities in Local Resilience" hosted by the National Council for Science and the Environment and the University of Colorado.

an unanticipated takeaway from the exhibit was the displacement of and distraction from wider considerations of decarbonization and system change.

Social science and humanities research has illuminated how individuals and groups view climate risks and how these perspectives shape potential responses. In 1982, Mary Douglas and Aaron Wildovsky coauthored the seminal work *Risk and Culture* that laid out how risk perception was a function of psychological, ethical, political and cultural factors as well as technical assessments. Before considering climate risks in particular, they mapped out a "cultural theory of risk." This was a theory that explained patterns of how people conceive of, perseverate over, or ignore various environmental risks based on how individuals, groups or societies see themselves in relation to others in society. A categorization into four groups is then contingent on orientation/preferences for "social regulation" and valuation of a "social contract" (see Chapter 6 for more about a "social contract" with society).

By "social regulation," they meant the extent to which people viewed rules as needed to guide/govern behavior; by "social contract," they meant the extent to which people are individual-oriented or group-oriented. Consequently, four categories were mapped:

- "Fatalists" (high social regulation/low social contract)
- "Hierarchalists" (high social regulation/high social contract)
- "Individualists" (low social regulation/low social contract)
- "Egalitarians" (low social regulation/high social contract)

These were then posited as the general ways in which individuals or groups viewed a relationship between individuals and society, and then mapped them onto cultural perceptions of risk (Chiao et al., 2009). For example, an egalitarian is seen to be one who values collective efforts, who is group-oriented and who shares a sense of solidarity in common struggles with other sectors of society while an individualist is soothed through separation with other segments of society or with a collective. However, as both have "low social regulation" characteristics, they share a view of low need for social structuring through rules and hierarchies. As another example, a "hierarchalist" may be someone who adopts a philosophy in which order and conservative/authoritarian rule is favored over liberty.[4]

---

[4] This perspective has been traced back to a Hobbesian philosophy, articulated in his book *Leviathan*, where he takes up a brash Darwinian approach. Hobbes stakes out the perspective that individuals will always be at war with one another, and "The life of man (is) solitary, poor, nasty, brutish and short." As a consequence, there is less concern for future generations and nonhuman life (Heilbroner, 2011).

Considering communication strategies that could resonate with these different perspectives, it can be useful to combine this theory with "social dominance theory" developed by Felicia Pratto, Jim Sidanius and Shana Levin (Pratto et al., 2006) in order to understand how varying personality traits can predict attitudes about and engagements with climate change. Social Dominance Theory (SDT) is a way to assess individual preferences for hierarchy within social systems, and Social Dominance Ordering (SDO) is the operationalization or an attitudinal aspect of SDT (Milfont et al., 2013). Those who score high on SDO measures are typically individuals or groups who value hierarchy, dominance, and power in intergroup relationships (Pratto et al., 2006; Sidanius and Pratto, 2001). These preferences then have consequences for attitudes toward social systems and institutions, as well as dimensions of equality and common purpose.

Social science research into relationships between cultural theory of risk and social dominance theory has been revealing (Amel et al., 2009; Barbaro and Pickett, 2015). For instance, Taciano Milfont and Chris Sibley (2014) found that a person or group with high SDO will prioritize human over environmental needs when faced with that trade-off. In another study, Angelo Panno and colleagues confirmed the relationship between pro-environmental behavior and egalitarianism, while also finding a positive associationwith Buddhist practices of mindfulness (Panno et al., 2018). They stated, "Understanding those aspects of mindfulness that are capable of promoting egalitarianism (thus reducing Social Dominance Orientation) could then be particularly interesting in addressing environmental issues from a psychological point of view" (Panno et al., 2018, p. 866). Of note, more work could usefully be done to examine how levels of SDO map onto communities of contrarianism (see Chapter 3).

Therefore, when developing creative communications about climate change, it behooves us to consider an individual's or group's orientation to the world around them (see Chapter 1 for more on audiences). Per Espen Stoknes (2015) has pointed out that we human animals are evolutionarily driven by ancestral forces including self-interest, social imitation and short-termism. However, he points out that "genes are not destiny" and "their current expression is now shaped by language, technology and culture" (p. 34). In other words, through resonant communications we can open up discussions that catalyze new perspectives and engagements.

In a report to the American Academy for the Advancement of Science, Matt Nisbet (2018b) posited, "A first step towards improved relations, the goal of dialogue-based science communication, may be to simply recognize and affirm shared values, beliefs, and goals" (p. 36). It can therefore be seen that communications that foster pathways to understand other points of view and mindfulness can effectively meet people where they are on climate change. Through this perspective, I consider an example of the "More than Scientists" project that has sought to humanize climate researchers in order to inspire mindfulness, empathy and a recognition of intersecting values and shared risks. Meeting the scale of the sustainability challenge means recognizing our connectedness, and that we are more than consumers (we are also citizens and stewards).

They're More than That

The More Than Scientists (MTS) project is an example of an initiative that deliberately frames these issues at the individual level. MTS is a collaborative activity of the nonprofit Climate Change Education Project organization, led by Eric Michelman as director. The campaign works to profile motivations, hopes, concerns and aspirations that prop up the work of prominent climate researchers around the world by publishing short videos about the human dimensions of their professional work. Eric Michelman has identified his target audience as the "movable middle," or those who may be concerned, cautious, disengaged or even doubtful, but open-minded on climate change. MTS effectively then sheds the professional lab coat to spotlight these researchers as citizens, parents and community members sharing their perspectives on the past, present and future in a changing climate. For example, in a video interview Mark Serreze – Director of the National Snow and Ice Data Center – discussed his first research trip to the Arctic in 1982 on northern Ellesmere Island. Professor Serreze recounted, "It was within one minute that I stepped outside the plane I knew this is what I wanted to do."

MTS has drawn on the communication power of Facebook and Twitter to share assembled humanizing profiles of climate researchers. As such, the interviews have gained further traction in the public arena through views, shares and retweets in new and social media. This project has capitalized on new developments in tools – such as through new and social media – that have recalibrated who has a say and how claims circulate (Cacciatore et al., 2012; Schäfer, 2012; Graham et al., 2013). Ashley Anderson (2017) has observed that there are "several positive impacts, with social media encouraging greater knowledge of climate change, mobilization of climate change activists, space for discussing the issue with others, and online discussions that frame climate change as a negative for society" (p. 496). On these social media platforms, MTS has utilized opportunities for content production to attempt to productively influence climate considerations in the public sphere.

Over a three-year period, MTS worked with Inside the Greenhouse (ITG) (see the Preface for more about ITG), and ITG students (in groups of two to four people) produced forty-eight MTS videos. Student content producers initially received guidance from MTS director Eric Michelman (either through a Skype chat or in person) and were shown previous MTS videos. Through the process of brainstorming, storyboarding, coordination with the interviewee and conducting and editing the interview, students gained confidence and competence in creative climate communication through production of an approximately three-minute MTS video. While the process was therefore as important as the products themselves, many of the students produced high-quality compositions that portrayed human dimensions of the interviewee's work and motivations behind their professional commitments to climate research.[5]

For example, on a snowy winter afternoon in Boulder Canyon near Boulder, Colorado, three students – Samantha Szabian, Alex Doyle and Trevor Bishop – hiked to the base of an icy canyon wall to capture footage of interviewee Dr. Ryan Vachon ice climbing. Ryan is both a renowned ice climber and climate scientist associated with the Institute for Arctic and Alpine Research at the University of Colorado Boulder (Rom, 2017). In this shoot, he swung his ice pick into the frozen waterfall of ice, pulled himself up, and then dug the short blade extending from the toe of his crampons into the ice to push up from one foot and then the next, repeating this process as ice chips flew. The students recorded his progress. Trevor Bishop, also an experienced ice climber, traversed up the canyon wall using ropes and tackle secured by Ryan after his first ascent to video record Ryan climbing the wall again, this time from a top view looking down (see Figure 5.1).

Other interviewees included climate researchers from the natural and social sciences as well as humanities who work in institutions such as the National Center for Atmospheric Research (NCAR), the National Oceanic and Atmospheric Administration (NOAA), the Cooperative Institute for Research in Environmental Sciences (CIRES), the Institute for Arctic and Alpine Research (INSTAAR), Western Water Assessment (WWA), the National Snow and Ice Data Center (NSIDC) and the Laboratory for Atmospheric and Space Physics (LASP) (see Figure 5.2 for examples). For example, one video profiled Waleed Abdalati (CIRES director).[6] In the interview, they asked him why he does the work that he does on glacial melt in Greenland and the Arctic. Professor Abdalati

---

[5] These videos can be seen here: http://insidethegreenhouse.org/project/inside-greenhouse-more-scientists-collaboration

[6] This video was produced by Greg Chancellor, Reghan Gillman, Luca Delpiccolo and Andrew Linenfelser.

Figure 5.1 (Left) Trevor Bishop, Alex Doyle and Samantha Szabian (from left) set up the shoot with interviewee Dr. Ryan Vachon (on right). (Right) Student interviewers Trevor Bishop (top) and Alex Doyle setting up for b-roll footage of Dr. Ryan Vachon to climb for the MTS video interview. (Photo credits: Beth Osnes)

brought climate change "home" when he responded, "We all care about our kids. How do we ensure the best future?" He then showed a "Best Dad Ever!" poster made for him by his two daughters that he has kept beside his desk as motivation.

Beth Osnes, Rebecca Safran and I analyzed these interviews through a Critical Discourse Analysis (CDA) approach.[7] CDA is among approaches that help examine conscious and unconscious framing of climate narratives in a spatial and temporal context, as well as account for subtle factors that shape their storytelling practices in context (Fairclough, 1995). CDA accounts for how meanings are partially fixed as well as negotiated as they are constructed over time (Laclau and Mouffe, 2001). We therefore examined the context underlying these interviews while paying attention to framing, salience, ideology, tone and tenor in the interviews themselves (Carvalho, 2007).

The approach helped to capture how representations contribute to discursive narratives that – while anchored to social, economic and cultural norms – dynamically shape ongoing considerations and actions (Phillips and Hardy, 2002; Jones and Peterson, 2017). Anabela Carvalho (2007) has pointed out that CDA "allows for a richer examination of the resource used in any type of text for producing meaning. It shares with framing analysis an interest in the variable social construction of the world but puts a stronger emphasis on language and on the relation between discourse and particular social, political, and cultural contexts" (p. 227). Through these analyses, we traced the contours of what has been referred to as an

---

[7] For more details about this collaboration and our findings, see Osnes et al. (2017) and Boykoff et al. (2018).

Figure 5.2 Interviewees have included Cassandra Brooks (Assistant Professor in Environmental Studies at University of Colorado Boulder) (top left), Peter Newton (Assistant Professor in Environmental Studies at University of Colorado Boulder) (top middle), Atreyee Bhattacharya (Instructor in Environmental Studies at University of Colorado Boulder) (top right), Kristopher Karnauskas (Fellow of the Cooperative Institute for Research in Environmental Sciences and Associate Professor in Atmospheric and Oceanic Sciences at University of Colorado Boulder) (middle left), Phaedra Pezzullo (Associate Professor in Communication at University of Colorado Boulder) (middle), Jota Samper (Assistant Professor of Environmental Design at University of Colorado Boulder) (middle right), Michelle Gabrieloff-Parrish (Energy & Climate Justice Manager at University of Colorado Boulder's Environmental Center) (bottom left), Ben Livneh (Assistant Professor in Civil, Environmental and Architectural Engineering at University of Colorado Boulder) (bottom middle) and Amanda Carrico (Assistant Professor in Environmental Studies at University of Colorado Boulder) (bottom right).

"engagement gap" in climate change advocacy as it relates to natural and social scientists as well as humanities scholars (see Chapter 6 for more). Variants of this general notion have also been called a research–practice gap (Han and Stenhouse, 2014) or a "science–action gap" (Moser and Dilling, 2011).

Our analyses revealed mechanics of how interviewees chose storytelling and personal narratives to describe how they view the significance of their professional work. For example, interviewee social scientist Dr. Paty Romero-Lankao from the National Renewable Energy Laboratory (NREL) talked about why she conducts her work on preparedness and resilience of cities around the world dealing with the effects of climate change. The video portrayed the Mexican-born and -educated scientist as she expressed concern for social justice, where "safety nets are being eroded as we face cascading effects from climate and other challenges."[8] This humanizing portrayal effectively linked scientific, social, political and cultural contexts through the storyline that was developed.

As another example, Dr. Jim White (Dean of Arts and Sciences at the University of Colorado) spoke of his motivation by "being a grandfather three times over" as he reflected on why he has devoted his life to studying the Earth's climate, and how he feels about leaving it to his grandkids' generation.[9] The interview earned tens of thousands of views on Facebook. Comments included this from Anne Sondgrath from Salem, Oregon, who commented, "The evidence is around those of us who have lived for many years and have seen the climate change." This intergenerational framing resonated with research from Lisa Zaval, Ezra Markowitz and Elke Weber (2015), who have found that motivation to leave behind positive legacies is associated with pro-environmental claims and behaviors. Charles Mann (2014) at *The Atlantic* has also recognized the importance of "legacy" in terms of what we leave for our grandchildren (see Chapter 7 for more).

Other videos drew from exemplification theory by showing how prominent natural and social scientists "walk the walk" in bridging their personal and professional commitments in regards to climate change. For instance, research glaciologist Mike MacFerrin in CIRES talked about extraordinary ice melt in Greenland.[10] He exclaimed, "I wish I could just scream and let everybody know how big this is!" Mike went on to say, "Climate change is important to me because it affects so much of what I do and what I love." This video about the motivations behind Mike's work struck a resonant tone and earned tens of thousands of views on Facebook. Among user comments, Rob Healy (a geography teacher in Wagga Wagga, Australia) wrote, "Big thanks to all the scientists who post this evidence. Believe me all of us Geography teachers use this in our classes, the next generation are on your side."

The MTS project has deliberately sought to focus on the intersecting ways in which experiential, emotional and aesthetic motivations and perspectives inform scientific ways of researching and knowing about climate change at the individual level. Content producers have engaged in careful framing of content in order to find common ground in the context of contemporary climate change. For instance, a video interview of Lisa Dilling (Environmental Studies

---

[8] This video was produced by Alex Nimkoff, Ross Matsumoto and Ned Chambers.
[9] This video was produced by Ben Crawford and Ciara Green.
[10] This video was produced by Barbara MacFerrin.

professor at the University of Colorado Boulder) drew out intergenerational considerations along with this element of bridge building.[11] In the video, Professor Dilling pondered, "Will our kids get to see the beauty of our planet? . . . Let's not treat climate change as a big fight. Let's look for opportunities to speak across world views and look for common ground."

In some ways, the project design has followed the logic that individuals do remain the most commonly perceived locus of agency, as described earlier. MTS videos have mindfully built awareness of climate change as selected individuals have spoken from a platform of her or his given agency, institution or center. Consequently, these videos have centered on heartfelt stories of personal motivations seeking to activate engagement by that movable middle of the public citizenry. However, through this focus on the individual, challenges lurk as political economic dimensions are often lost. As Jo Littler (2009) has pointed out, these atomizing representational practices can distract from important political economic realities of carbon-based industry resistance to engagement. While videos like these may appear to open up space for the development of more (pro)activist stances through their reach on Facebook and Twitter, this individualization risks reducing collective action problems to individual acts of caring, commitment and kindness.

Moreover, these can appeal more immediately to the concerned or cautious, more than disengaged or doubtful, even despite Eric Michelman's and MTS's best intentions. And those pathways are littered with further barriers to engagement and action. To that point, Hanrie Han, Aaron Sparks and Nate Towery (2017) have commented that paths of engagement often "are a function not only of whether people are willing and able to get involved, but also whether the structure of opportunities offered by a group is appealing to the people the group seeks to engage" (p. 22). As I mentioned earlier, when climate change action becomes the sole responsibility of the individual advocate *in place of* governments or regulators who might affect significant policy changes through altering production and distribution, such advocacy – though potentially sincere and heroic – is effectively problematic when seeking to effectively confront the scale of the climate challenge (see Chapter 6 for more).

Yet, by no means do these pitfalls mean that initiatives like these are not worth pursuing. To the contrary: MTS has been an example of thoughtful work done to find common ground rather than placing misguided investments into winning an argument. The MTS project has worked to tell these human stories behind climate change research. Rather than providing stimuli that trigger defense mechanisms, the strategies deployed by MTS have shown these videos

[11] This video was produced by Elana Selinger, Katelynne Knight, Alex Posen and Mariel Kramer.

to effectively lower defenses and provide beginnings to potential ongoing and authentic conversations. These have then had the potential to induce a sharing of varied perspectives and an active practice of listening openly to other points of view. In turn, initiatives like MTS have fed into important dimensions of trust (see Chapter 1 for more) and have fought effectively against a prevailing myth that "there's nothing I can do about this big issue of climate change" (Hayhoe, 2018). As projects like these have permeated our everyday awareness, actions and activities, they have succeeded in terms of promoting mindfulness (thereby helping to reduce SDO), shared values, shared struggles.

Felicia Pratto, Jim Sidanius and colleagues have offered psychological reasons why mindfulness can promote empathy, compassion and openness while staving off self-interest and self-centered indulgence (see also Shapiro et al., 1998; Dekeyser et al., 2008; Kemeny et al., 2012). Among them, it has been found to promote altruism and other prosocial behaviors (Condon et al., 2013; Lim et al., 2015), reduce hostility toward other points of view (Kemeny et al., 2012) and enhance ethical awareness (Ruedy and Schweitzer, 2010). For example, Elizabeth Rush has examined the efficacy of burden sharing narratives through her creative nonfiction book *Rising: Dispatches from the New American Shore*. She portrayed life in a changing climate, bringing us in – experientially, scientifically, emotionally, aesthetically, viscerally – to US communities grappling with the realities of human-induced climate change at the water's edge on the coasts. Through the power of Rush's storytelling by way of nonexploitative testimonials, readers gained insights into and empathy for climate challenges experienced by people living in these places. Through the tragic beauty surrounding these stories told, readers could better understand attachment to place as well as attachment to the past. In so doing, creative approaches to storytelling like this have helped readers more mindfully and compassionately understand present-day dimensions of climate change through stories of people who are struggling to remake their realities in the face of worldly change.

These micro-level focused creative approaches help to illuminate narrative strategies that seek to more effectively motivate public citizens about aspects of climate change by sharing the human side of climate change. Katharine Hayhoe has argued that the most powerful thing that individuals can do to confront climate change is to talk to others about it (see Chapter 1 for "climate silence"). These approaches and interventions are usefully considered in the context of ongoing creative climate communications approaches that seek to help people effectively access and address "matters of concern" (Latour, 2004). Clearly, these efforts will not provide silver bullet solutions across multiple audiences and contexts, but they illustrate critical dimensions of silver buckshot approaches that can engage segments of the public in novel and productive deliberations (see Chapter 7 for more).

97% Consensus, So There (?)

It is useful to then scale up from individuals as the locus of agency to considerations of collective perspectives. It so doing, it can be illustrative to turn to a common expression of convergent agreement articulated through scientific and fact-based ways of knowing that the climate is changing and that humans contribute to those changes. The shortcut term is "consensus." Seth Darling and Douglas Sisterson (2014) have defined consensus in this way: "Over time, the community tests, and tests, and tests each of these ideas, most of which turn out to be wrong. In science, you know you have consensus when most scientists simply stop arguing against the emerging consensus point of view because the evidence supporting it is too compelling to disagree with anymore (p. 3).

Over decades, despite lingering uncertainties in other areas of climate science, peer-reviewed scientific reports and findings have signaled a broad scientific consensus that human activity has significantly driven climate changes in the past two centuries, and that climate change since the Industrial Revolution has not been merely the result of natural fluctuations (Tett et al., 1999; Allen et al., 2000). This storyline of consensus regarding anthropogenic climate change has evolved over time. Its evolution is evident through considerations of IPCC reports and outputs from correlated scientific bodies like the National Academies of Sciences. These top-level assessments have emerged from the aggregation of numerous peer-reviewed climate studies and reports (Argrawala, 1998).

In the case of anthropogenic climate change, IPCC statements have achieved greater clarity and detail regarding human influences on the global climate. For instance, the IPCC Second Assessment Report in 1995 made this detection and attribution work clear through the statement, "The balance of evidence suggests that there is a discernible human influence on the global climate" (Houghton et al., 1995, p. 4). This convergent view strengthened further in the years that followed. Prominently, the Third Assessment Report in 2001 contained the statement that "There is new and stronger evidence that most of the warming observed over the last 50 years is attributable to human activities" (Houghton et al., 2001, p. 10). This consensus was made even more detailed and explicit in the 2007 IPCC Fourth

Assessment Report statement that "Most of the observed increase in globally averaged temperatures since the mid-20th century is *very likely* due to the observed increase in anthropogenic greenhouse gas concentrations" (Solomon et al., 2007).

Clearly, relevant experts across the science–policy interface can and do continue to disagree on a range of topics associated with climate change: rates of change, climate sensitivity/transient climate response to emissions of greenhouse gases, the wisdom of a carbon tax and so on.

From the perspective of everyday people, the contours and specifics of these disagreements can be hard to make sense of and adjudicate (Dunwoody and Kohl, 2017). Survey research by Brandon Johnson and Nathan Dieckmann (2017) found that everyday people in the US fail to sort out more nuanced explanations behind what may be perceived as scientific disputes. For instance, members of the general public did not distinguish between deliberate bias (interests) and implicit bias (values) that may lurk behind an apparent debate. In other words, these debates all just look like generalized scientific disagreement without more textured considerations as to why.

As such, it is much easier to pollute the waters of public understanding than to keep them clean, productive and useful in a twenty-first-century communications ecosystem. It does not take much to create a communications environment of misunderstanding and mistrust (see Chapter 3 for more). So from a communications standpoint, ways to effectively express the extent of this convergent agreement through "consensus" have provided an appealing shortcut. It has then helped various audiences to detect the signal of our understandings about climate change in the noise of the everyday.

In this way, a prominent climate communication strategy that has emerged in recent years has been the "97% consensus" trope. This essentially points out that 97% of relevant expert scientists agree that humans contribute to climate change. By extension, this communication device is meant to enhance collective acknowledgment of the issue (and to then inspire action).

This 97% figure has been arrived at through a number of studies that have examined the extent to which relevant peer-reviewed literature agrees that humans contribute to climate change (Oreskes, 2004b; Anderegg et al., 2010; Cook et al., 2013; Verheggen et al., 2014). These research projects have converged on the finding that approximately 97% of relevant experts agree on this point. The first of these studies was published in 2004 by Naomi Oreskes in the peer-reviewed journal *Science*. Back then she assessed nearly 1,000 peer-reviewed research papers up to that point addressing this question of a human influence and found unanimous agreement over the decade-long study period (1993–2003).[12] She concluded, "Politicians, economists, journalists, and others may have the impressions of confusion, disagreement, or discord among climate scientists, but that impression is incorrect" (Oreskes, 2004b, p. 1686). Reflecting on this work a decade and a half later, Oreskes (2018) wrote, "To deny that global warming is real is to deny that humans have become geological agents, changing the most basic physical processes of the earth and therefore to deny that we bear responsibility for adverse changes that are taking place around us" (p. 57).

As another example, in a follow-up study John Cook and colleagues examined nearly 12,000 peer-reviewed journal abstracts published over a twenty-year period (1991–2011) and found just twenty-six rejections of the scientific consensus that humans contribute to climate change (Cook et al., 2013). Joe Romm (2018) offered an analogy: "We are as certain that humans are responsible for recent climate change as we are that cigarettes are dangerous to your health."

As this "97% consensus" meme emerged, it gained more widespread notoriety in the public sphere. For example, a comedy segment by John Oliver on his show *Last Week Tonight* in May 2014 represented this convergent agreement. To do so, he staged a "debate" on anthropogenic climate change initially pitting Bill Nye the Science Guy against a stereotypical climate contrarian, calling the segment "a statistically representative climate change debate." John Oliver prefaced the debate by stating

> You don't need people's opinions on a fact. You may as well have a poll on which number is bigger, 15 or 5, or do owls exist . . . there is a mountain of research on this topic [of climate change] . . . the only accurate way to report that one out of four Americans are sceptical about global warming is to say 'a poll says that one out of four Americans are wrong about something'. Because a survey of thousands of scientific papers that took a position on climate change found that 97% endorsed the position that humans are causing global warming. And I think I know why people still think this issue is open to debate. Because on TV it is, and it is always one person for, one person against. When you look at the screen it's fifty-fifty, which is inherently misleading.

After introducing the two debaters, John Oliver interjected into Bill Nye's opening statement that humans contribute to climate change, saying, "Wait, before we begin, in the interest of mathematical balance I'm going to bring out two people who agree with the climate skeptic and ninety-six other scientists who agree with Bill Nye. It's a little unwieldly but this is the only way we can

---

[12] In particular, Naomi Oreskes found that three quarters accepted the consensus view that humans contribute to climate change, while the authors of the remaining quarter of the papers took no position.

actually have a representative discussion … this is going to make the debate difficult, we should not really be having it in the first place."

As a crowd of caricatured climate scientists filed in with lab coats, glasses and clip boards, John Oliver got up on his desk to see out over the large group. Amid this scene of many scientists swarming three contrarians, John Oliver welcomed the cliché dissenting contrarian to make his case. With a caricatured southern drawl, the contrarian commented, "I just don't think the science is in yet or settled." From atop his desk, John Oliver then asked, "What is the overwhelming view of the entire scientific community?" As all the scientists spoke at once, the noise became overwhelming. Yelling over it all, John Oliver closed the segment by shouting, "This whole debate should not have happened, I apologize to everyone." In playing out this parody, John Oliver very effectively illustrated the ridiculous nature of these ongoing debates – also dubbed "false balance" – played out on television and elsewhere.

As John Oliver pointed out in this segment, even though there is this clear and convergent agreement within the relevant expert science community, there has remained low public awareness of this in the general public around the world (Skuce et al., 2016). In the USA, polling by Anthony Leiserowitz and colleagues has revealed that only about one in eight people (12%) correctly estimate the strength of this 97% consensus (Leiserowitz et al., 2017). Consequently, social sciences researchers have examined the efficacy of discussing convergent agreement in this way. There has been research into differences between people who accept or reject it (Solomon, 1993; Lewandowsky, 2013) and how this has mapped onto political ideology (Dunlap and McCright, 2008; Hart and Nisbet, 2012; Kahan et al., 2012; Nisbet et al., 2015).

These research endeavors have generated productive discussions about whether this trope has provided a wise and effective pathway forward at this scale in catalyzing support for public action and policy prioritization (e.g., van der Linden et al., 2015; Cook et al., 2016; Kahan, 2016). For example, Sander van der Linden, Anthony Leiserowitz, Geoffrey Feinberg and Ed Maibach (2014) conducted experimental research in the USA about how consensus messaging may strengthen perspectives among the public that climate change is happening, humans contribute to it, there is cause for concern and action must be taken to address it. They have called this the "Gateway Belief Model." (see Figure 5.3)

Through analysis of this nationally representative sample they posited that "perceived scientific agreement is an important gateway belief, ultimately influencing public responses to climate change" (p. 1). Moreover, they found that "Consensus-messaging does not increase political polarization on the issue (perhaps due to the neutral scientific character of the message) and shifts the

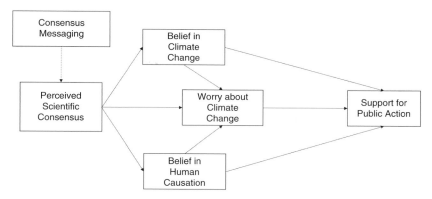

Figure 5.3 Gateway Belief Model (van der Linden et al., 2017). This model posits that perceived scientific agreement as a "gateway belief" then contributes to other beliefs about climate change and together influences indirect support for public action.[13]

opinions of both [US] Democrats and Republicans in directions consistent with the conclusions of climate science" (van der Linden et al., 2014, p. 6).[14] They concluded, "Repeated exposure to simple messages that correctly state the actual scientific consensus on human-caused climate change is a strategy likely to help counter the concerted efforts to misinform the public. Effectively communicating the scientific consensus can also help move the issue of climate change forward on the public policy agenda" (p. 7).

Digging more deeply into this, in research that related scientific consensus to perceptions of scientific (un)certainty, social science survey work has found that "Communicating higher levels of consensus increases perceptions of scientific certainty, which is associated with greater personal agreement and policy support for non-political issues" (Chinn et al., 2018, p. 1).

Meanwhile, research by Baobao Zhang and colleagues provided additional texture. They found that responsiveness to consensus messaging is highest in more conservative parts of the USA, and they attribute it in part to the fact that "conservatives value conformity to consensus and authority more than liberals" (Zhang et al., 2018, p. 373). They concluded that this research shows possibilities for pathways to "national convergence in perceptions of the climate

---

[13] Ed Maibach and Sander van der Linden (2016) have also advocated for the communication of normative agreement among experts in the case of conspiracy theories about the existence of a secret, large-scale atmospheric spraying program (SLAP) despite that 99% of surveyed experts have not found evidence to support SLAP concerns.

[14] See Chapter 3 for more on this communication strategy in relation to information-deficit model logic and contrarianism.

science consensus across diverse political geographies" (Zhang et al., 2018, p. 370). That effects were strongest in rural areas where consensus was initially reported as low has shown that this "97% consensus" meme may be a sharp tool for popping "filter bubbles" that inhibit more widespread engagement with climate change (Brugger, 2018a).

However, Sedona Chinn, Daniel Lane and P. Sol Hart (2018) also found that trust in science mediated this finding, where "Those with low trust in science fail to perceive higher agreement as indicative of greater scientific certainty" (p. 1). Along those lines, research by Dan Kahan (2015, 2017) has contested 97% consensus messaging approaches. He has argued that people's predisposition to motivated reasoning prevents consensus messaging to make much of a difference to their views. This might be called the "Don't confuse me with consensus I've got my mind made up already" perspective.

Kahan (2017) has therefore recommended that other communication strategies – not the "consensus strategy" as a gateway belief to public action approach – are wiser investments of time and resources. Furthermore, Warren Pearce, Reiner Grundmann, Mike Hulme, Sujatha Ramn, Eleanor Hadley Kershaw and Judith Tsouvalis (2017a) have argued that "centering on consensus about climate science in public debates does little to resolve the most pressing questions in climate policy design and implementation" (p. 728). They argued, "It distracts attention away from important practical challenges that highlight the need to negotiate between different scales of concern and action rather than box them into a linear relationship between scientific consensus and political action" (p. 728).

Moreover, other social science and humanities researchers have pointed out that there is potential for this messaging to backfire in a variety of ways (Hamilton, 2011; Cook and Lewandowsky, 2016; Pasek, 2017) including devaluing disagreement or respectful dissent (see Chapter 3 for more). This research instead has shown these approaches have limited or no effect (Deryugina and Shurchkov, 2016; Dixon et al., 2017; Kobayashi, 2018). For example, research by Adam Corner, Olga Roberts and Agathe Pellisier (2014) in the UK found that 18- to 25-year-old participants viewed a "97% consensus" among scientists as compelling, but "not necessarily enough on its own to inspire action-oriented response among young people" (p. 5).

What's more, these Gateway Belief Model and 97% consensus strategies have been found to have favored ways of knowing that can also build insider–outsider discourses. In other words, there may be seduction or alienation that comes from whether someone feels he or she is in line with this 97% or not. As was mentioned in the Chapter 4, Josh Pasek's (2017) survey research illuminated some of these potential pitfalls (see also Chapter 1 for more about trust).

He found that a questioning of scientific consensus emerged from identities associated with religion and political ideology. He also then favored motivated reasoning as a key driver of whether this consensus is or is "not my consensus" (Pasek, 2017).

Further critiques of this Gateway Belief Model include that it can be a simple appeal to facts, when the realities of the many ways of learning and knowing about climate change paint a more complex picture of understanding and engagement (Pearce et al., 2017). Yet, Naomi Oreskes (2017) has commented that this is an effective way to simply dispute a contrarian talking point that there is "no consensus." She argued, "In a political environment where contra-rians have repeatedly misrepresented scientific consensus in a deliberate attempt to influence public policy, it is both reasonable and necessary for scholars to participate in attempting to clarify what scientists believe that they have established" (p. 731).

John Cook has noted that an either/or consideration that may appear to emerge is unnecessary. Whether or not to invoke consensus in climate communications represents a false dichotomy. He commented, "Establishing expert consensus on human-caused global warming is a stepping stone leading to discussion of mitigation and adaptation policies ... failure to address misconceptions about consensus enables the persistence of distractions that can delay substantive policy discussion" (Cook, 2017, p. 733). Elsewhere, Dana Nuccitelli, John Cook, Sander van der Linden, Anthony Leiserowitz and Ed Maibach (2017) have claimed, "No one argues that the scientific consensus is the one and only message that needs to be communicated ... but informing the public about the scientific consensus that the problem actually exists can support the discussion on how to best solve the problem. The argument that we have to choose between debunking the climate policy opponent's disinformation about the scientific consensus or engage in meaningful policy dialogue is a false choice." Tom Toles has creatively pointed out some of these unnecessary contradictions (see Figure 5.4).

There are other ways to address these issues while more smartly meeting people where they are. As I mentioned in Chapter 2, sometimes avoiding explicit mentions of climate change can help find common ground more effectively than an insistence on the climate change frame. For example, in a study of political ideology and framing in communications in the UK, Lorraine Whitmarsh and Adam Corner examined climate change messaging of climate justice and energy savings. While the first message of climate justice elicited less engagement from right-of-center respondents, messages of energy savings found common ground across the political spectrum through shared values of conservation and waste

Figure 5.4 Tom Toles cartoon regarding contradictions and problems that lurk behind narrow pursuits of certainty and consensus.

avoidance (Whitmarsh and Corner, 2017). Of note, explicit mention of climate change ultimately was found to be less effective than its implicit presence in calls for energy conservation. They concluded that this research "offers clear implications for how climate change communicators can move beyond preaching to the converted and initiate constructive dialogue about climate change with traditionally disengaged audiences" (Whitmarsh and Corner, 2017, p. 122).

In 2018 polling research, Abel Gustafson and Matthew Goldberg explored audience segmentation of perceptions about scientific consensus in the USA, building on the "Six Americas" work first developed in 2009 (see Chapter 1 for more). They found that all groups underestimated the 97% convergent agreement among relevant-expert climate researchers that humans contribute to climate change. Among those who offered guesses about the level of scientific agreement, the "alarmed" group provided the closest estimate (84%) and the "dismissive group" provided the lowest estimate (44%) (see Figure 5.5). These findings

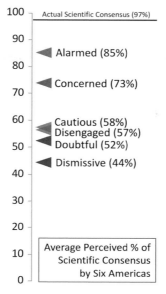

Figure 5.5 Findings by Abel Gustafson and Matthew Goldberg at the Yale Program on Climate Change Communication (2018).

provided some possible insights into both climate literacy (see Chapter 3 for more) and motivated reasoning (see Chapter 2 for more).

On one hand this may have seemed like a daunting and depressing set of results. However, on the other hand these findings revealed opportunities for climate communicators to further examine how trust, framing, communicators, social norms, values and other factors shape stated knowledge about levels of consensus on anthropogenic climate change.[15]

Consensus messaging or "Gateway Belief Model" may then be seen to be a useful tool or shortcut in relevant and appropriate contexts. This 97% meme can build support and greater clarity in the public sphere (e.g., in rural and conservative areas). It is a deliberate articulation characterizing a collective perspective, and by extension it is meant to then facilitate a progression of collective considerations into engagement and action in the public sphere in the face of this evidence of convergent agreement.

However, it is not a silver bullet strategy, especially with particular audiences (e.g., those with low trust in science). It is also important to note here too

---

[15] More information can be found here: http://campaign.r20.constantcontact.com/render?m=1102608159466&ca=c536a9ef-2fb9-4893-bf62-8144021fc2ef

that these scientific ways of knowing about climate change through consensus are steeped in social, political, cultural and ideological histories.

To illustrate, in the USA Louisiana congressman Clay Higgins had some choice words to offer in his remarks at the March 2018 "Expanding Global Gas Infrastructure seminar."[16] He said, "Welcome to the war for the future of our planet ... My role as your representative is to be not just your ally, but your warrior. Please allow the service of my office to represent the point of the spear that you wield. We'll knock down every bureaucratic wall. We'll kick down every federal barrier. We'll work with you. We'll work for you ... some of you know, some of you perhaps do not, that the Revolutionary soldiers that gave birth to this nation, represented only three percent of the populace, the colonists. Some of us refer to ourselves as 'Three Percenters'" (Kelly, 2018). Perhaps also making an oblique reference to those standing outside the "97% consensus" regarding human contributions to climate change, Higgins declared to the gathering of representatives from fossil fuel and liquefied natural gas companies, "You, ladies and gentlemen, are the Three Percenters of the modern era, where wars are fought with monies and strategies and energy."[17]

As such, 97% consensus does not represent the end of the tale, but rather a period in the ongoing story of collective (in)action.

Creative Communications with Each Other (at Scales ...)

In this quest for more resonant and meaningful conversations about and engagements with climate change, it is important to recognize that as inhabitants of planet Earth we are all responsible for the changing climate. Candis Callison (2014) has commented that "climate change challenges people to see themselves as part of global environmental, industrial and capital systems, and in many ways it demands a co-articulation of how to locate oneself in a larger collective" (p. 23).

Yet, it is critical to keep in mind that some are also much more culpable than others (Agyeman et al., 2007). Therefore, keeping the status quo through inaction (as a form of action) is then a choice to leave some people behind, and to allow others to suffer in the face of climate change. Another one of the

---

[16] More about the seminar agenda can be found here: www.ourenergypolicy.org/event/expanding-global-gas-infrastructure/

[17] An undercurrent here too are historical connections between a 3% movement and right-wing militia movements and white supremacism. It was unclear whether Clay Higgins was aware of these connections in making these comments but they have been made explicit elsewhere (e.g., Leber, 2018b).

cruel paradoxes mentioned in Chapter 1 is a paradox that people at the forefront of climate impacts are rarely the people who have contributed much to climate change problems.

Susanne Moser and Carol Berzonsky (2015) have offered three motivations to work systematically to overcome apparent cultural divides when confronting climate change awareness and action in the everyday. They outlined necessity, addressing the movable middle (and not providing outsized attention to extreme views) and attending to moral dimensions of these issues. Through authentic dialogue, they argued that these can help with collective efforts to achieve deeper engagement on these issues. In the context of climate adaptation, they wrote, "The promise and track record of dialogue to engage people, initiated by smart, courageous and trustworthy leaders who are willing to face the adaptive challenge before us offers a true and promising alternative" (Moser and Berzonsky, 2015, p. 306).

Elsewhere, Mike Goodman and I have examined how credible and trusted messengers have become a new form of "charismatic megafauna" (Boykoff and Goodman, 2009), displacing images of distant polar bears and melting glaciers that have been found to occupy the imaginaries of the public minds in the issue of climate change over the years (Leiserowitz, 2006) (see Chapter 1 for more). In so doing, these kinds of efforts can more effectively bring climate change "home" through resonant communication strategies (Slocum, 2004; Gold et al., 2015).

These findings and recommendations cohere with a range of social science and humanities research on climate change communication over time (e.g., Cormick et al., 2014; Smith and Leiserowitz, 2014). These amount to a call to not accept communications conditions and environments as they are, but rather to meld them into more productive spaces of multiscale engagement both over time and from place to place. This instead is a recognition that circumstances are made, not found; therefore, they can be unmade or made differently (Rutherford, 2007). Together, these elements can then reveal numerous temporal and spatial challenges as well as contradictions.

In terms of the temporal scale challenges, Andrew Revkin (2007) has called climate change a commonly perceived "classic incremental story." This runs up against our "finite pool of worry" (Linnville and Fisher, 1991) and our "caring capacities" amid many other burdens of daily life. Frankly, day to day most of us are inundated by immediate challenges surrounding us, and we often may feel that our actions are shaped more by external circumstances than by internal passions and desires.

Also, in terms of climate change issues themselves, frequent attention is often paid to average temperature and sea level rise over decades, which translates to communication challenges through their implicit minimization of potential nonlinear and abrupt climate changes (Mastrandrea and Schneider, 2004). Research in risk communication by Granger Morgan, Baruch Fischhoff, Ann Bostrom, and Cynthia Atman (2002) has informed how a combination of three main factors often inhibits everyday action on climate change:

- First, it takes a substantial time commitment to significantly confront climate change (long residence time of greenhouse gases in the atmosphere; e.g., carbon dioxide stays in the atmosphere on average between 50 and 200 years so emissions from the Model T Ford in 1911 are being confronted today).
- Second, many resources are required and at stake in confronting climate change (associated decarbonization of industry and society impacts every element of our lives, e.g., how we get from place to place via personal transportation to how our food and amenities are transported around the globe).
- Third, collective action on climate change demands overcoming cultural and political distance between interests, organizations and actors involved (considering how climate change manifests in challenges very differently in various places around the globe, e.g., desertification issues in sub-Saharan Africa vis-à-vis sea level rise and storm surge in island nations).

Moreover, John McPhee (1998) expressed another dimension to the challenges of communicating long-term climate changes when he wrote, "[T]he human mind may not have evolved enough to be able to comprehend deep time"(see Chapter 8 for more regarding temporal considerations through a discussions of future generations).

In terms of engagements from place to place, Adam Corner and Jamie Clarke (2017) have called for a scaling up to collective climate engagements "anchored in deeper notions of identity, values and citizenship if they are to have meaningful influence on promoting a proportional response to climate change" (p. 85). Robert Benford and David Snow (2000) have pointed out the critical importance of effective discursive framing for mobilizing social movements. In her book *Communicating Climate Change: The Path Forward*, Susanna Priest (2016) issued the challenge to move beyond individual conceptions of action to consider the power of social movements, to move beyond merely communicating problems into mobilizing informed and wise collective actions.

For instance, the #WeAreStillIn campaign has brought climate change "home" by scaling it down to local climate change engagement in the absence of a US policy void (e.g., *New York Times* editors, 2016). Efforts like these point to the importance of the individual locus of agency embedded in a larger collective

Figure 5.6 A sampling of Science March protest signs from 2017 and 2018.

action challenge. This more effectively confronts and reduces tensions that may thread through these multiscale challenges (see Chapter 4 for more).

The "March for Science" has been another useful example of these intersections in relation to creative climate communication. These marches, taking place in April 2017 and April 2018, were a coordinated set of rallies held near Earth Day (April 22). These were organized amid a backdrop of increased mobilizations in the USA and around the world (like the January 2017 "Women's March"). Other satellite events included a "Rally to Stand Up for Science" outside the 2017 American Association for the Advancement of Science (AAAS) annual meeting. Climate researchers who participated in these marches for science took "steps" from talk to action, and these have long been uncomfortable and unclear demarcations. They also demonstrated commitment to the general public while the rally raised the issue in the public view (Makri, 2017).

At the 2018 March for Science, Suan Svrluga from the *Washington Post* reported "A few people chanted 'Science is real. It's not how you feel,' beating a tempo on buckets, but mostly the mass of people marched through Washington quietly Saturday, letting their homemade signs show their support for empirical research" (Svrluga, 2018). Through the power of social media, messaging from these events (see Figure 5.6 for examples) crossed scales in creative communications around the globe.

These were marches not necessarily organized for a specific cause, but for advocacy for the integrity of scientific inquiry (see Chapter 6 for more on advocacy). However, these events rose up partly in response to US Trump administration actions that were seen to threaten science, scientists and scientific ways of knowing about contemporary environmental issues. Calls for a return to evidence-based policymaking and funding for scientific research moved at times from general statements and signs to explicit linkages to the Trump administration's suppression and side lining of science. Survey work on the marches and marchers by Teresa Myers, John Kotcher, John Cook, Lindsey Beall and Ed Maibach (2018) found that 89% marched because they wanted more evidence in policy decisions, and most scientist participants stated that they planned to engage in other science advocacy following the march (e.g., contacting elected officials). Organizers of the events insisted that the marches were nonpartisan celebrations of the importance of science in our everyday lives.[18] However, these marches inevitably mapped onto a politically charged and partisan cultural milieu regarding the intersections of science, policy and society (see Chapter 6 for more).

Critiques of these intersections and movements followed from many different perspectives. As examples, sociologist Robert Brulle (2018a) argued that by placing climate scientists as leading spokespeople for climate change action, "it fed into and exacerbated the existing polarized divide" rather than bridging it (p. 3). Meanwhile, physicist Jim Gates opined that "such a politically-charged event might send a message to the public that scientists are driven by ideology more than by evidence" (Flam, 2017). Communications scholar Matt Nisbet (2017) argued that "in the long run, the March for Science may have only deepened partisan differences, while jeopardizing trust in the impartiality and credibility of scientists" (p. 17).

However, numerous scientists and science advocates countered these critiques. Among them, AAAS leader Rush Holt has commented, "There's nothing philosophically incompatible with science and politics ... but there seem to be psychological hurdles that scientists have to overcome. One of them is this misconception that science is pure and politics is dirty. Neither side of that is completely true. It's not a good justification, but it may be something of an explanation. Of course, science is hard work, and people are busy with their laboratory work, or their writing, or their meetings, or whatever. The ethic in the profession is that you stick to your science, and if you're interested in how science affects public policy or public questions, just let the facts speak for themselves. Of course, there's a fallacy there, too. Facts are, by themselves, voiceless." (Roberts, 2017).

Exemplification theory suggests that concrete cases of influential actors grappling with issues such as climate change can significantly influence citizens' awareness and inclination to act themselves (Gibson and Zillman, 1994). This is the case because such exertions have been found to lower the psychological barriers to engagement (Zillman, 2006). Pro-environmental and pro-social behavioral engagement through inspirational leadership has been

---

[18] See https://web.archive.org/web/20170318183426/ and www.marchforscience.com/mission-and-vision for more.

evidenced in numerous studies (e.g., Sun et al., 2008; Lin, 2013). For example, through their studies of pro-environmental messaging in Singapore and in the USA, Sonny Rosenthal and Michael Dahlstrom (2017) recommended that "environmental communicators should leverage features of a message that engender first-person perception, with the goal of increasing its reach via audience members who feel inclined by their perceptions to promote it" (p. 14).[19] As such, these actions can catalyze others to follow, where the "Do what your neighbours do" mentality has a powerful effect.[20]

Moreover, Per Espen Stoknes (2015) has cited the power of positive and infectious action in the public view, stating

> The more people see happy others conserve energy, install solar rooftops, recycle, shop green, and drive electric cars, the more they are inclined to support ambitious climate policies on local, state and national levels. By seeing solutions in action, their feelings of helplessness and of being overwhelmed by global climate facts are eased. By seeing and believing that others – neighbors and friends – are taking action on the climate message, they start perceiving it as more personal, nearer and more urgent, too, counteracting the barriers. (p. 109)

These trends also interact with findings from humanities and social science research that our views and behaviors are influenced by our perspectives on what "tribe" we feel we belong to and align with (e.g., Kahan et al., 2012). So for example, if our in-group friends (or neighbors or family) are all recycling and switching to renewable energy in their households we feel compelled to also take action.

Furthermore, it has been found that researchers' understandings and engagements are strong predictors of their legitimacy in the public sphere (Gauchat et al., 2017). Gordon Gauchat, Timothy O'Brien, and Oriol Mirosa (2017) argued, "Given that polarization extends beyond claims about climate change to perceptions of the scientists themselves, science communication strategies ... may require direct engagement by *scientists themselves* rather than relying on the surrogates in the news media or elected officials" (p. 303, italics in the original). These efforts also have cohered with survey research by Steven Brieger (2018), who has found (across a thirty-eight-country study involving more than 30,000 respondents) that awareness and connectivity with social units across scale like community, nation and world "strengthen

---

[19] Moreover, researchers have found that barriers can be lowered by introducing specific ways that people can take action. For example, watching a short YouTube video with a specific pledge and plan for action could be seen to be effective here ... This can be seen to lower the barriers to pro-environmental actions after an engaging communication (Lokhorst et al., 2013).

[20] University of Colorado Boulder Mechanical Engineering Professor Shelly Miller has discussed this in the context of addressing indoor air pollution concerns in households.

in group solidarity and empathy and, in consequence, readiness to protect the environment benefiting the in-group's welfare" (p. 1). Together, creative communications that capture active leadership can arouse first-person perceptions that, in turn, help to build connectivity that sparks conversations, discussions, reflections and meaningful engagements across scales.

Conclusions

In this chapter, I have shown that context matters and discourses must be considered in context in order to be effective with target audiences. To draw these arguments out, I have examined conscious and unconscious framing approaches and multiscale initiatives taking place where climate communications seek to create change. Research findings in these areas have been brought to bear on questions as to how to effectively mobilize engagement by way of authentic connections through shared values. Through a wider and context-sensitive lens of cultural politics (see Chapter 2 for more), these explorations have sought to help guide effort to achieve greater success in efforts to meet people where they are. These are spaces of meaning construction that involved nested questions of how people make sense of, value and feel they could and should act in the world. This work then links how discourses (the ways we talk about climate change) contribute to social and material practices (the ways we respond to climate challenges). This can be seen productively as part of a silver buckshot approach, rather than adherence to notions of one-size-fits-all or magic bullet ways to communicate about climate change (see Chapter 7 for more).

These many critiques may seem like they amount to a great deal of finger pointing. However, precisely the opposite is true. Rather than naming and shaming and demonizing others, the embedded assertion here is that this is a collective action problem, and responsibility rests with all of us. Going forward with this mentality, consequent undertakings still face formidable – and sometimes fundamental – challenges.

Rather than viewing this rightly as work to "smarten up" and increase sophistication to produce resonant communications, some mistakenly view these endeavors as exercises in "dumbing down" information for public audiences. Furthermore, rather than working to meet people where they are, some erroneously concoct visions of where people *should* be, engaging in practices of naming and shaming instead of bridge-building through shared values. Together, these contradictions and competing approaches can generate attention but also alienate people from intended engagement with the issues raised.

There are many instances of these opportunities for connection going awry in the public arena over time. As one example, at the launch of his "culture strategy," then–London mayor Boris Johnson embodied the promises and perils of such a misguided perspective. In his remarks, Johnson argued, "Arts chiefs must stop dumbing down culture for young people." According to reporter Ian Drury (2008), "The mayor of London pledged to stop targeting [young people] with hip-hop music and movies, and instead encourage them to enjoy opera and ballet." It may be laudably aspirational from Boris Johnson to think that young people could enjoy opera and ballet like they do hip-hop and movies, but his approach failed to meet London youth where they were as he instead declared where youth *should* be, embracing and engaging with culture (Boykoff, 2011).

As another example, in 2011 the legislature in the US state of North Carolina proposed a bill (S.L. 2012–201) to ban state and local agencies from basing their coastal policies on scientific models indicating accelerating sea level rise (Harish, 2012). The bill (which became law in 2012) called on agencies to instead rely on historical linear predictions. Effectively this meant that a forecasted sea level rise was revised down through this policy to just eight inches, thereby enabling more coastal property development than would have otherwise been legally possible. Motivated by this drive for lucrative property development on the coastline, the bill was supported by a business-backed consortium of North Carolina property owners known as NC-20.[21] It was essentially a short-termism "hear no evil, sea no evil" policy proposal, in the view of satirist Stephen Colbert.[22] The bill also mandated that the Coastal Resources Commission and the Division of Coastal Management could not redefine this sea level change before July 2016.

Geologist Stan Riggs served on the North Carolina science panel that recommended the original thirty-nine-inch forecast. On the passage of this bill into law, Riggs quit the panel and commented, "the state is completely not dealing with this . . . [it] is just a short-term answer" (Leavenworth, 2017). As a result of these special interests shaping North Carolina coastal development policy, adaptation and resilience work was effectively limited during that five-year period of 2012–2016 as this law enabled overdevelopment in threatened areas over this period of time. That discursive missed connection has had many material effects in the public arena. Most notably, it is both difficult to pin down but impossible to rule out how this law had retarded measures for protection, thereby exacerbating damages caused by hurricane Florence that led to

---

[21] The bill can be viewed here: www.ncleg.net/Sessions/2011/Bills/House/PDF/H819v4.pdf
[22] The segment called "Sink or Swim" from The Colbert Report can be viewed here www.cc.com /video-clips/w6itwj/the-colbert-report-the-word–sink-or-swim

1.7 million North Carolinians being evacuated as the storm came on land in September 2018 (Kristof, 2018; Zezima, 2018).

Amid instances like these of "missed connections," there has been good news: more organizations, institutions, scholars and practitioners around the world are increasingly confronting the twin challenges that more research must be done to examine creative and effective projects and practices on climate change, and that more efforts must be made to build capacities and provide feedback to practitioners and everyday citizen communities to enhance climate awareness and engagement (see Chapter 7 for more). In other words, creativity (as applied imagination) is expanding rather than retracting from this core challenge of meeting people where they are on climate change in the twenty-first century.

6

Academic Climate Advocacy and Activism

"Evidence-based," "science-based," "vulnerable," "entitlement," "diversity," "transgender," and "fetus." These are the words that the Donald J. Trump administration's Department of Health and Human Services (HHS) were reported to have forbidden the Centers for Disease Control and Prevention (CDC) from using in their budget request for fiscal year 2019 in the United States (Anapol, 2017). As part of the US budget formulation process, in December 2017 news first broke in the *Washington Post* (Sun and Eilperin, 2017) that the HHS recommended that the CDC avoid this language in order to maintain support from Republicans as a budget moved to Congressional approval in 2018 (Cohen, 2017). Sheila Kaplan and Donald McNeil, Jr. (2017) from the *New York Times* reported, "The news set off an uproar among advocacy groups and some Democratic officials, who denounced any efforts to muzzle federal agencies or censor their language."

Reasons behind the apparent censorship and consequent backlash pointed to the highly politicized tenor of interactions at the contemporary science–policy interface. It also punctuated how dog-whistle discourses have lurked just beneath stances taken for evidence- and science-based decisions. In part, these contentious circumstances have been fueled by the funding from carbon-based industry interests that put seemingly benign advocacy for science and evidence on edge (Brulle, 2014, 2018a) (see Chapter 3 for more).

This has also illustrated how present conditions are a product of past trends as well: over time, socially constructed narratives about climate "advocacy" along with "activism" have woven their way into culture, ideology, politics and society (Hammond, 2017; Kukkonen et al., 2018; Nissan and Conway, 2018), where science is often seen to be a left-wing conspiracy (Lewandowsky et al., 2012). Furthermore, these influences have inhabited our psychology, where many relevant expert researchers choose to "self-silence" (Brysse et al., 2013; Lewandowsky et al., 2015). Self-silencing about climate change has also been found among everyday citizens as well (Geiger and Swim, 2016) (see Chapter 1 for more). Together, we have arrived at a state of affairs in which researchers are wary of discussing climate change so as not to "rock the boat," not to jeopardize their career advancement and/or to not potentially sacrifice their self-perception of credibility, authority and competence.[1]

The term "advocacy" comes from late Middle English and Medieval Latin. The Cambridge Dictionary defines advocacy of 'public support of an idea, plan or way of doing something' (Cambridge University Press, 2008) while the Oxford Dictionary defines advocacy as "public support for or recommendation or a particular cause or policy" (*Oxford English Dictionary*, 2016). By extension, an "advocate" is defined respectively as someone or something that "speaks in support of an idea or course of action" and who "publicly supports or recommends a particular cause or policy." While these standard definitions of advocacy and advocates help to corral the nature of the actions and interventions, there remains a great deal of wriggle room between advocacy of a "way of doing something" and a "particular cause or policy."

As such, while an advocate can be seen from one perspective as someone providing "policy relevant" information, he or she can also be perceived from other perspectives as offering (either overtly or stealthily) "policy prescriptive" information. This conflation is compounded by the ways in which "advocacy" and "advocates" have been invoked synonymously with "activism" and "activist" over time (Han et al., 2017). "Activism" has been defined as "the use of direct and public methods to try to bring about (especially social and political) changes that you and others want" (Cambridge University Press, 2008), and "campaigning to bring about political or social change" (*Oxford English Dictionary*, 2016). How academics productively contribute to science–policy interactions is often unclear (Fähnrich, 2017).

Simon Donner has pointed out that there are many risks and responsibilities associated with maintaining and enhancing reputation and power of scientists' voices in the public arena (Donner, 2014, 2017). Organizationally, the UN Intergovernmental Panel on Climate Change (IPCC) has taken the tactic to provide "policy relevant, not policy prescriptive" information about climate change (IPCC, 2010, p. 1). But such prudence has the potential of squandering obligations of engagement with public citizens, leaving the IPCC isolated in proverbial "ivory towers."

Journalist Chris Mooney of the *Washington Post* has observed that these factors have contributed to a "vicious cycle" that has effectively dissuaded

---

[1] This latter condition is referred to in psychology as "pluralistic ignorance" by Nathaniel Geiger and Janet Swim (2016).

people – among them climate researchers – from needed everyday discussions and engagements with climate change (Marshall, 2014; Mooney, 2016;). In the public sphere, it is important though to remember that scientists are not the only authority on science (Scheitle et al., 2018). For example, Mike Goodman and I have mapped out how cultural celebrities have shaped public understanding and engagement at the climate science–policy interface (Boykoff and Goodman, 2009) (see Chapter 1 for more).

In this chapter, I delve into the richness of various flavors of climate advocacy and activism through a focus on academic researchers. I ponder questions about whether academic climate advocacy should be celebrated and pursued, or derided and eschewed. I explore conditions in which some in academic communities facilitate various forms of engagement relating to their research while others respond to pressure from one's scientific peer communities to not overextend their outreach beyond their own specialized research programs. I also consider how some just simply do not have an interest in being engaged with the outputs of their work while others find it to be woven into the fabric of their scientific responsibilities in society.

Like in Chapter 3, I focus attention primarily on dynamics in the US context, but lessons learned can provide insights into cultural and political contexts in other places and scales.[2] I interrogate different approaches to climate advocacy, and animate their promises and pitfalls in the context of highly politicized deliberations on climate change in the public sphere. Nested in here, I argue that the conflation of these advocacy approaches between Type I advocacy (advocacy for [scientific] evidence) and Type II advocacy (advocacy for policy outcomes) contributes to confusion, apolitical intellectualism and excessive restraint that then increases the prevalence of Type 0 advocacy. Type 0 advocacy as inaction is then effectively a form of action. These explorations help to work through fundamental considerations of what is a social contract for science through engagement of scientists and academic researchers in society.

The Politicized Climate Science–Policy Landscape

The climate science–policy landscape has been littered by politicization for many decades (Boykoff, 2011). Consequent pollution of a communications environment has had pervasive and enduring effects. In this regard, even when

---

[2] For instance, Kirils Makarovs and Peter Achterberg (2018) have examined relationships between democratization, public engagement and citizen participation in science–policy issues in thirty-two European countries. Further analyses of conceptions and practices of climate advocacy could deepen understanding of these interactions.

focus is placed (here and elsewhere) on the contemporary context, the structures and conditions that have given rise to these interactions cannot be overlooked. As such, the adage "Don't hate the player, hate the game" can be an appropriate description of this state of affairs.

Amid this multidecade history, one of the long shadows over climate advocacy has been has been cast by former US vice president Al Gore. While he had worked tirelessly for decades to promote scientific evidence as well as policy engagement on climate change, segments of the public citizenry have conflated these two approaches and have consequently viewed his initial recognition through politics as an elected official as a way that climate change has been cast as a left-of-center belief system. For instance, in a *New York Times* article about how farming communities navigate questions of weather and climate change, Georgia farmer Casey Cox opined, "I really wish that Al Gore hadn't been the messenger, it just turned everybody off. It allowed people to say that it was just a liberal thing" (Thrush, 2018, p. A10).

Particularly since the advent of the US Donald J. Trump administration, there has been increased scrutiny regarding what may be appropriate roles for climate researchers to play in society, as potential "advocates" and "activists" (e.g., Oransky and Marcu, 2017). For example, climate researchers have stepped into the political fray not just on climate science in society but on other issues such as immigration policy. Climate science research articulated connections between these issues because they argued that a more draconian immigration policy then stifled a free exchange of scientific knowledge through collaborations and international conferences and workshop and meeting participation (Marshall, 2017).

The American Association for the Advancement of Science (AAAS) meeting in early 2017 saw many scientists step away from the conference for a short demonstration in Boston's Copley Square to advocate for evidence-based science (see Chapter 5 for links to the March for Science) (see Figure 6.1). Karen Weintraub (2017) reported, "Many scientists view political activism as a potential taint or threat to the absolute empiricism that science strives for – or simply feel they cannot afford to take time away from their work. But several said . . . they believe they no longer have the luxury of remaining in their labs. Instead . . . they felt compelled to speak out against the new Trump administration's use of 'alternative facts', climate change denial and restrictions on immigrants – many of whom work in medicine and science."

Later that year, at the American Geophysical Union meeting, increased attention was paid to these dimensions in panel presentations and discussions. For example, one session called "Legal Advice for Scientists Interested in Activism" featured a talk by Lauren Kurtz from the Climate Science Legal

Figure 6.1 A scientists-led demonstration advocating for evidence-based science in Boston's Copley Square outside the American Association for the Advancement of Science (AAAS) meeting in early 2017.

Defense Fund and Lee Rowland from the American Civil Liberties Union (Kaplan, 2017b). And in the 2018 midterm elections, in part as a result of the efforts of the "STEM the Divide" initiative – part of the 314 Action[3] – more people with backgrounds in science, technology, engineering and math (STEM) were campaigning to run for an elected political office at various levels of governance in the USA (Harmon and Fountain, 2017; Kaplan, 2017a).

But with this swirl of activity there is a substantial amount of confusion and conflation within the academic community about different points of entry into this world of "advocacy" and "activism." Mixed in here are also normative ingredients about what may be the "right" or "appropriate" place for academic researchers to enter these worlds. What results is often anxiety and reticence about how to navigate these often high-profile, high-stakes and highly politi-cized spaces of engagement at the science–policy interface and in the public sphere.

To illustrate, Luis Hestres (2018) studied email advocacy. He defined advo-cacy emails as "those addressed to a national audience whose primary purpose was to elicit political action from the recipient, or were part of an ongoing

[3] 314 refers to the first three digits of the Greek letter pi.

stream of action-oriented communications" (p. 8). The study provided some insights into links between online advocacy and engagement, but its analyses vacillated indiscriminately between advocacy for evidence-based climate science information and advocacy for policy outcomes (Hestres, 2018). But the failure to sufficiently distinguish between types of advocacy limited the explanatory power of the findings.

What has been needed is a basic taxonomy that helps distinguish between different manifestations of climate advocacy. This can then provide more solid ground for climate researchers to stand on when considering their stances in the public sphere. These varying types can help climate researchers in particular when they navigate the choppy waters of climate discourse in the public sphere through communications. While there is no particularly easy sailing when temporal and spatial context both matter for conditions, key features of advocacy can help to map these spaces of engagement. Among them, there are three clear advocacy approaches where distinctions between them are important:

- **Type 0 advocacy** = those who choose to stay away from any semblance of advocacy, sometimes because of disinterest in engagement, and sometimes owing to confusion and conflation of perceptions of academic advocacy in the public sphere; this appearance of inaction is in fact a choice or action.
- **Type I advocacy** = advocacy for (scientific) evidence: this approach advocates for the intersecting ways in which experiential, emotional and aesthetic information informs scientific ways of knowing about climate change.
- **Type II advocacy** = advocacy for policy outcomes: this approach promotes particular decisions (e.g., environmental policies or legislation) based on evidence ascertained in various forms to know about climate change; one strain of this type of advocacy may then involve advocacy for particular political parties that advance preferred policies or decisions.

This categorization of Type 0, Type I and Type II advocacy should not be interpreted as a binary or blunt interpretation of varied stakes and conditions. Rather, these "types" represent three distinct nodes across a spectrum of chosen engagements. This categorization also does not suggest that academic researchers will slot into one node or the other. Instead, there is dynamism in these flavors of engagement (see Figure 6.2 for examples) across issues and over time, along with a range from low- to high-stakes situations, all possibly experienced by the same academic researcher. Moreover, this is not just about frequency of advocacy; it is also it is more about efficacy. John Besley and Anthony Dudo (2017) have pointed out that there is a need to "evolve the focus from helping scientists engage with the

Figure 6.2 Prominent Type I advocates may include Katharine Hayhoe (left), and Neil DeGrasse Tyson (center); prominent Type II advocates may include James Hansen (right).

public *more often* to helping them engage *more effectively*" (p. 410, emphasis in the original).

This set of considerations is focused as well on academic researchers as advocates. Here I argue that confusion and conflation of these distinct modes of engagement has led to reticence and shying away from engagement in the public sphere. Elsewhere, other researchers such as Lydia Messling, John Kotcher and Shahzeen Attari have examined advocacy more widely.

Here, I argue that because of this confusion and conflation there is a consequent shying away from Type I advocacy into Type 0 advocacy precisely at the time when publicly engaged scientists are needed to confront these collective action challenges at the human–environment interface and combat post-truth conditions polluting a productive communications ecosystem.

Approaches must be keenly aware of the thickets of potential conflation between portrayal of varied approaches and advocacy for any one response in particular. In championing or chastising individuals as advocates, there is also a danger of overindividualizing issues that are fundamentally failures of collective action (see Chapter 5 for more). There are also pitfalls in terms of potentially reducing public engagement with science to a limited group of influential academics and scientists, effectively excluding other voices (Tudge, 1997). In 2013, Amy Luers interviewed climate advocates, academics and foundation funders to better understand the landscape of climate advocacy in the USA. She found that "climate advocates have focused too narrowly on specific policy goals and insufficiently on influencing the larger political landscape" (Luers, 2013, p. 13).

Amid these risks, advocacy and activism often flow into varieties of public intellectualism. Declan Fahy (2015) has mapped out a four-step process whereby scientists and researchers effectively become public intellectuals.

This involves becoming expert in their specialist subject, accessing media channels to speak more broadly to general audiences, expressing added-value opinions and views on a range of relevant issues and establishing a reputation in the public sphere.[4] Fahy surmised that "public intellectuals do not work solely within a *professional culture* of other credentialed experts. They also work within a broader *public culture* that includes experts from other fields, journalists, writers, critics and citizens. Scientific public intellectuals, like other public intellectuals, have a powerful cultural role" (p. 12, emphasis in the original). In the twenty-first-century of climate science–policy–society interactions, there is an unavoidable blurring of roles in the public arena.

Declan Fahy (2015) has also noted that "scientists who become public intellectuals are brilliant synthesizers of science" (p. 215). He has also called them "emblems of a new era of science, one embedded in the dynamics of media, the needs of celebrity culture, and the vicissitudes of public life." He continued, "They are powerful figures, who influence the public understanding of science *and* the future direction of society *and* the inner workings of science. They are science's new influential leaders, the ultramodern scientific elite" (p. 219).

One such influential leader in science is Katharine Hayhoe. She herself has mapped associated considerations of "outreach" onto a spectrum, noting that outreach is not the primary role of academic climate researchers. She characterized outreach choices across this range: on one end there is "outreach" in publications in peer-reviewed specialist journals like *Journal of Geophysical Research*; on the other end is running for elected office "because you've sold out and you're doing 100% outreach and you have 0% time to do research." Katharine Hayhoe importantly has pointed out that where you engage on the spectrum is a personal choice. Moreover, she opined that the goal is to determine where academic climate researchers can play to their strengths, their interests and where they "can uniquely contribute most effectively."[5]

In this context, it is also worth considering the intersections with the "four idealized roles of science in policy and politics" that Roger Pielke, Jr. mapped out in his thought-provoking 2007 book. In various science–policy topics, Pielke argued that the realm of "pure science" has been a working fallacy onto which scientists mistakenly cling, while "science arbiters" (who weigh in

[4] Fahy introduces a more linear view through these four steps than I promote here, but I value the elements as they work in tandem to produce academic public intellectuals.
[5] Katharine Hayhoe's full talk where she discussed this on September 6, 2018 in the Cooperative Institute for Research in Environmental Sciences (CIRES) Distinguished Lecture Series at the University of Colorado Boulder can be viewed here: www.youtube.com/watch?time_continue=2158&v=114RNExgwkg

on narrowly defined scientifically testable questions) and "issue advocates" (who argue for specific courses of action) suboptimally navigate paths of responsible engagement by relevant expert scientists. Then the optimal role is the "honest broker." According to Pielke (2007), "The defining characteristic of the honest broker of policy alternatives is an effort to expand (or at least clarify) the scope of choice for decision-making in a way that allows for the decision-maker to reduce choice based on his or her own preferences and values" (p. 3).

While Pielke rightly argued that when it comes to climate science, policy and society, things are much more complex, his "honest broker of policy alternatives" positioning still could have further dwelt in the dynamic contours of climate advocacy in US science, values and political culture. Pielke provided a useful set of starting points. Yet, the nuances of these roles can still be developed further to map onto the promises and perils of academic climate advocacy in contemporary US science, politics, culture and society.

In these movements of advocacy to activism, between evidence-based and policy prescriptive recommendations by relevant-experts, inevitably politics, values and culture pervade.[6] In addition, these choices and positionalities have the potential to change day to day, week to week or job to job, depending on a number of contextual factors. These factors range from one's mood and blood-sugar levels to deep-seated ethics or conceptions of the audience with whom they are engaging.

Overall, involvement of academic climate researchers as advocates and activists inevitably increases the scope of potential engagement in the public sphere (Bucchi, 2015). Engagement of academic researchers as public intellectuals "catalyzes wider discussions of science" so that members of the public "are brought into a wider cultural conversation about the impact of scientific discoveries, the justification for scientific work, and the value of science in society" (Fahy, 2015, p. 219). These are dynamic and contested spaces where various influencers construct and negotiate meaning while they shape and challenge public understanding and engagement. These are places where formal climate science, policy and politics operating at multiple scales permeate the spaces of the "everyday."

More than twenty years ago, then–American Association for the Advancement of Science (AAAS) president Jane Lubchenco posited that scientists needed to rethink their roles and responsibilities within society in order to capably confront environmental challenges as they related to climate change, public health, security, social justice and the economy (Lubchenco,

---

[6] For example, Theda Skocpol (2013) has lamented a perceived passivity in climate activism, and has called for more grassroots mobilization around climate change.

1998). In this call for a "rethink" there remained some dangers of being pulled into comfortable platitudes (see Preface for more), rather than pushing for radical change through specific actions. Nonetheless, Lubchenco revisited her call for a realignment of scientific engagement to meet the needs of society in 2017 when she reflected on seven reasons why she viewed academics to be ambivalent about public engagement (Lubchenco, 2017). Among these reasons she argued, "We fear our colleagues will criticize us for seeking glory by having our names in the media or label us as (the dreaded) Advocate (That's spelled with a scarlet letter A!)" (Lubchenco, 2017, p. 100). Distinguishing between Type I and Type II advocacy, she continued:

> Scientists are conflicted on the topic of advocacy. On the one hand, they feel a moral obligation to help society deal with important issues but are simultaneously cautioned that tainting science with bias will undermine the credibility of science. I can tell you that many scientists feel that they are not only scientists, but citizens, and that they have a right as citizens to express their opinions about the solutions that they think are the right ones based on both their information, but also their values. They say that they can do so in a way that's not confusing . . . Other scientists say that any scientists who voice their own opinions undermine the credibility of all scientists. They believe that any advocacy will compromise all of science. I would note that physicians are routinely advocates, and are expected to be, but do not lose their credibility in the process. Recommending that people not smoke or that they exercise does not seem to make physicians less credible. But the dialogue in the environmental science arena seems to have different rules. This is a very rich dialogue for which there is no one single answer for all scientists or all academics. Many scientists choose a middle ground in which scientists offer useful, actionable input to policymakers without making overt recommendations. For example, one can say about climate change, "This is what we know, and based on our understanding of what we know, if we choose this path, this is the likely outcome. If we choose a different path, this is the likely outcome." So you can frame answers in the fashion of choices with consequences in which you are not making overt recommendation but are focusing mostly on the scientific understanding. (Lubchenco, 2017, pp. 106–107)

Some Causes and Consequences of Academic Climate Advocacy

In a 2018 study David Oonk and I explored some of the causes and consequences of academic climate advocacy in contemporary times by surveying academic researchers and scholars (e.g., professors, postdoctoral researchers, graduate students) as we focused on Type I advocacy (advocacy for [scientific] evidence) (Boykoff and Oonk, 2019).[7] We asked them about their views of climate advocacy, soliciting responses based on the extent to which they agreed or disagreed with numerous statements. Our research built upon work by Declan Fahy (2015); Shahzeen Attari, David Krantz and Elke Weber (2016); and John Kotcher, Teresa Myers, Emily Vraga, Neil Stenhouse and Ed Maibach (2017).

Statements in our online survey included, "Academic climate advocacy *should* be carried out without endangering long-term credibility and integrity of the academic climate advocates" and "I find it appropriate for relevant expert academic researchers to engage in advocacy for evidence-based climate science" (see the Appendix for a full list of survey prompts).[8] Following these responses, respondents provided additional comments that they had in relation to these issues. Through our research we tested for main effects and interactive effects with gender, age categories and self-identification as social or natural scientists.

Among our findings, we detected broad agreement among respondents that climate change is a pressing issue, that climate advocacy should not be criticized, that it is appropriate for academic researchers to advocate for specific policies and that small carbon footprints are associated with a positive effect on influencing change. There were differences in level of agreement based on gender and between natural and social scientists. In particular, we found that women were more likely to agree that advocacy was important. This finding was consistent with previous survey work of female ecologists and higher associations with conservation (Reiners et al., 2013). More widely, this could also link to research that those with perspectives labeled "dismissive" are more frequently men (63%) (Leiserowitz et al., 2015).

Furthermore, we found that younger respondents were more swayed by advocacy from someone with a small carbon footprint than older respondents were. We also found that social scientists were more likely than natural scientists to be moved to act by someone with a small carbon footprint. This

[7] We assembled our sample similarly to approaches by John Besley, Anthony Dudo and Shupei Yuan (2018). In particular, John Besley and colleagues invited participation from five scientific organizational groups: a geological, geophysical, ecological, biological and a more general scientific society. We drew our convenience sample from societies that support both natural and social scientists, including the Association of American Geographers, the Anthropology and Environment Society of the American Anthropological Association, the Political Ecology Society, the Association for Environmental Studies and Sciences and the American Institute of Physics.

[8] These are 7-point semantic differential statements or fixed choice response formats to measure attitudes and opinions. This is adapted from Likert (1932), Bandium and Davis (1972), Paulhus (1984), Bowling (1997), Burns and Grove (1997), McCroskey and Teven (1999) and Kotcher et al. (2017).

did not support notions that seniority (and by extension possibly job security and tenure) has lower barriers and therefore acted as an accelerant for greater engagement in the public sphere. The low-carbon-footprint effect was more pronounced for social scientists than for natural scientists and for younger respondents than for older respondents.

These results provided evidence of confirmation biases, where a small carbon footprint is powerful for those who are inclined to change, and a large footprint is justification for inaction for those motivated not to change. They also were consistent with findings from Shahzeen Attari et al. (2016), who examined personal choices by use of public transportation (not intentions to fly or home energy conservation) and found that "differences in perceived credibility strongly affect participants' reported intentions to change personal energy consumption" (p. 325). While self-assessments of behavior may be lacking among climate researchers at times (Bleys et al., 2018), our research contributed to social science evidence that external assessments of how relevant-expert climate researchers taking up advocacy roles have remained critically important.

When we dug into the qualitative comments, we found some remarkable differences between respondents regarding how they viewed advocacy. For instance, some took up the perspective that advocacy was outside of their professional responsibilities, while others viewed it as integrated into their research. To illustrate, one respondent mused, "If you aren't willing to have your research used for advocacy and practical application, then why are you doing the research?" Another observed that even determined advocates for evidence-based climate science "can make matters worse if we bring our evidence-based climate science to the general public and do a poor job communicating." Yet another respondent noted, "I'd enjoy being blunt about reality, but I'd rather moderate, listen more, and try to plant seeds of information and doubt about the disinformation and try to more effectively change thinking than the stupid academic approach of winning the argument as if some cosmic scorekeeper were awarding points, instead of making some progress."

Amid these qualitative results, a respondent called for people to "listen with open minds to evidence" while another commented, "I'd much rather have expert academic researchers influencing climate policy rather than career politicians who lack expertise in climate science." These findings are consistent with the work of John Kotcher, Teresa Myers, Emily Vraga, Neil Stenhouse and Ed Maibach (2017), who found that "Climate scientists can safely engage in public dialogue about policy matters" ... "and in certain forms of advocacy without directly harming their credibility or the credibility of the scientific community" (p. 9). These findings with David Oonk also supported the assertion of John Kotcher et al., "It is a mistake to assume that *all* normative statements made by scientists are detrimental to their credibility" (pp. 9–10) and that "climate scientists advocating for action broadly may not harm their credibility" (p. 12).

Comments also delved into the importance of effective public engagement by considering target audiences and communication goals, through Type 0, Type I and Type II advocacy. One respondent noted, "It is challenging to think through the nuance here . . . I'd wonder if climate scientists are more likely to distance themselves from advocacy than the energy forecasters" while another commented, "In addition to relevant expertise, skill at advocacy (and communication) is important." Along these lines, another wrote, "I think that those who are advocating need not only be experts but also be good communicators. Scientists need training to speak with lay audiences." Meanwhile, a respondent to this survey commented, "I do not believe that those who are academic researchers should avoid advocacy for particular policies. In some cases, this is very fitting to the science itself, while others are an expressions of values. Where it overlaps the expertise, it is fine to mobilize that expertise. What is not fine is to conflate one's values with one's expertise in the public mind – so it would be critical to be clear when one is advocating for a particular policy because of one's personal values, not professional expertise."

Qualitative responses included a comment from a twenty-something-year-old that "Understanding basic climate science and the social, economic, and environmental impacts of climate change is not 'advocacy.' It's our job." This comment was consistent with views expressed by Andrew Hoffman (2016) that "young scholars are seeking more impact from their work" (p. 89). It also supported Jane Lubchenco's (2017) remark, "As a senior scientist, I don't believe that my students should follow the path that I took: establish your scientific credentials first and then begin to be more public" (p. 108). These insights could lend evidence to a notion that a deeper paradigm shift may be underfoot regarding definitions of academics and scientists contributing through their research, advocacy and public engagement in society.

Survey research into scientists' willingness to engage in the public sphere (through face-to-face encounters, speaking through the news media and professional engagement online) by John Besley, Anthony Dudo, Shupei Yuan and Frank Lawrence (2018) found, "The most consistent predictors of willingness to take part in engagement activities with the public are a belief that she or he will enjoy the experience (attitude), make a difference through engagement (response efficacy), and has the time to engage" (p. 559). In a related study, Shupei Yuan, John Besley and Anthony Dudo compared scientists' and communication scholars' views of public engagement activities by scientists. They found that while both communities viewed communication objectives similarly, scientists rated their engagement as generally high while communication scholars (working on issues relating to science, health and risk communication) viewed scientists as underperforming in their professional engagement online (Yuan et al., 2018). In that online communications world, the *Nature Geoscience* editors (2018) pointed out, "Scientists have many options for engaging with the scientific community online. But even for those who prefer not to actively engage in science communication with the broader public, there is much to be gained professionally and scientifically from an online presence" (p. 701).

Mind the Engagement Gap – But How?

Many members of academic communities have entered into various forms of engagement relating to their research every day. They have engaged in advocacy in part because they view it as part of their responsibility as contemporary climate researchers. Others have engaged because they seek to shift public conversations (Schifeling and Hoffman, 2017). However, many also shy away from (creative) applications of their work, in part owing to fear of being labeled an "advocate" (Type 0 advocacy). This chapter has sought to provoke and inspire productive deliberations on how one might navigate these fears and challenges associated with advocacy at the science–policy interface and in the public arena.

In this contemporary context, particularly on this subject of climate change, choices are made because of a number of interlinked reasons. Among them, many – who already feel inundated and/or overwhelmed by the time pressures involved in other aspects of their roles as researchers – deprioritize the importance of communication efforts. Others do so at the sacrifice of their core duties of research (Nelson and Vucetich, 2009). However, as many relevant-expert researchers calculate both the time investment and value into reaching beyond their academic circles to communicate the relevance and importance of their climate change research, there are consequential and often deleterious impacts on wider public understanding and engagement.

Andrew Hoffman (2016) has argued, "Increased engagement is unavoidable in an emerging educational context where the calibre of public discourse has become so degraded and social media is changing the nature of science and scientific discourse within society" (p. 77). Moreover, Adam Corner et al. have commented that "accepting that scientists are inevitably advocates for their work helps humanize them. Bringing science out of its academic bubble and into the public discourse allows the people in lab coats and behind data sets to be seen and heard directly; a vital step for rebuilding trust and understanding across society" (Corner et al., 2018, p. 25; see also Corner and van Eck, 2014).

Nonetheless, reticence can be explained in part by the confusion that surrounds advocacy in academia because of a failure to distinguish between and understand different types of engagement. To date, some universities have innovated through new guides and expectations relating to public engagement. However, this has entered into codified promotion and tenure incentives in only limited ways (Alperin et al., 2019). In an editorial in *Scientific American* (2018b) called "Universities Should Encourage Scientists to Speak Out about Public Issues" the editors commented that antiquated tenure and promotion policies constrained the ability of relevant experts to advocate meaningfully evidence-based science for fear of either reprisal or lack of valuation.

Yet, some faculty see this push to public engagement part of an ongoing process of "responsibilization" wherein academic researchers-as-communicators become responsible for tasks that were previously the duties of others (Goodman, 2013; Doyle et al., 2017). Director of the Hayden Planetarium in New York City Neil deGrasse Tyson has commented, "Everything I do that is not in a laboratory does not accrue to me professionally in academia" (Patel, 2018). Though lamenting that the twentieth-century tenure process generally does not account for twenty-first-century communications environments, he acknowledged some progress. He noted, "In my field now, we have come far enough that what does not accrue to my professional standing at least does not subtract from it" (Patel, 2018).

Many within and outside of relevant-expert academic research communities are puzzled by rules and norms as they relate to individual stances and their representation of their research groups or universities. As a result, they often choose to avoid the treacherous waters of advocacy, broadly construed, for fear of undertow. Others still respond to pressure from one's scientific peer communities not to overextend their outreach beyond their own specialized research programs and risk compromising their scientific credibility. They therefore largely avoid advocacy entanglements, especially with this highly contentious and highly politicized issue in the USA. Moreover, even some just simply do not have an interest in being engaged with the outputs of their work and therefore shun advocacy, often treating it as a dirty word. Hence, an "engagement gap" has emerged. Demonstrating bewilderment and concern with conflation, confusion, befuddlement and hesitancy a scholar has observed, "I think that the relationship between science and policy advocacy is incredibly sticky and fraught" (Boykoff and Oonk, 2019, p. 10).

Despite these pitfalls, in 2014 AAAS produced a report entitled "What We Know" to articulate where there are areas of common evidence-based understandings about climate change (Molina et al., 2014). When this report was released, *New York Times* journalist Justin Gillis called the report a "sharper, clearer and more accessible" explanation of climate change "than perhaps anything the scientific community has put out to date" (Mock, 2014). Five years later, they followed up with a "How We Respond" communications initiative focusing on how communities are responding to climate change through adaptation and mitigation strategies.[9]

Roger Pielke, Jr.'s "honest broker of policy alternatives" at the science–policy interface, along with the taxonomy of climate advocacy offered earlier, can be helpful guidance for practical cases where many value judgments and criticisms are often shepherded into assessments of scientific excellence. To illustrate, in February 2018 the AAAS bestowed Penn State University professor and climate scientist Michael E. Mann with the AAAS Public Engagement with Science award. According to Stephen Waldron (2018) from AAAS, the honor recognized Mann's "tireless efforts to communicate the science of climate change to the media, public and policymakers." His numerous public talks, media interviews, op-eds and commentaries along with his testimony to the US Congress and his participation in the 2017 "March for Science" in Washington, DC (see Chapter 5 for more about the "March for Science") were cited as reasons for his selection as the awardee.

While praise poured in for his advocacy for science, some backlash was also immediate. For example, in a post entitled "Michael Mann gets award for climate activism," Anthony Watts from the active right-wing blog "Watts Up with That" opined, "I guess if you are a 'loud mouthed climate activist' that equates to good science today" (Watts, 2018).

Meanwhile, Roger Pielke Jr. (2018) commented on his own "Honest Broker" blog in a post called "AAAS sends a message." He outlined a number of transgressions by Professor Mann from his perspective, before commenting, "Mann is fully entitled to his views and the use of whatever techniques he thinks appropriate to advance his extreme politics . . . my issue here is not with Mann. It is with the decision by AAAS to single out Mann as someone who embodies our highest values of the scientific community: a role model to emulate who engages in behaviours to celebrate. With this award, what message is AAAS sending to the scientific community and to the public? The AAAS is telling us that engaging in hyper-partisan, gutter politics, targeted against Republicans and colleagues you disagree with, using unethical tactics, will be rewarded by leaders in the scientific community."

While the AAAS touted Mann's evidence-based science engagement (Type I advocacy) in the public space, Pielke and Watts interpreted this as an endorsement of a particular partisan political stance (Type II advocacy). These interpretations represented ongoing confusion that often arises between advocacy for evidence-based climate science and advocacy for particular policy outcomes. Varied reactions also demonstrated how recognition for academic climate advocacy gives way to a spectrum ranging from consternation to congratulation.

This spat can reveal another set of challenges: differing perspectives on the extent to which one can and should be active (but not necessarily activists) and political (but not necessarily politicized). As another example, Timothy Sullivan et al. (2018) reviewed effective roles of policy in air quality management in the USA. Despite explicit objectives to enhance public understanding and discourse of the value of evidence-based interventions (through data gathering and analysis) over "partisan opinion and ideology" (p. 69) they found these challenges of active/activism and political/politicized. They concluded, "We increasingly hear public narratives that appear to be grounded in a post-truth world, where empirical evidence and science take a back seat to ideology . . . perhaps nowhere is this disregard for facts more evident than in discourse on environmental policy" (p. 72) (see Chapter 1 for more on post-truth or selective truth). While independence is a worthy goal, in the next section I explore how objectivity is illusory and efforts to be apolitical can be damaging.

Judge Not, Lest Ye Be Judged

Amid these many examples, however, challenges remain regarding how to decide what positions are appropriately staked out and to grapple with how they may be viewed (and judged) from these different perspectives. Again, in these spaces it is critical to acknowledge that our values imbue our considerations and interactions.[10] As was mentioned in Chapter 3, our ideologies penetrate our culturally infused perceptions as we seek to understand climate change. Rather

---

[10] For example, during a time of deep social, ideological, political and cultural struggles in Central America in the 1960s, Guatemalan poet Otto Rene Castillo captured the danger of "apolitical intellectualism" through a poem containing these lines: "One day the apolitical intellectuals of my country will be interrogated by the simplest of our people . . . And they'll ask, 'What did you do when the poor suffered, when tenderness and life burned out in them?'"

than a nuisance, considerations of how ideologies shape communication and engagement help fortify pathways to knowing about climate change. We frankly cannot "objectively" assess climate change outside these filters of understanding (Carvalho, 2007).

Moreover, it is important to recognize that these are politically saturated spaces and to pretend otherwise is damaging. This process of depoliticization refers to "any communication practice that misrecognizes or conceals the ideological values, perspectives, and choices at work in, for instance, a given policy decision or framework, thereby shielding the latter from contestation and closing democratic discussion" (Peppermans and Maeseele, 2018). Depoliticization then has a tendency to reduce climate change engagement to market-driven technocratic innovations or ongoing (albeit "green") consumption (Swyngedouw, 2010). It can also then paper over important considerations of representation and associated struggles for justice (Asher and Wainwright, 2018). This depoliticization process therefore masks the effects of embedded political economic structures and constellations of power as well as knowledge formations that shape social practices (Foucault, 1980) that are at the root of climate challenges (Carvalho, 2010; Maeseele, 2015).

Through hyper-focus on individual attitudes and actions (see Chapter 5 for more) ascertained from polling or from a moralization frame (where emissions reductions are individual responsibilities), there can also be a danger of naturalizing knowledge production and maintenance through various "fixes" (Barnes, 2001; Bok, 2018). Moreover, there is a danger of concealing textured political struggles as simplified and individualized matters of "right versus wrong" or "good versus evil" (Mouffe, 2005). Trevor Barnes (1991) has cautioned that we "must continually think critically about the metaphors we use" in order to avoid naturalizing a depoliticized terrain of engagement and advocacy (p. 118). Multiscalar considerations – steeped in politics, values and culture – then help to more capably and realistically confront present-day climate challenges through mindful advocacy and activism (Peppermans and Maeseele, 2016) and a tussling with different visions and alternatives for the future (Kenis and Lievens, 2015).

Navigating these dangers of depoliticization and overpoliticization of climate science can feel like walking a tightrope. For example, Chris Reddy and David Valentine (2017) penned an opinion piece about the role of scientists shaping policy responses to the 2010 Deepwater Horizon oil spill in the Gulf of Mexico. They observed, "In our current polarized political climate, people increasingly are predisposed to conclusions before hearing the evidence. Or they ignore, spin, or debunk evidence, and sometimes create unsubstantiated facts, to buttress their positions. Against this backdrop, many scientists feel that their voices are being drowned out, and that is stirring more scientists to step beyond their role as evidence collectors to become activists." They then argued that "We do need to be more active about communicating our findings to help policymakers make the wisest decisions. But as scientists, we believe our greatest value lies in holding – not crossing or blurring – the line that separates science from activism. If we fail to remain independent, objective and apolitical, we risk attacks that will further mute the collective voice of all scientists." They then concluded, "If we fail to remain independent, objective and apolitical, we risk attacks that will further mute the collective voice of all scientists." They then effectively call for Type 0 advocacy in order to avoid the dangers of too much politicization.

Despite their pleas, dangers of associations with overt political motivations nonetheless lurk. To illustrate, in the fall of 2018 US journalist Leslie Stahl interviewed President Donald Trump on the television program *60 Minutes*. During this conversation, she asked him, "Do you still think climate change is a hoax?" referring to a well-known Tweet of his from 2012. He answered, "I think something's happening. Something's changing and it'll change back again. I don't think it's a hoax, I think there's probably a difference. But I don't know that it's manmade." Leslie Stahl then pointed out to the president that his views are out of step with those of the US government and relevant expert climate scientists. He replied, "You'd have to show me the scientists because they have a very big political agenda."[11]

Executive Director Keith Seitter from the American Meteorological Society (AMS) responded to these comments by the president by way of an open letter dated October 16, 2018.[12] The response was a form of Type I advocacy but was encased in information-deficit model logics (see Chapter 3 for more). In the letter, he sincerely (but perhaps decidedly naïvely) pointed out to him that there is "an overwhelming body of scientific evidence that shows that the warming global climate we have been experiencing in recent decades is primarily caused by human activity and that the current long-term warming trends cannot be expected to reverse if no action is taken." Seitter also then wrote, "You also said that scientists are making this political, which is misleading and very damaging."

To be fair, from his positionality as AMS executive director this reply emphasizing scientific ways of knowing is both logical and understandable. And his expression of concern about politically motivated scientists is indeed

[11] For more, see www.cbsnews.com/news/donald-trump-interview-60-minutes-full-transcript-lesley-stahl-jamal-khashoggi-james-mattis-brett-kavanaugh-vladimir-putin-2018–10-14/

[12] The letter can be found here: www.ametsoc.org/index.cfm/ams/about-ams/ams-position-letters/letter-to-president-trump-on-climate-change/

troubling. Along these lines, Gretchen Goldman et al. (2017) have pointed out that "The public will suffer if politicization of science is normalized" (p. 697). However, to suggest that this approach might have made the president change his stance, or even take note, borders on daft. Alternative misguided approaches from academic researchers may also be to point out the logical fallacies of such a set of arguments, or to dismiss them as "deniers" or "contrarians" (see Chapter 3 for more). Yet performances of oppositions like these are all that many academic researchers generally feel they have as tools at their disposal.

This limited range of perceived responses shows that there are many challenges to confront when academic climate advocates of various hues engage in the public sphere. Among them, rather than viewing advocacy as work to "smarten up" and increase sophistication to produce resonant communications, some mistakenly view these endeavors as exercises in "dumbing down" information for public audiences. Together, these contradictions and competing approaches can then generate attention but also alienate people from intended engagement with the issues raised. Moreover, there is the challenge of erroneously concocting visions of advocacy as an inappropriate exercise of telling others where this *should* be rather than treating advocacy as a vehicle to meet people where they are. As an outgrowth of this misgiving, there is then a tendency to engage in practices of naming, critiquing and shaming instead of bridge-building through diversity and shared values.[13]

In Chapter 3 and elsewhere (e.g., Boykoff, 2013), I have argued that labels like "contrarian" and "denier" have the propensity to obscure rather than illuminate the contours of one's argument(s). In turn, I have pointed out that these labels have an obfuscating influence on public discussions (Boykoff, 2013). Saffron O'Neill and I have also cautioned that continued indiscriminate use of terms like "denier" has more potential to catalyze polarization of climate change discussions than to facilitate two-way discussions and understanding (O'Neill and Boykoff, 2010). Similarly, Candice Howarth and Amelia Sharman (2017) have argued that labels "can be used to mask particular points of view ... can be used in a pejorative and derogatory manner ... only identify those at the polarized extremes ... [and] act as fixed markers for opinion" (p. 12). Elsewhere, they have argued that labels "also serve to demonize individual debate participants" (p. 780). Furthermore, Matt Nisbet (2018a) has pointed out, "The war of labels

---

[13] See Kennedy et al. (2018) for an exploration of inclusive discursive bridge-building engagement through science festivals.

cultivates a discourse culture where protecting one's own identity, group, and preferred political narrative takes priority over constructive consideration of knowledge and evidence." Again referring to Kyle Powys-Whyte's comment (in Chapter 3) that "dissent is not the problem, it is an indicator of a problem," disagreement must be contained respectfully in order to promote productive dialogue and discourse.

"Go Public or Perish"

In February 2018, the editors of *Scientific American* (2018a) penned an opinion piece entitled "Go Public or Perish." In it, they made the observation, "If citizens never hear from legitimate experts, no one can blame them for indifference to fake-science tweets, decisions by politicians that ignore facts, or cuts to federal agencies that are supposed to be built on sound science." Moreover, Gretchen Goldman, Emily Berman, Michael Halpern, Charise Johnson, Yogin Kathari, Genna Reed and Andrew Rosenberg (2017) have forcefully stated that "the scientific community should be prepared to defend the scientific basis for public protections ... [and] the process by which scientists engage with decision makers" (p. 697).

In today's society, outreach and engagement flow through considerations of academic responsibility. Michael Nelson and John Vucetich (2009) have argued, "Advocacy is nearly unavoidable, and scientists, by virtue of being citizens first and scientists second, have a responsibility to advocate to the best of their abilities, to improve their advocacy abilities, and to advocate in a justified and transparent manner" (p. 1090). These ruminations have played out in other critical issues in the past, like advocacy and engagement from scientists who were concerned about implications of their atomic research on nuclear proliferation (Hart and Victor, 1993). These are vexing challenges that endure at the interface of science and society.

By extension, communications are critical parts of a researcher's job to share his or her findings beyond the fire-walled journals themselves. Long-time journalist Andrew Revkin (2007) has commented that the more that scientists work to communicate in the public sphere, "the more likely it is that the public ... will appreciate what science can (and cannot) offer as society grapples with difficult questions about how to invest scarce resources. An intensified dialogue of this sort is becoming ever more important as science and technology increasingly underpin daily life and the progress of modern civilization" (p. 158).

That said, these engagements are not for everyone. Some see these endeavors as new and extra burdens on an already demanding job as an academic climate researcher. Others may feel that these commitments pull time and energy from their core passions of research (and possibly also teaching). Moreover, some climate scientists may simply be bad communicators. For these reasons and others, many natural and social scientists who conduct research on topics associated with climate change may then not want to step into these dynamic and contentious arenas and effectively choose Type 0 advocacy. And that is their prerogative. While institutions and universities begin to address advocacy as a subset of public engagement, these complexities must be taken into account.

Academic researchers still have much to learn about the thickets of advocacy and activism (Green, 2017). There remain many vagaries in the realm of value–action gaps. Returning to Jane Lubchenco (2017), she closed her "Delivering on Science's Social Contract" argument commenting, "Which level of engagement you choose is a personal choice, and that you need to think deeply about the issues and make a conscious decision" (p. 107). Similarly, Karen McLeod (2017) has posited, "Figuring out where you fall on the advocacy spectrum is a personal choice – it's not a matter of right or wrong." In his call for a new social contract for global change research, Noel Castree (2016) has pointed to "the need to connect directly with non-academic constituencies" (p. 331).[14]

On the topic of climate change, how to determine what are the (acceptable) elements of a "conscious decision," what is adequately "justified and transparent" or when personal values problematically overlap with professional expertise remain open and contested concepts going forward in the USA and around the world. But engaging in these messy questions about climate advocacy by relevant-expert researchers in contemporary is both warranted and worthwhile. As Michael Nelson and John Vucetich (2009) have pointed out, "our goal here should not be simplicity but rather the betterment of society" (p. 1100). If relevant-expert climate researchers are not stepping into this public arena, then others do instead. For instance, in a 2013 publication with Shawn Olson-Hazboun, we documented that CATO Institute Senior Fellow Patrick Michaels was paid – largely by carbon-based industry interests – to spend up to 40% of his working hours doing media outreach in the public arena (Boykoff and Olson, 2013). This type of commitment to communication can therefore serve to amplify certain outlier perspectives on key climate science–policy

---

[14] David Demeritt (2000) has also called for a new social contract for science through the dimensions of accountability, relevance and value.

issues and shape public understanding, discourse and engagement in the public sphere.

As much as academic researchers are trained extensively in the scientific process, communication and engagement are often still seen as an extension of their responsibilities and competencies rather than components of it. Yet, generationally, social science research has revealed some trends that views of academic climate advocacy and activism may be changing (e.g., Boykoff and Oonk, 2019). Furthermore, there are institutional signs of support for a generational conversion toward engagement and trainings (see Chapter 2 for more).

More widely, Nobel Prize–winning physicist Max Planck (1936) argued, "Important scientific innovations rarely make their way by gradually winning over and converting their opponents ... what does happen is that their opponents gradually die out and the growing generation is familiarized with the idea from the beginning" (p. 97). Planck took up a decidedly optimistic view that overcomes the effects of repeated and reinscribed practices of the proverbial hiding away in ivory towers.

Questions about when and how to advocate as well as what one advocates for (from evidence-based science to particular actions, or Type I and Type II advocacy) remain contentious and challenging. For instance, advocacy for climate researchers to "walk the walk" by no longer engaging in air travel has been a contentious (and possibly taboo) topic that has stirred and unsettled public discussions and considerations of actions (Dolšak and Prakash, 2018; Knox, 2018). In an essay entitled "No One's Perfect," Shahzeen Attari (2018) has commented that conserving energy in our own lives is a way to stave off ad hominem attacks about contradictions like flying and eating meat while advocating for climate policy action. She added, "We need to think about shifting practice, people, politics and policy" to thereby "inspire others to change and support needed climate policies" in order to ultimately build "political will and public support to decarbonize our energy system."

This chapter has sought to better understand the contours of what may be a social contract for communication about climate change in the twenty-first century. As climate change cuts to the heart of how we live, work, play and relax in modern life, engagement through research and through communications entail reflection on how our personal lives mesh with our professional ones. Through these explorations in Chapter 6, I have worked to more effectively map out how advocacy may be pursued in various forms as it operates in this high-stakes, high-profile and highly contested milieu. By distinguishing between Type 0, Type I and Type II advocacy, I have aimed to provide firmer footing for academic climate researchers to stand, without risk of sliding down a slippery slope to banishment from a constructive communications environment. With this framework in mind, it becomes more difficult to dismiss advocates for scientific evidence as some caricature like 'left-wing socialists seeking to dismantle US power around the globe'. This framework also helps to strengthen rather than tarnish the reputation of science through politically relevant advocacy and activism.

There are many reasons why academic researchers and scholars may be reticent to engage in Type I and Type II climate advocacy. However, as these folks are unwilling to invest time and value into reaching beyond their academic circles to communicate the relevance and importance of their work in fear of being labeled an "advocate" or "activist," the fact is that they effectively risk impairing public understanding and engagement when they squander their opportunities to share their relevant insights. This is then effectively Type 0 advocacy-as-inaction. When those recoiling from spaces of advocacy for evidence-based climate research (characterizing Type 0 and Type I advocacy) are the relevant experts who hold insights for useful and informed commentary, I ultimately argue that they should be viewed as missed opportunities to attend to their present-day responsibilities of meeting people where they are on climate change. This chapter has interrogated varied approaches to climate advocacy in today's highly politicized environment in order to more effectively trace a current engagement gap plaguing ongoing climate communications. The next two chapters describe ways in which these considerations can be usefully addressed going forward.

Appendix

List of survey prompts and dependent variables from Boykoff and Oonk (2019).

1. Climate change is an important issue in 2018.
2. Academic climate advocacy *can* be carried out while ensuring optimal use of scientific knowledge in policy.
3. Academic climate advocacy *can* be carried out without distorting truth(s)
4. Academic climate advocacy *can* be carried out without endangering long-term credibility and integrity of the academic climate advocates
5. Academic climate advocacy *should* be carried out while ensuring optimal use of scientific knowledge in policy.
6. Academic climate advocacy *should* be carried out without distorting truth(s).

7. Academic climate advocacy *should* be carried out without endangering long-term credibility and integrity of the academic climate advocates.
8. I would take advocacy advice from a climate researcher even if climate researchers themselves personally had a *large* carbon footprint.
9. I would take advocacy advice from a climate researcher if climate researchers themselves personally had a *small* carbon footprint.
10. Advocacy from climate researchers with *large* carbon footprints compel me to make changes in my own energy consumption.
11. Advocacy from climate researchers with *small* carbon footprints compel me to make changes in my own energy consumption.
12. Academic climate advocacy should be *criticized* in these contemporary times.
13. It is a bad thing that many academic researchers shy away from applications of their work, at times in fear of being labeled an "advocate"
14. Men are *more likely* than women to advocate for evidence-based climate science.
15. Men are *more likely* than women to advocate for particular climate policies and actions.
16. White climate researchers are *more likely* than minority climate researchers to advocate for evidence-based climate science.
17. White climate researchers are *more likely* than minority climate researchers to advocate for particular climate policies and actions.
18. Older climate researchers are *more likely* than younger climate researchers to advocate for evidence-based climate science.
19. Older climate researchers are *more likely* than younger climate researchers to advocate for particular climate policies and actions.
20. When those recoiling from communications and engagement are the relevant experts who hold insights for useful and informed commentary, avoided advocacy is a critical mistake.
21. I find it appropriate for relevant expert academic researchers to engage in advocacy for evidence-based climate science.
22. I find it appropriate for relevant expert academic researchers to engage in advocacy for particular policies and actions.

7

Silver Buckshot

Know thy audience, know thyself, know thy stuff.
–*Stephen H. Schneider*[1]

This adage has been a useful guide for effective communications about climate change. In this chapter, I highlight selected multimodal works to illustrate the many ways that climate change is being represented, portrayed and communicated in the public sphere. These artefacts have manifested through fine arts, documentaries and science fiction films, videos, theatre, photography, art installations, imagery, internet-based projects, sonification, creative writings and more. I then draw together themes, topics and illustrations from the previous chapters and assemble features on a "road map" along with "rules of the road" for ongoing creative climate communications. I pursue these elements in the spirit of bringing research to bear on social change while accounting for the ongoing importance of temporal and spatial context.

Archaic views are ones that take up a perspective that communication and engagement in the public sphere are outside of one's responsibilities; current views are that they are integrated. While many individuals are increasingly acknowledging and adopting a more integrated perspective, many structures and institutions are still working to still catch up. While connections between information sharing and attitudes, perspectives, intentions, decision-making, and behavioral change are far from uncomplicated, experimentation is critical to successful communications about climate change going forward.

Alexander Jutkowitz (2017) has commented, "Innovation and creativity are the defining words of this collective moment" (p. x). Experimentation has been moving forward with a recognition that progress does not happen on its own: rather, it requires multiscale – from individual to collective – struggles along

---

[1] A reference to this adage can be found here: www.sciencemag.org/news/2010/02/know-thy-audience-know-thy-self-know-thy-stuff

with sustained endeavor. It also requires empowerment of experimentation through philanthropic support (Luers, 2013), investments in science and society, and knowledge production through social science research in creative climate communications as it relates in our personal and collective futures. These efforts defy linear and modernist ideologies of progress. These spaces of innovation, creativity and experimentation necessarily then embrace co-temporal, nonlinear and open view knowledge generation and learning (see Chapter 3 for more).

In terms of experimentation, Ezra Markowitz and Meaghan Guckian (2018) have highlighted "the critical need for communicators and issues advocates to continue experimenting with techniques and approaches that may be able to engage multiple, diverse audiences simultaneously" (p. 51). David Roberts (2017b) has posited, "it seems the wise course of action on climate communications is to encourage diversity, experimentation, and most of all, a spirit of charity and the assumption of good faith toward others who are attempting to tell the same story in different ways" Susanne Moser (2017) has challenged researchers and practitioners to consider how "competencies, resources, institutional support, and interactions with each other" can help effectively connect on climate change in our current communication ecosystem (p. 361). Moreover, Amy Luers (2013) has called for greater experimentation and more coordination between academic researchers and practitioners "to build a culture of learning" (p. 13).

One of the leading examples of practitioner groups seeking climate change engagement through creative forms of communication is a group called Connect4Climate. From their positioning within the World Bank, Connect4Climate have described themselves as "a campaign, a coalition, and a community that cares about climate change." Their stated goal has been to "create a participatory, open knowledge platform that engages the global community in climate change conversation to drive local action through advocacy, operational support, research, and capacity building." This initiative was first established through a novel partnership between the World Bank and the Ministry of Environment and Territory of the Government of Italy. Together, they first established the Communication for Climate Change Multi-Donor Trust Fund at the World Bank in January 2009 to raise awareness about climate change at multiple scales, to promote multisector collaborations and partnerships and to build wider coalitions for climate action. This partnership was designed to advance work in communication for development by building climate change awareness, and to enhance initiatives and efforts seeking to improve climate adaptation.

The Connect4Climate initiative began in September 2011, through the initiative and commitment of Lucia Grenna. Along with her team, Lucia sought to pursue shared goals of improved climate communications, enhanced culturally sensitive knowledge sharing and more effective coalition-building on climate between policy actors, grassroots communities and members of the private sector. Bridging from 2009 to 2001, a key *World Development Report: Development and Climate Change* from the World Bank in 2010 demonstrated budding institutional support for these endeavors as it emphasized critical influences of strategic communications programs in accelerating climate action (or overcoming barriers to engagement). As Connect4Climate formed, a three-pronged approach was articulated to achieve these stated goals, within the context of World Bank institutional architectures: (1) support wider World Bank and World Bank partner operations, (2) offer research and capacity building within the World Bank and between the World Bank and its partners and (3) advocate for the leveraging of funds to support to ongoing and future initiatives.

The first project of Connect4Climate in 2011 centered on drawing attention to the voices of African youth (ages 13–35) in ongoing climate discussions in the public arena. Leading up to the United Nations Conference of Parties meeting that year in Durban, South Africa (COP17), Connect4Climate solicited personal stories on agriculture, energy, forests, gender, health and water. An impressive total of more than 700 entries were received, and submissions were entered from all the recognized sovereign nations of Africa at that time.

From the successes of that initial photo and video competition, Connect4Climate went on to engage more widely with the global citizenry primarily through popular culture. For example, multiple projects connected climate themes with production, distribution and consumption dimensions of fashion. The first of these projects was "Sustainability Dialogues in the Design Industry," mobilized during Milan Design Week. Another example has been that of connections to climate change through sport. #Sport4Climate activities and partnerships – such as with Instituto E and Fluminense football club in Rio de Janeiro (launched at World Environment Day events of June 2014) – have engaged with novel and varied segments of the private and public spheres.

Despite the untimely passing of Lucia Grenna in 2017, a leadership team including Giulia Camilla Braga, Max Thabiso Edkins[2] and Francis Dobbs has continued to move forward with an ambitious portfolio of projects including fashion, sport, music, film, banking and financial services and youth

[2] After the book's writing, Max tragically died along with 156 others in the Ethiopian Air disaster on March 10, 2019. www.connect4climate.org/initiative/loving-memory-max-thabiso-edkins-champion-grand-transformation

empowerment. Together, these projects have taken advantage of their influence within the World Bank institution and have engaged with influential partners across the public, policymakers and the private sector to find traction on climate change in the public sphere.

As another leading example of innovative avenues for enhanced climate change engagement through creative forms of communication, through research and experimentation, "serious games" have developed to creatively respond to and support climate risk management and collective adaptation efforts. "Serious games" refer to a set of initiatives and experiential learning activities designed to help confront climate-related challenges (Suarez et al., 2013). These strategies translate complex and formal climate science–policy concepts in ways that resonate and "stick" with everyday citizens (Suarez, 2009).

"Serious games" span a range of digital and analog modes with the intention of sparking learning, dialogue, participation, engagement, trust and empathy (e.g., Gordon and Schirra, 2011; Suarez et al., 2015; Gordon et al., 2017). These games help make climate challenges more meaningful for typically under-served populations at the forefront of climate impacts by integrating participatory learning "whereby knowledge is created through transformation of experience" (Kolb, 1984, p. 38). "Serious games" are founded on principles from decision sciences that have found that experiential learning is an effective means of learning and knowledge-sharing (Kemmis and McTaggart, 2000; Lieberman, 2006, Mitgutsch and Alvarado, 2012). Learning therefore emerges from participatory endeavors that have been crafted to invoke an understanding of opposing ways of seeing and dealing with the world (Osnes, 2012; Mendler de Suarez et al., 2012).

"Serious games" align with many of the communications objectives sought through an approach that has been referred to as community-based adaptation (CBA). Both endeavors pursue multimodal, collaborative, inclusive approaches to increasing resilience and anticipatory capacity (McGonigal 2011; Koelle and Annecke, 2014). Through these activities, participants become agents in their own education (Winn, 2009). Serious games and CBA approaches are also grounded in participatory learning and action (PLA) methods (Wadsworth, 2001; Koelle and Oettle, 2009). These nonhierarchical processes work with participants, rather than merely delivering information to them. PLA methods also value multiple pathways of learning and knowing about climate change and other associated issues.

From their respective vantage points, serious games, CBA and PLA initiatives and projects very deliberately seek to reach particular audiences through creative, resonant and carefully constructed communications. These are prominent examples amid burgeoning spaces of experimentation that, in the aggregate, help to better understand how and why creative (climate) communications work in certain contexts and conditions. Additional examples of multimodal initiatives illustrate the expanding spaces of creative climate communications in recent years.

Show and Tell

Presented somewhat chronologically and somewhat modally, these selections are nowhere near an exhaustive set of samples that show and tell creative communication approaches. These illustrations help profile innovative individuals and teams,[3] and indicate the varied multimodal ways that they creatively promote learning and knowing about climate change in today's public sphere.[4] Again, these creative (climate) communications pursue a range of goals and objectives, from enhancing individual- to collective-scale awareness raising and behavior change to improving education and literacy as well as catalyzing cultural change.

Some influential early forays into these spaces of creative communications about climate change have helped to set the stage for creative works that have followed. For a first example, painter Alexis Rockman is an artist who has long addressed scientific themes in climate change and genetic engineering through his depictions of future North American landscapes. He first waded into these spaces two decades ago, through his 1999 oil painting "Manifest Destiny." This is a 24-foot-long exploration of a setting of the Brooklyn waterfront after land-based glaciers melt. The piece now has been on display in the Smithsonian Museum of American Art.[5] When asked about how he researches these connections as he creates works of art communicating about climate change, Rockman responded, "I start like a journalist. I ask questions, and I figure out what is the interesting thing going on. Then I find out who knows about it, and then I try to talk to them. Before I begin, I'll go out to see things first hand" (Dreifus, 2016).

---

[3] In 2018, *New York Times* reporter Tatiana Schlossberg profiled innovators from varied sectors – including Kim Cobb from Georgia Tech working on communications through the climate sciences and Kendra Krul from Opus 12 working on turning carbon dioxide emissions into fuels – confronting climate challenges.

[4] This also highlights English-language and North American or European communication ecosystems and biomes, owing primarily to the positionality of myself as the author. That by no means is meant to diminish or distract from creative communications in other parts of the globe in other languages as well.

[5] More about Rockman's work can be found here: www.honoringthefuture.org/climate-smarts /artist-to-know/alexis-rockman/

As a second example, "Cape Farewell'" was created in 2001. First started by artist David Buckland, this initiative has grown into an international program that has engaged with influential artists to help them develop projects that inspire, transform and engage society. The group has hosted exhibitions (such as Carbon 12, Carbon 13 and Carbon 14 in Paris, Texas and Toronto, Canada); "encounters" with prominent "creatives, scientists and informers" discussing various aspects of a changing climate (at Oxford University); and expeditions to the High Arctic and to the Peruvian Andes.[6] These activities have been devised to inspire and activate influential actors to respond creatively to the climate challenges in the twenty-first century.[7]

As a third early illustration of creative (climate) communications, in 2003 Kate Evans created the climate comic book *Funny Weather: Everything You Didn't Want to Know About Climate Change but Probably Should Find Out*. With an introduction by George Monbiot, this was an approachable illustrated story of a teenage girl, a big businessman and a quirky scientist that provided both entertainment and instructional value about issues from the greenhouse effect to carbon sequestration.

Documentaries as well as **science fiction films** have sought to open up ways of learning and knowing about climate change. For example, in 2004 the science fiction disaster film *The Day After Tomorrow* was released in theatres. Directed by Roland Emmerich, the lead protagonist (played by Dennis Quaid) was a paleoclimatologist named Jack Hall. Jack conducted ice core research on the Larsen B ice shelf for the National Oceanographic and Atmospheric Administration (NOAA) and also spoke at the 2002 United Nations climate talks in Delhi, India (COP8). From there, as the result of a (scientifically inaccurate) rapid shutting down of the North Atlantic Oscillation, catastrophic events ensued. Of note, these events were part of a sudden (and very unrealistic) global cooling in North America and a swiftly developing ice age. Despite scientific shortcomings, the film did well at the proverbial box office.

Two years later, in 2006 *An Inconvenient Truth* debuted. This was a documentary film directed by Davis Guggenheim that followed former US vice president Al Gore's multiyear and multicountry efforts to raise awareness and engagement about climate change. Tethered to his slide show presentation, the film also captured causes and consequences of climate change around planet Earth. As a relatively successful documentary at the box office, the film sparked many discussions about climate change at the

---

[6] Disclosure that I was on an organizing team for encounters at Oxford University in 2007–2008.
[7] More information about the Cape Farewell project can be found here: https://capefarewell.com /about.html

time that the Intergovernmental Panel on Climate Change (IPCC) released its Fourth Assessment Report (AR4) (Solomon et al., 2007). In part because of the success of the film and the added attention paid to climate change, in 2007 Al Gore and the IPCC were awarded the Nobel Peace Prize. The Nobel Committee stated that the award was earned "for their efforts to build up and disseminate greater knowledge about man-made climate change, and to lay the foundations for the measures that are needed to counteract such change."[8]

As another example, in 2009 in the UK, through innovative crowd-funding and distribution the dystopian science fiction thriller *The Age of Stupid* was released with a good amount of British fanfare. This was an imaginative apocalyptic film set in the year 2055, on a planet (Earth) wrought with the impacts of climate change. The protagonist was an archivist who was tasked with saving records from the arts and sciences of the past. Directed by Franny Armstrong, it communicated a dire prognosis for the future unless significant shifts were made regarding human activities contributing to negative climate change reverberations.

In 2016, the documentary film *Before the Flood* was released in theatres and then on the National Geographic Channel. Co-created by Fisher Stevens and Leonardo DiCaprio, this film documented the causes and consequences of climate change through travels around the world. The film also delved into policy tools such as a carbon tax, as well as efforts (and barriers) to decarbonization. Then, just over ten years after the release of *An Inconvenient Truth*, Directors Bonni Cohen and Jon Shenk released *An Inconvenient Sequel: Truth to Power.*[9] This 2017 documentary depicted Al Gore's continuing efforts to help confront climate change through his ongoing talks and endeavors. It also portrayed multiscale struggles to engage with climate change causes and consequences through culture, politics, economics and society. The film also gave space to Gore to reflect on his involve-ment in the UN Conference of Parties meeting in Paris, France in December 2015 that led to the Paris Climate Agreement.

Approaches have manifested as well through **videos** and **shows** (on television or on the internet) that have also endeavored to spark ways of learning and knowing about climate change. In 2010, Tom Friedman popularized the term "global weirding" in a *New York Times* opinion

---

[8] More information about the award can be found here: www.nobelprize.org/prizes/peace/2007/summary/

[9] Jon Shenk also directed the 2012 documentary film *Island President*, featuring then–Maldives president Mohamed Nasheed. This was a film that traced Nasheed's efforts to raise awareness about threats of sea level rise and storm surge on the residents of the Maldives as well as other island nations.

piece.[10] He argued, "Avoid the term 'global warming'." I prefer the term "global weirding," because that is what actually happens as global temperatures rise and the climate changes. The weather gets weird. The hots are expected to get hotter, the wets wetter, the dries drier and the most violent storms more numerous" (Friedman, 2010, p. A23). In the years that followed, Katharine Hayhoe built upon this notion by establishing a series of short YouTube videos answering questions about global warming and climate change. This project, entitled "Global Weirding," has been produced by KTTZ Texas Public Media and distributed by PBS Digital Studios (Peach, 2016). These short videos have garnered attention in the public sphere by clearly confronting many questions about climate change that members of the general public may have in mind but are potentially hesitant to ask, like "If I just explain the facts, they'll get it right?" (see Chapter 3 for more on this subject)[11] and "Developing countries need fossil fuels to reach the standard of living we enjoy, right?"[12] These six- to ten-minute videos have earned thousands to tens of thousands of views on YouTube.[13]

In 2013, Beth Osnes created a series of short live performances in a project she called "Striking the Match." Originally performed live, she then recreated the performances as stand-alone videos posted on the internet to illuminate issues of sustainability (Osnes and Gammon, 2013). The project comprised twenty-six short segments, all performed by Beth Osnes herself. As one example of a piece, Beth played a doctor in "You've Got the Big CC." In this one-minute video, Beth wore a white lab coat while she was seated in a doctor's office. She looked directly into the camera to tell a patient the results of the patient's medical tests. With a heartfelt expression, she said, "This is the worst part of my job. I'm really sorry to tell you this ... but you've got the big CC. You've got climate change. All the tests came back positive. I had the chance to share the results of your tests with scientists across the world, and 97% confirmed that you've got climate change." She paused again and then disclosed, "What do I think you should do about it? Nothing, just go about your daily life as though nothing is wrong. Go on, have fun. Just don't let it bother you." This dry humor helped to creatively draw out the contradictions between the diagnosis facing humanity and the actual business-as-usual response.[14]

---

[10] Hunter Lovins, co-founder of the Rocky Mountain Institute, has been credited with coining "global weirding." This term first emerged around 2007 as a way to describe how extreme weather can be explained by human contributions to climate change.
[11] This can be viewed here: www.youtube.com/watch?v=nkMIjbDtdo0
[12] This can be viewed here: www.youtube.com/watch?v=h687bvUB5jI&t=23s
[13] See www.pbs.org/show/global-weirding/ for more.
[14] See this performance piece here: www.youtube.com/watch?v=uBFaiQgeMiU

In 2014 Joel Bach and David Gelber created "Years of Living Dangerously" (YOLD). This was a documentary television series that ran for two seasons. YOLD depicted scientific, economic, ecological, political and cultural dimensions of climate change. The first season aired on Showtime while the second season appeared on the National Geographic Channel. Supported by executive producers James Cameron and Arnold Schwarzenegger, YOLD sought to overcome political partisanship and polarization as it explored climate impacts on everyday people around the globe. Through a telling of human stories, YOLD episodes found traction on common and often otherwise missing middle grounds. The show creators uniquely deployed cultural celebrities to serve as de facto investigative journalists interrogating how climate change has manifested in threats such as sea level rise, storm surge, desertification, deforestation, ocean acidification and species extinction. With hosts including Sigourney Weaver, Harrison Ford, Jessica Alba, David Letterman, Don Cheadle, Olivia Munn, Jack Black, Matt Damon, Aasif Mandvi and America Farrera, the episodes took viewers from palm oil production in Indonesia to coal burning and air toxics near the US city of Chicago.

As another example, John Oliver aired a segment on his show *Last Week Tonight with John Oliver* in 2017 where he accused Robert Murray – CEO of Murray Energy Corporation – of fighting coal industry regulations meant to protect workers (Wang, 2017). John Oliver recounted the 2007 collapse at the Crandall Canyon coal mine in Utah that killed six miners. A government investigation found that the collapse was caused by unauthorized mining practices. John Oliver called Murray a "geriatric Dr. Evil" as Murray claimed that the accident was instead caused by an earthquake. After the show's airing, Murray promptly filed a defamation lawsuit. This suit was later dismissed (Brown, 2017).

Scripted television shows have increasingly incorporated themes associated with climate change (Smith et al., 2018). For example, the seventh season of *Game of Thrones* (aired in 2017) contained the storyline of the need for widespread action in the face of threats of White Walkers destroying the seven kingdoms. Speculation for some time was that this was a subtle parable for the collective action problem associated with climate change (Skibell, 2017b) but *Game of Thrones* book author George R. R. Martin cleared it up in an interview, stating that there have been deliberate parallels (Sims, 2018, p. A2). He commented that his characters

> are fighting their individual battles over power and status and wealth. And those are so distracting [for] them that they're ignoring the threat of 'winter is coming,' which has the potential to destroy all of them and to destroy their world . . . there is a great parallel there to, I think, what I see this planet doing here, where we're fighting our

own battles. We're fighting over issues, important issues, mind you – foreign policy, domestic policy, civil rights, social responsibility, social justice. All of these things are important. But while we're tearing ourselves apart over this and expending so much energy, there exists this threat of climate change, which, to my mind, is conclusively proved by most of the data and 99.9% of the scientific community. And it really has the potential to destroy our world . . . really, climate change should be the number one priority for any politician who is capable of looking past the next election. But unfortunately, there are only a handful of those. (Sims, 2018, p. A2)

Numerous **online media projects** have sprung up to creatively contend with climate challenges in the public sphere. For example, the "Creative Climate Diary Project" was a relatively early foray into these multimodal spaces of creative (climate) communications. It was started in in 2009 by Joe Smith and others at Open University in the UK, to create a living archive of online entries from individuals up to 2020 in order to better understand how people apprehend climate and environmental change from varying perspectives. Juxtaposing perspectives from scientists, artists and religious leaders, this project has intended to provide a compelling account of how humans are confronting climate change.[15] As another online media tool, Mark Trexler created the decision support tool called "The Climate Web" in 2010. This was designed to help crowd-source knowledge to address climate change by providing open access to information from relevant climate experts around the globe.[16] As a third example, "Dear Tomorrow" has been a living and growing digital archive of "personal, hopeful messages about climate change to inspire new thinking and action."[17] This has been a deliberate effort to bridge political, social and cultural divides through connecting frames of love, family and legacy. The project co-creators, Dr. Trisha Shrum and Jill Kubit, began "Dear Tomorrow" in 2015 as a way to address future-looking intergenerational considerations on climate change. They have sought to provide pathways for physically distant but emotionally adjacent concerns about climate change threats in today's society by utilizing trusted messengers and multimodal narrative storytelling. They have set a goal of reaching 20 million people by 2020 with an enduring archive of more than 10,000 personal messages about climate change, through letters, videos and photos intended for future generations or one's future self. They have solicited public-facing entries that address why climate change is important to future inhabitants of this planet, in order to

---

[15] The project can be viewed here: www.open.edu/openlearn/nature-environment/the-environment/creative-climate
[16] More information about the Climate Web can be found here: www.theclimateweb.com/
[17] This website can be found here: www.deartomorrow.org/

Figure 7.1 Logos from a selection of the many alternative-to-government-agencies Twitter accounts that sprung up in early 2017 to promote fact-based communication.

prompt heartfelt commentaries and resonant pleas that can be shared further through new and social media channels.

Through online media in the wake of the 2016 US presidential election, a number of creative Twitter accounts sprung up in the early months of 2017. These alternative Twitter accounts included groups like @RogueNASA, @altUSEPA, @AltForestServ @altNOAA and @alt_fda (see Figure 7.1). The first @AltNatParkSer tweet read "Hello, we are the Alternative National Park Service Twitter Account activated in time of war and censorship to ensure fact-based education." Journalist Jessica Goldstein (2017) reported, "The tone is tongue-in-cheek, but the messaging is clear: Climate change is real. Facts are facts. The public has the right to know the truth. Science should not be dictated by politics. Attempts to silence official means of communication will only spark alternative means of communication."

In 2017, artists Marina Zurkow (behind the "Dear Climate" project from 2012 described later) and Viniyata Pany along with others at New York University's Tisch School of the Arts launched "Climojis" (see Figure 7.2). These were made available for iPhones and Androids, meant to "connect audiences to difficult issues worldlessly, emotionally and with humor as grave as the issues may be" (Cusick, 2018).[18]

In the gaming world, Angela Watercutter (2018) from *Wired Magazine* covered a development where climate science has deliberately been communicated through the popular online game *Fortnite* (first released by Epic Games in 2017). She wrote, "That's how science goes on ClimateFortnite, a channel full of climate scientists who discuss issues of global warming while playing Fortnite on Twitch, hoping the platform's massive reach will get their message in front of more eyeballs."[19] In the piece, she quoted climate scientist

---

[18] See https://climoji.org/ for more.
[19] See www.twitch.tv/climatefortnite for more.

Figure 7.2 Examples of Climojis created in 2017, made available for texting communications.

Andrew Dessler, who observed, "Scientists do a good job of communicating via traditional routes – talking to journalists and policymakers and writing op-eds. But not everyone listens to policymakers or reads op-eds or follows scientists on Twitter. ClimateFortnite is a great way to reach people who don't get news from traditional sources" (Watercutter, 2018).

Climate themes have also been emergent through creativity in **theatre**, **photography** and **art installations**. For example, in 2009 artist and playwright Chantal Biladeau founded the Artists and Climate Change blog and began a "Climate Change Theatre Action" initiative that included a project called "The Arctic Cycle" to accelerate confrontations with a collective future in the face of climate challenges. Also in 2009, scientist Gavin Schmidt and photographer Josh Wolfe teamed up to publish *Climate Change: Picturing the Science*. This was a creative collection of essays written by climate researchers along with visuals such as photographs, figures and schematics to draw out stories of a changing climate through different ways of knowing about the issues.

Around that same time period, a network of ten arts organizations in the European Union began "Imagine2020." Groups included Artsadmin (from London, England), Bunker (based in Ljublijana, Slovenia), Coalition pour l'art et le développement durable (COA) (from Paris, France), Teato Maria Matos (from Lisbon, Portugal), Kaaitheater (based in Brussels, Belgium), Kampnagel Center for the Finer Arts (from Hamburg, Germany), The Rotterdam Theatre (from Rotterdam, Netherlands) and the New Teatre Institute (based in Riga, Latvia). This network has sought to raise ecological awareness and engagement through the arts, by funding works that creatively communicate about climate change.[20]

In 2012, Marina Zurkow, Una Chaudhuri, Fritz Ertl and Oliver Kelihammer created "Dear Climate" (see Figure 7.3). This has been a growing set of posters and installations to help people get more intimately acquainted with dimensions of climate change. With the tagline "Meet the climate, befriend the climate, become the climate" the creators have framed these creative climate communications through a mix of counter-intuition, humor and jest to garner attention and trigger "affection" for various climate-related issues.

In 2014 (through 2015) in Taipei, Taiwan, an exhibition including performances, conferences and publications called *The Great Acceleration* was a part of larger Biennial celebrations there. The title denoted the Anthropocene era and featured representations of challenges at the human–environment interface. Focusing on the interactions of the arts, machines and nature, the activities centered on the idea that humans have been in fact just one element among

---

[20] More about these efforts can be found here: www.imagine2020.eu/home/

Figure 7.3 Some artefacts from the "Dear Climate" project, created in 2012 by Marina Zurkow, Una Chaudhuri, Fritz Ertl and Oliver Kelihammer.

others networked across planet Earth. This was a provocative Biennial that enrolled more than fifty artists and collectives from many different countries including Brazil, Japan, Germany, Poland, Argentina Thailand, Sierra Leone, Netherlands, Korea, Iceland Belgium, France, Algeria, the UK, USA, Taiwan and China.[21]

Also, ART+CLIMATE=CHANGE festivals have been staged in 2015 and 2017 in Melbourne and Victoria, Australia. These festivals – comprising more than two dozen exhibitions as museums in the region – have recruited participation from scientific researchers, policy decision-makers and artists as they have sought to promote pro-environmental actions related to climate change, and efforts to achieve net zero emissions by mid-century. These have been led by Guy Abrahams and Bronwyn Johnson, with the support of Bank Australia, the University of Melbourne and the Victorian government.[22]

The American Geophysical Union (AGU) meeting in December 2017 included a creative form of climate communications by way of a contribution to the poster session by Miles Traer, Ryan Haupt and Emily Grubert. Entitled "Stop Saving the Planet! Carbon Accounting of Superheroes and Their Impacts on Climate Change," this represented research of carbon footprinting for various superheroes (Showstack, 2017). For example, they found that Batman's annual carbon footprint was calculated to be 5.5 million pounds,

[21] More information about the Great Acceleration can be found here: www.e-flux.com/announce ments/30740/taipei-biennial-2014-the-great-acceleration/

[22] More information about ART+CLIMATE=CHANGE can be found here: www .artclimatechange.org/about-us/

Figure 7.4   A construction billboard – eleven feet by seven feet in size – among ten total that were placed in public locations around New York City by artist Justin Brice Guariglia. (Photo by Lisa Goulet)

where his bat wing airplane accounts for 31% of his footprint if he flew 48 hours a year while his batmobile accounted for just 1% if he only drives 20,000 miles a year at 8 miles per gallon. It was an effort to draw creatively on science fiction in order to make real science points. Traer commented to journalist Randy Showstack (2017) of *Eos* that linking science to popular culture "was a great way of getting people to engage with the science … It's a way of tricking people into learning."

In fall 2018, there was also a provocative exhibition that popped up around the boroughs of New York City through the work of artist Justin Brice Guariglia and the Climate Museum (see Figure 7.4). Guariglia repurposed construction semaphores to communicate climate-related messages through the "Climate Signals" project. Ten solar-powered highway signs flashing text with climate messages like "Climate denial kills" and "Climate change at work" in order to garner attention and spark conversations about climate change. This was not a renegade exhibition; rather it was installed in partnership with the Climate Policy and Programs unit of the New York City Mayor's Office as well as with eighteen additional local programmatic partners. Devi Lockwood (2018) wrote about the project, noting that "the Climate Museum aims to humanize and personalize climate change."

And in 2019, British artist John Akomfrah installed a six-channel immersive video display called "Purple" in the Boston Harbor Shipyard. This effort built on the works from the Arctic Cycle and the Climate Museum to foster visceral ways of knowing about climate change. This artwork has woven together video footage of climate impacts on people and landscapes in regions around the globe (Kagubare, 2018).

Moving to sound, **sonification** of climate data has blossomed in the past decade (see Chapter 4 for more). Sonification is defined as the use of music and other auditory techniques to convey information or data. For example, Julio Friedmann (chief energy technologist at Lawrence Livermore National Laboratory) and Michael Loomis (a Lawrence Livermore National Laboratory programmer) created "Fiddling a Warming Tune" in 2013. This was the sonification of 600 years of global temperature data (1400–2012) (Revkin, 2013).

As another example, Marco Tedesco (a marine geologist and geophysicist) and Jonathan Perl (a musicologist) at the City College of New York created "Greenland Melt Music." This was sonified data of changing reflectivity of ice and snow (called "albedo") and the melting of the Greenland ice sheet. This project was said to have come about through Tedesco's interest in representing melting ice in far-flung places like Greenland (where he has done fieldwork) in ways that are accessible and visceral for everyday people (Fitzgerald, 2013). Jonathan Perl helped with the project, commenting "When you take the time to actually listen to it, you experience what's actually happening in a way that is more visceral, and has a different kind of impact than if you just look at a graph." Reporter Emma Fitzgerald (2013) noted, "there are thousands of global warming graphs and statistics out there, but given what 2012 sounded like, hearing climate change may bring the message home."

As a third example, Daniel Crawford at the University of Minnesota wrote "A Song of Our Warming Planet" which was analog data sonification of annual global temperatures from 1880 to 2012 (Revkin, 2013). Also, two years later with Julian Maddox, Jason Shu, Alastair Witherspoon and Nigel Witherspoon, Daniel composed and performed "Planetary Bands, Warming World," that tracked and musically represented rising Northern Hemisphere temperatures. Daniel Crawford commented, "Music is an important tool because it acts to bridge the divide between logic and emotion . . . through music, we can convey the data in a different way, which draws on the science of the numbers and also the emotional power of hearing sound" (Hansman, 2015).

Creative written works have increasingly grappled with climate change through creative non-fiction, climate fiction ("cli-fi") and children's books, among others. For example, Amitav Ghosh assembled the 2016 book *The Great Derangement: Climate Change and the Unthinkable* from a set of lectures he delivered at the University of Chicago. In this set of writings, he challenged readers to draw on creativity to collectively confront climate change and overcome failures of imagination to date. As another example, in 2016 Dennis Meadows, Linda Booth Sweeney and Gillian Martin Mehers (2016) co-wrote *The Climate Change Playbook*. This assembled almost two dozen creative group activities called "games or strategic exercises" to enhance learning about climate change through multiple learning pathways (p. x). Through experiential, tactile, critical analytical and interactive approaches, these projects were designed to help participants better understand dimensions of climate change while dispelling various misconceptions about climate change. Brian Foo also uniquely assembled the *Climate Change Coloring Book* with activities to explore and sit with climate research. Through a successful crowd-sourced Kickstarter campaign in 2017, he translated research from places like the US Department of Energy (DOE) and the National Aeronautics and Space Administration (NASA) to create approachable and creative ways to engage with complex data. He commented, "Coloring helps people slow down, which is not just good for relaxing, it's good for absorbing some critical facts about our warming planet" (Harrington, 2017).

Cli-fi literature has emerged over the past two decades with books such as Barbara Kingsolver's *Flight Behaviour* (2001), Ian McEwan's *Solar* (2010) and Alice Robinson's *Anchor Point* (2015). As a result of these developments, several schools and colleges around the world are increasingly incorporating climate fiction into their curricula, as a pathway to understanding human–environment interactions in a changing climate (Ring, 2016).

In the first empirical study from the social sciences and humanities on the efficacy of cli-fi, Matthew Schneider-Mayerson found that "'cli-fi' reminds concerned readers of the severity of climate change while impelling them to imagine environmental futures and consider the impact of climate change on human and nonhuman life" but "the affective responses of many readers suggest that most works of climate fiction are leading readers to associate climate change with intensely negative emotions, which could prove counterproductive to efforts at environmental engagement or persuasion" (Schneider-Mayerson, 2018, p. 473).

And as an example of children's literature, in 2018 Megan Herbert and Michael Mann published a (carbon-neutral) kids' book to bridge what can be a difficult subject of threats associated with climate change and positive action. Engaging through an empathetic story, the authors seek to empower young readers through their introduction of a "World Saving Action Plan."

Together, creative communications have increasingly pervaded high-profile spaces like the UN climate talks in recent years. For instance, in

Figure 7.5 The image on the left (credit Lila Howard) captures some of the 20,000 pairs of shoes placed in the Place de la Republique in Paris at the 2015 "Silent March" preceding the UN climate talks. The image on the right (credit Avaaz) shows one of the notable contributors of shoes to the protest, Pope Francis.

December 2015 many creative climate communications surrounded and swirled in the United Nations Conference on Climate Change (COP21) in Paris, France. As an example, there was a powerful "silent march" organized by Avaaz and other organizations. There, 20,000 pairs of shoes were laid out in Place de la République to provide symbolic public pressure for climate policy action at COP21 (see Figure 7.5). This silent march took place near the memorial to victims of the Paris terrorist attacks that took place approximately one month earlier, leading to a ban on large gatherings during these talks. As another example, the "ArtCOP21" – in Gaîté Lyrique and connected to others around the world in solidarity – convened many creative communicators through a global programme of more than 500 plays, talks and exhibitions fusing science and culture. Through their activities, they made the case that "climate is culture."[23] A third example was on display through the subversive work of UK-based "Brandalism." During these climate talks, they sought to call attention to corporate influence in mainstream discussions of options for climate action through their 600 fake outdoor advertisements throughout central Paris (see Figure 7.6).[24]

These examples have illustrated auspicious experiments and avenues for creative and effective communication about climate change. However, these multimodal communications generally have remained piecemeal strategies or approaches groping for resonance in the face of the climate engagement challenges. To make sense of promises and pitfalls associated with creative

[23] ArtCOP21 works can be seen here: www.artcop21.com/
[24] In addition to those in Figure 7.6, other fake advertisements can be seen here: www
.brandalism.org.uk/

Figure 7.6 Two examples of the 600 in situ hangings and displays around Paris during the December 2015, UN Conference of Parties meeting (COP21).

(climate) communication, "road rules" along with features on a "road map" of context-, audience- and time-sensitive tenets of successful ways to communicate about climate change are vital. These help with ongoing efforts to meet people where they are and to activate policy engagements as well as public participation. This guidance can help maximize effectiveness and opportunities and minimize mistakes and dead ends in a resource-, energy- and time-constrained environment.

Ezra Markowitz and Meaghan Guckian (2018) have commented, "Climate change communication is hard to do well and easy to do poorly" (p. 53). Therefore, a "road map" along with "rules of the road" can help as researchers and practitioners proceed with both ambition and caution into struggles to find meaningful common ground on topics associated with climate change in the twenty-first century.

Five Rules of the Road and Five Features on a Road Map

Lessons from creative (climate) communication experiments have been sprinkled throughout the book. Here I pull together these lessons into a more cogent yet nonexhaustive list. I do so to help guide a path toward more creative and ultimately effective (climate) communication to inspire people to pay attention and take action. This is an ongoing process of aggregating lessons learned from research and practice, in order to collectively integrate them into creative (climate) communications planning going forward. I hope this then helps researchers, practitioners and students move to more concrete and actionable spaces of engagement as they work to confront persistent climate communication challenges and improve climate communication outcomes.

While I call this "rules of the road" and "features on a road map," I am careful to not call it a disembodied "recipe book," "set of directions," or a "tool kit." Unfortunately, there is no clear and simple path to get from here to there. Philip Smith and Nicolas Howe (2015) have noted that "'tool kits' are only part of the equation" (p. 29). Successful and creative climate communications strategies must be tailored to particular epistemic communities, by way of perceived and intended audiences and through relations of trust. Spatial and temporal context is critical. Cultural, political, social, environmental, economic, ideological and psychological conditions matter.

With this in mind, I preface the features on a road map with a set of "rules of the road" to increase opportunities to arrive at a place of effective (climate) communications. These may seem obvious. But many communications efforts often fall short of their goals and objectives by violating these basic road rules. Lessons learned from social science and humanities research as well as experimentation and practices can be integrated into plans to effectively map how people thoughtfully consider, productively discuss and substantively engage with climate change in current times. It can then help find common ground while closing an engagement gap.

First, **be authentic**. Authentic and effective engagement requires connection to audience in ways that reflect their values and lived experiences while acknowledging many ways of knowing (Pérez and Simon, 2017). This is captured in the "know thy self" portion of the Schneider quote beginning this chapter. For example, if you care deeply about the plight of those at the forefront of climate impacts, communicate authentically about your passions and concerns. As another example, if you are a naturally snippy and cynical person who responds to denigrating humor, be authentic about that. Disingenuousness, deceitful and phony approaches prove ineffective when working to legitimately establish relations of trust (see Chapter 1 for more) and connect with others (Neffinger and Kohut, 2013). For instance, an insincere recycling of hollow clichés about taking action to address climate change might seem like useful communication but it proves tiresome, bogus and unsuccessful over time. Moreover, pretending to be in allegiance with a group when you authentically are not ultimately proves unconvincing and ineffectual.

Second, **be aware**. This can manifest in many ways, through motivated mindfulness of varied audiences, and attentiveness to what people you want to reach actually care about (see Chapter 1 for more). This essentially addresses the "know thy audience" part of the Schneider quote that opens this chapter. Being aware here also encompasses respect, humility and awareness of effective scales of engagement and interdependencies therein (see Chapter 5 for more). This is essentially about remaining mindful of meeting the scale of challenges with an effective scale of response(s). Respect and mindfulness here also involve active empathy and work to understand how values and ethics guide our lives (in different ways and through other points of view) (Schwartz, 1992) (see Chapter 4 for more).

Third, **be accurate**. This is embodied in the "know thy stuff" excerpt of Stephen Schneider's statement at the beginning of this chapter. This brings in elements of making it/keeping it real (see Chapter 2 for more) as well as awareness of conditions and contexts (see Chapter 5 for more). This can be bolstered by elements of climate literacy (see Chapter 3 for more). This can involve accurate representations of scientific ways of knowing, as well as accurate descriptions of conditions and contexts and cultural and political dimensions shaping climate change engagements in the public sphere. This can then bolster elements of legitimacy and trust (see Chapter 5 for more).

Fourth, **be imaginative**. Deploying creativity and innovation through experimentation helps to unlock many different answers and insights in the face of climate challenges. Imagination involves a nimbleness, agility and ability to adapt to changing communication contexts. Imagination also involves a willingness to take risks, potentially make mistakes and to suspend (self-) stigmatism along the way. Imagination can lead to inspiration and resonances (see Chapter 2 for more). This imaginativeness applied to climate challenges often involves interdisciplinary approaches (see Chapter 2 for more). Imagination then marshals resources for a "smartening up" of communications that then more effectively makes connections to issues, people and things that everyday citizens care about.

Fifth, **be bold**. Coordinated, inspiring and honest experimentation can cat- alyze segments of a broader "public" to engage more constructively in shared climate challenges. This "exemplification" can boost other people's inclination to recognize a common struggle, and to take action themselves through your leadership and action. It can also lower psychological barriers to engagement by engendering first-person perceptions that help build connectivity. This connectivity can strengthen collective solidarity and induce empathy. It can then spark conversations, discussions, reflections and a readiness to engage across scales (see Chapter 5 for more). John Dryzek and Alex Lo (2015) have commented, "What we need is not just Medicare for the Climate – but Mandela (or multiple Mandelas) for the climate," referring to leadership like that of Nelson Mandela in today's creative (climate) communication environment (p. 15).

Considering these tenets together can increase possibilities for creative (climate) communications success once you hit the road. Katharine Hayhoe (2018) has commented, "We all live on the same planet, and we all want the same things. By connecting our heads to our hearts, we all can talk about – and tackle – the problem of climate change together" (p. 943). Features on a road map can be helpful then when seeking to navigate toward resonant and effective communications. These five features – elaborated on in other parts of this book – can then help to meet people where they are in their everyday realities while opening up spaces for productive discussions and deliberations about climate change. These are clearly not mutually exclusive features, but rather overlapping elements that guide a process of creative (climate) communications.

First, **<u>find common ground</u> and meet people where they are on climate change**. Social science and humanities research has found communication strategies and tactics that value experiential, visceral, affective, visual, tactile, aesthetic and scientific pathways of knowing help open people up to other perspectives. This then involves knowing your audience: some people are swayed by personal stories while others just want "the cold hard facts." There are many ways of learning about, knowing and researching creative (climate) communications. If it cannot be measured (e.g., through polling data) it still counts (see Preface for more). Insights can be gained through both quantitative (e.g., surveys) and qualitative (e.g., interviews) approaches. They both require active awareness and valuing of other points of view. Andrew Hoffman (2015) has pointed out that "We cannot scold, lecture or treat people with disrespect if we are going to gain their trust . . . trust will not be gained by bludgeoning those we engage" (p. 82). This may be about dialogue and two-way communications as much as talking at people. Social change often begins as conversations. Kersty Hobson (2008) has commented, "Positive change is not about obtaining or instilling a new set of ethics or addressing public ignorance with information campaigns. Rather . . . we can 'start from where we are', to open up and explore the parts of us that want to live differently – the already ethical parts of us that are open to and already do think sustainably" (p. 11) (see Chapter 2 for more).

Second, **emphasize how climate change affects us here and now, in our everyday lives.** This work to enhance relevancy then creates openings for discussions and actions. It can be in part a process of considering the "here and now" (see Chapter 1 for more) in "bringing climate change home" (see Chapter 2 for more) and connecting with place by overcoming perceptions of distant climate impacts in far-flung places and considering present, local and personal risk (see Chapter 4 for more). Social science and humanities research outlined in this volume has confirmed the notion that people prioritize events near to them in time and space (e.g., Leiserowitz, 2006; Trope et al., 2007; Gifford and Comeau, 2011). Furthermore, social science and humanities research has revealed there is great value of overcoming psychological as well as physical distance when seeking to enhance climate awareness and action (Markowitz et al., 2014; Stoknes, 2015). This can also be in part a process of overcoming abstraction by tapping into visual, visceral, experiential and other ways of knowing through multimodal communications (see Chapter 4 for more). This can also involve a process of connecting economic concerns that more clearly impact people each day to climate concerns that are a slow moving threat unfolding in real time (see Chapter 5 for more). Taken together, Susanne Moser (2010) has called for climate communication work to make the invisible visible and distant consequences resonant in our everyday lives.

Third, **focus on how climate change engagement makes our lives and livelihoods better**. Clearly this must particularly be accurate and authentic. Social science and humanities research to date has illuminated how communications successes can derive from a focus on "a better life" through climate engagements. Part of this can involve framing communications with a focus on "opportunities" and "benefits" (Markowitz et al., 2014) and desired futures that can be achieved as engaged ecological citizens (Dobson, 2003). For example, Good Life Goals were designed as deliberately positive and action-oriented framings for effective communications about climate and wider environmental issues (see Chapter 8 for more). These have sought to creatively help promote agency and empowerment in the face of these global challenges. Part of effective climate communications can involve forgiveness of mistakes, be they big ones (e.g., long-held contrarians' viewpoints [see Chapter 3 for more]) or small ones (e.g., "crossing the line" with comedy [see Chapter 4 for more]). In this context, disagreements must be able to be on the table, along with the fact that realities of differential climate impacts and vulnerabilities must still be part of conversations, handled consistently, authentically and carefully (see Chapter 2 for more). As such, this continued mindfulness comes with an awareness that these framing choices do not change behavioral realities. In other words, this is not merely about reframing of issues, as recognition of the linkages between discourse and our material practices is vital to sustained success. But, there is a gravitation pull toward enjoyable experiences and the good things in life (Osnes, 2014).

Fourth, **creatively empower people to take meaningful and purposeful action on climate change**. This approach is in part about illuminating (and creating) pathways to make commitments habitual, easy, productive and fun. This is also in part about deploying motivating language and tactics to inspire engagement. While there is a time and place for everything (see Chapter 2 for more), this is also, in part, about provoking informed agitation (Sen, 2013), where creative (climate) communications can mobilize intersubjective meaning systems in order to value our differences (as they are situated in cultural processes in political economic spaces). Social science and humanities research has shown that this involves methodical and deliberate multimodal (see Chapter 4 for more) and multiscale (see Chapter 5 for more) efforts to confront individual as well as collective searches for meaning in our shared society, both in the present as well as into the future (see Chapter 8 for more). As such, this reinforces findings from both researchers and practitioners that "binding communications" lead to resonant and pro-environmental behavioral action (e.g., Dryzek and Lo, 2015). These might be communications that combine a persuasive message and potential solution (Parant et al., 2017).[25] It also promotes Katharine Hayhoe's three-part approach to communication through (1) "bonding" or finding common ground, (2) '"connecting" or finding ways to show that it matters and (3) "inspiring" or showing how solutions are all around (Frey, 2018).

Fifth, **"smarten up" communications about climate change to match the demands of a twenty-first-century communications environment.** Social science and humanities research has provided guidance to help improve efforts to help bridge divides that impede progress on climate change. This work can also inform science and policy amidst our shared and distinct contemporary realities. However, while lessons have been learned, work remains regarding integration of these lessons into our ongoing communication practices. Part of this integration involves shedding information-deficit model communication logics, thereby more effectively addressing (mis)perceptions and (mis)understandings and no longer being duped by many counterproductive, destructive and outlier voices (see Chapter 3 for more). Part of this involves navigating the challenges of both depoliticization and overpoliticization while more capably distinguishing among Type 0, Type I and Type II advocacy (see Chapter 6 for more). This also partly means great accounting for the power of social norms and burden- as well as risk-sharing in fueling inspiration to collaborate across scales to confront this collective action challenge (see Chapter 5 for more). And this may partly comprise contemplation about how aspects like humor (Boykoff and Osnes, 2019) legacy (Zaval et al., 2015), patriotism (Gabehart, 2015), humanization (Boykoff et al., 2018), long-term time horizons and desires to "be on the right side of history" can effectively be deployed with varied audiences in order to garner engagement and action. This is about "smartly" connecting content in

---

[25] They call this a "preparatory act" and this addition "increases the efficiency of the persuasive message" (Parant et al., 2017, p. 349).

communications with high-quality form, and working carefully to make these work well together. Together, this "smartened up" approach can help to carefully and honestly assess how creative (climate) communications can effectively shape the spectrum of possibility for meaningful, substantive and sustained responses to contemporary climate challenges.

Together, these five rules of the road and five features on a road map fortify observations from Julia Corbett and Brett Clark (2017), who cite "storytelling, corporally sensed and felt experiences, interdependency with the world, engaged emotions, and connection with place" as key factors shaping engagement in the public sphere (p. 48). They also bolster insights from research by Sander van der Linden, Ed Maibach and Anthony Leiserowitz (2015), who mapped out five "best practices" to improve public engagement with climate change. They wrote, "Instead of a future, distant, global, nonpersonal, and analytical risk that is often framed as an overt loss for society, we argue . . . [to] (a) emphasize climate change as a present, local, and personal risk, (b) facilitate more affective and experiential engagement, (c) leverage relevant social group norms, (d) frame policy solutions in terms of what can be gained from immediate action, and (e) appeal to intrinsically valued long-term environmental goals and outcomes" (p. 758). The road map and rules of the road also cohere with work by Ezra Markowitz et al. (2014), who have called for communicators to carefully consider audience, emphasize solutions, bring climate impacts "home" and tell meaningful stories (see Chapter 5 for more). Together, creative (climate) communications – through various media channels and through participatory and in-person routes – all play distinct roles in shaping politically, culturally, environmentally and socially infused attitudes and behaviors (Gavin, 2018).

These have been rules of the road and features of a road map focused on creative climate communications. But these can be transferable and applicable to creative communications about many other scientific, political, economic, cultural and social issues and challenges, irrespective of specific communication goals or objectives. Examples abound: genetically modified organisms, vaccinations, nuclear power, stem cell research, homelessness, transgender rights, healthcare reform, policy brutality, racial discrimination, abortion, student loan debt, immigration.

These rules of the road and features of a road map can also help guide ways to live more gracefully in a changing world. They can equip communication strategies to generate more signals of productive discussions and deliberations amid the noise of everyday life. With these road rules and road map we can more capably be nimble and agile as we mindfully, carefully and meaningfully move forward.

Work remains to systematically strengthen connections between these creative (climate) communication experiments and research on them. Susanne Moser (2017) has pointed, "Relatively few communication researchers ... actively, frequently, or on a sustained basis interact with those who do the lion's share of climate communication ... meanwhile, few practitioners have the time, inclination, or access to read ... and keep up with the ever increasing output from researchers in climate communication" (p. 356). While some bridging efforts are exceptions (e.g., Galindo, 2016; Svoboda, 2016; Tate, 2017) there remains a critical need for better coordination between these communities of research and practice and better integration of lessons learned.

Conclusions

Rather than pursuing a quixotic silver bullet that will compel people to resoundingly respond to climate change, researchers and practitioners have been finding common ground and inspiring awareness and engagement through creative climate communication strategies that can best be described as "silver buckshot." Silver buckshot approaches include scientific, experiential, visceral, emotional and aesthetic learning about climate change. Silver buckshot designs involve deliberate and multimodal communication endeavors that seek to meet people where they are, drawing varied constituents of the public citizenry into new considerations and behaviors. These are "all of the above" tactics. Silver buckshot plans are those creative strategies that open up spaces for inspired climate engagement and action.

It is important to therefore remember that some communication approaches work for some audiences and not for others. As such, not all experiments are always successful; there are no cookie-cutter approaches that work for all audiences. For example, some who communicate about the urgency of climate action have sought to impose notions of deadlines. These targets and timetables can motivate some people while paralyzing and even backfiring with others. For example, in 1988 the World Meteorological Organization (WMO) held an international conference called "Our Changing Atmosphere" in Toronto, Canada (Pearce, 1989). At this conference, 300 scientists and policymakers representing 46 countries convened, and from this meeting, participants called on countries for 20% CO_2 emissions reductions by 2005 (Gupta, 2001). These countries did not achieve these reductions by 2005, and the target moved. That moving target appeared to be a dose of realism for some, but an injection of deceit for others.

As a second example, in a 2017 issue of *Nature* magazine, Christiana Figueres, Hans Joachim Schellnhuber, Gail Whiteman, Johan Rockstrom, Anthony Hobley and Stefan Rahmstorf, set out a six-point plan for reducing carbon dioxide emissions into the atmosphere. Addressing energy, infrastructure, transport, land use, industry and finance, they dubbed the effort "Mission 2020" to raise ambition in the interim three years, arguing that 2020 "has more to do with physics than with politics" (p. 593). As a third example, the 2018 IPCC Special Report on 1.5°C discussed actions needed by 2030 to "avoid the worst impacts of climate change" (Allen et al., 2018). While these examples do not suggest that the world will implode if insufficient action is taken on the articulated timelines, this kind of language can fill some audiences with mixtures of fear, complacency and disregard rather than enthusiasm and interest. In particular, climate contrarians have at times met "deadline" language with ridicule. For example, in response to the statement in the 2018 IPCC Special Report on 1.5°C, Marc Morano from Climate Depot commented, "Climate tipping points have a long history of repetition, moved deadlines and utter failure."

Again, there is a time and place for everything. Places of worship, sporting events, concerts or movie theatres are the preferred gathering spots for some folks. Others may choose to gather in bars and cafés. Successful (climate) communications are ones that creatively and effectively frame messages through trusted messengers to reach particular audiences through resonant scientific, experiential, visceral, emotional and aesthetic registers.

Shawn Zheng Kai Tan and Jose Angelo Udal Perucho (2018) have argued for "bringing science to bars." They commented that "science communication programs should be developed around the locales of the target community as an effective strategy to counter the rising mistrust in science and scientists" (p. 819). Similarly, Daniel Clery (2003) has argued for bringing science to cafés in a French context.

Silver buckshot involves a variety of trusted leaders and messengers giving voice to many perspectives and ways of knowing about climate change (see Chapter 4 for more). Because climate change cuts to the heart of all our activities and actions – from choices about whether to have children (or how many) to the food we put on our plates – we need storytelling leadership from the sciences, from the policy arena, from medicine, from the business community, from the public sector, from social movements, from entertainers and more in order to help make sense of this collective action challenge in the twenty-first century.

Samuel Miller-McDonald (2018) has argued,

> If you're a writer, then you have to write about this. If you're an artist, then you have to make art about it. If you're an Uber driver, then talk to your passengers about it and every day bring someone on board. If you're an entrepreneur, then start a non-carbon company and infiltrate business associations, convince your customers to act . . . every single day, in our jobs and lives, we have to find creative ways to do something, anything, that can work towards not only avoiding collapse, but building a new world. This new world, one capable of equitably hosting life on earth, must be built on our heroism. It will only be birthed by the pangs of sacrifice. It will emerge out of the joys of mutual support and interdependence.

For example, some politically conservative leaders have stepped forward to overcome partisanship and polarization to address climate change. Prominently, the group RepublicEn was founded in 2012 by former South Carolina Republican Congressman Bob Inglis. Inglis was in the US Congress from 1993 to 1995 and again from 2005 to 2011. It was widely viewed that Inglis lost his reelection bid because of his strong stance that humans contribute to climate change and that action needed to be taken to address it. Seeking to reclaim a right-of-center engagement with climate change through market-based interventions, RepublicEn has promoted conservative approaches to climate action. Inglis has noted, "We're making progress. Conservatives are going to come around on climate. We're going to realized that we're all in this together, and the world is going to celebrate American leadership. This is our generation's moonshot" (in Roser-Renouf and Maibach, 2018).

Bob Inglis has also commented, "I know this is going to sound rather strange. We need a song. It's terribly frustrating to scientists; it's not the science that's going to do this. A poet is what we need. A poet, a priest or a rabbi or somebody who can summon the heart to do something. I'm completely convinced that the heart decides and the head justifies" (in Mathiesen, 2016, p. 981). There are clear needs for ongoing experimental work to access other ways of knowing and to activate engagement through authentic person-to-person connections through shared values. With these creative (climate) communications coursing through the veins of our shared society, Joanna Wolf and Susanne Moser (2011) have described success as taking place through "the mind, the heart and the hands" (p. 550). Through these pathways, the final chapter now turns to meaning making and prognoses for the future.

8

Search for Meaning

Viktor Frankl's *Man's Search for Meaning* was an influential book in my early years. I think I was roughly thirteen years old when my mother suggested I read the chronicle of Frankl's experiences in an Auschwitz concentration camp during World War II. His story of depravation and mistreatment during the Holocaust was stunning, particularly from my perspective in a comfortable life in Madison, Wisconsin. His emergent psychological theory of "logotherapy" fascinated me, even at that age. From the harshness of his living conditions during that uncertain period of time in captivity, Frankl managed to discover that, to him, the meaning of life was found through a combination of meaning itself and hope for the future. "Logos" comes from the Greek term referring to "word"; this stated the importance of both discursive and material realities that shape our existence. The theory of logotherapy developed as a theoretical perspective that one's freedom to find meaning in life is the primary and most powerful motivating force in human development. Progress along this pathway to meaning is then marked by understanding of (shared) values, (common) purpose and (mutual) respect for varied perspectives and differential suffering.

For many years that followed my initial reading of this book, the concepts and insights from it continued to bounce around in my head. When I began my own considerations of human–environment interactions and individual as well as collective searches for meaning in our shared yet differentiated society it pervaded my thoughts again. I admittedly dismissed logotherapy for a while as oversimplistic. But more recently I have found key parts of it to be resonant. Meaning can derive from fundamental desires to find common ground, to meet people where they are and to contend with our collective futures amid the various ways of knowing about the current state of climate change.

Lessons from the brutality of Frankl's experiences – and the millions of others who also endured suffering in these concentration camps – transfer in ways to many of today's ongoing struggles. In the face of contemporary climate change, many have argued that this is no less than an existential challenge. Furthermore, they argue that this is a vital communication and engagement challenge. In a 2018 opinion piece entitled "Stopping Climate Change Is Hopeless. Let's Do It," Auden Schendler and Andrew Jones (2018) noted, "Historically, we've tackled the biggest challenge – that of meaning, and the question of how to live a life – through the concept of 'practice' in the form of religion, cultural tradition or disciplines like yoga or martial arts" (p. SR10). Yet they pointed to engagement with climate challenges as still an under-developed vehicle to "endow our lives with some of the oldest and most numinous aspirations of humankind: leading a good life; treating our neighbors well; imbuing our short existence with timeless ideas like grace, dignity, respect, tolerance and love. The climate struggle embodies the essence of what it means to be human, which is that we strive for the divine" (p. SR10).

Creative (climate) communications can then foster pathways toward meaning and purpose in our shared lives. In the book *Why We Disagree about Climate Change* (mentioned in Chapter 7), Mike Hulme (2009) argued that instead of trying to "solve" climate change, "we need to see how we can use the idea of climate change – *the matrix of ecological functions, power relationships, cultural discourses and material flows that climate change reveals* – to rethink how we take forward our political, social, economic and personal projects over the decades to come" (p. 362, emphasis added). This then becomes about a changing attitude toward climate change, instead of stances that seek to merely conquer climate change and squelch dissent. These then become meaningful struggles that are incumbent on us all to communicate about and address in diverse ways.

With this in mind, in this final chapter I look at "kids today." I consider how youth can be empowered going forward in a twenty-first-century communications environment in order to combat what John Schellnhuber has referred to as the "dictatorship of now" in the context of creative climate communications (see Chapter 1). In taking up this set of considerations, I return to intergenerational/intragenerational equity questions about who has a voice and how, going forward in creative communications about climate change, decarbonization and sustainability.

The Kids Are Alright[1]

In February 2000, Angela Davis gave a lecture in Boulder, Colorado where she posed the question, "What will we say in 2030"? She commented, "I want you to think about your responsibility . . . what you tell people in 2030 is contingent on today."[2] While her talk focused on prisons, inequality, gender bias and capital punishment, it was a question that can usefully be extrapolated to dimensions of climate change. After all, 2030 is nearer to the present now than were the terrorist attacks on the World Trade Center and US Pentagon in 2001.

The 2018 IPCC Special Report warned that to avoid passing 1.5°C, emissions must drop 45% from 2010 levels by 2030, and must reach "net zero" by 2050 (Allen et al., 2018). It has struck me that nearly halving emissions by 2030 would create a much different world for my teenagers than the world we live in now. And in 2050 they will be roughly my current age. Clearly, much work remains to be done to overcome path dependency and political and institutional inertia as it relates to our relationships with carbon-based energy generation and consumption, along with associated contributions to climate change.

Meanwhile, there are commonly disparaging comments about Millennials or Generation Y (born between the mid-1980s and mid-1990s) and Generation Z (born in the mid-1990s through the early 2000s) as lazy (BBC, 2017; Smit, 2018), delicate, sensitive or fragile (Campbell and Manning, 2016; Skenazy and Haidt, 2017), narcissistic (Bergman et al., 2011), selfish (Keene and Handrich, 2010) and entitled (Alexander and Sysko, 2012; Berger, 2018). As such, they have denigratingly been called the "me me me" generation (Stein, 2013).

These comments then suggest that previous generations (e.g., baby boomers, Generation X) have been more motivated, more altruistic, more resilient. Those of us old enough to remember the "Calgon take me away" commercials from the 1970s and 1980s can relate in a benign way to such wistfulness. Those considering the more distressing undercurrents associated with the "Make America Great Again" slogan can also reflect on the power of such nods to nostalgia that can resonate with key segments of our shared society. (A survey by Morning Consult actually revealed a split regarding what nostalgic eras were conjured through "Make America Great Again." Responses ranged from no distinct pattern to 1955, 1960, 1970, 1985 and even 2000 [Sanger-Katz,

---

[1] This is a song originally created and performed by The Who on their "My Generation" album in 1965. It has been revamped and recreated most notably by the US band Pearl Jam since then.
[2] This talk can be found by David Barsamian's *Alternative Radio* recording from February 23 in the Boulder Theater here: www.alternativeradio.org/products/dava008

2016].) These cases point to nostalgia as symbolic expressions of romanticized times gone by.

Nostalgia can be a powerful intoxicant, a yearning for an idyllic or edenic past. Fred Davis (1979) has described nostalgia as "the means for holding onto and reaffirming identities which has been badly bruised by the turmoil of the times . . . nostalgia looks backward rather than forward, for the familiar rather than the novel, for certainty rather than discovery" (pp. 107–108). Gestures to nostalgia have historically been associated with times of social upheaval and rapid social change (Williams, 1973). The responses captured by the Morning Consult survey mentioned earlier in the context of "Make America Great Again" nonetheless point to times when minorities, women and people with disabilities were left out of huge parts of life, thus times marked with racism, sexism and inequities of many sorts. This shows that perspectives and ways of learning and knowing (see Chapter 4 for more) matter too.

Anders Hansen (2010) has examined nostalgia in relation to environmental awareness. He found that nostalgia is deployed particularly through nature imagery "to construct a mythical image of the past (including childhood) as a time of endless summers, sunny and orderly green landscapes, and, perhaps most importantly of all, as a time and place of community, belonging and well-defined identity" (pp. 148–149). In short, nostalgia may emanate from feelings that "things aren't what they used to be."

Yet, denigrating comments about Millennials, Generation Y or Generation Z have extended into their general comfort and engagement through online and social media. As such, they have been described as "slacktivists," where old-school activism is confined to armchairs in isolated comfort. So embedded are such critiques that the actions taken on the internet (like signing an online petition or tweeting with a favored hashtag) displace necessary physical activism (like marches) needed to bolster social movements. However, Shelley Boulianne and Yannis Theocharis reviewed more than 100 surveys about youth, civic and political engagement, and digital media activity. They found little evidence that online engagement has displaced physical civic engagement. In fact, they detected the opposite effect where digital activism catalyzed off-line political and social actions. They concluded that these findings "undermine claims of slacktivism among youth" (Boulianne and Theocharis, 2018, p. 1).

No doubt, things are not what they used to be. Millennials and Generation Zers are people who have generally been born into a world where they are comfortable with technology and social media because they were exposed to the internet at an early age. These are also generations who were born into a world where climate change as a science–policy and society challenge was already being discussed in the public sphere (Boykoff, 2011). In fact, most Millennials and all in Generation Z were born after December 1984, which was the last cooler than average month in global temperatures on planet Earth (Clark, 2018). It has been more than 400 months, nearing 35 years, suggesting that anyone born since then may consider these trends as a new (ab)normal compared to periods before then. Considered in this way, nostalgia is overrated and Millennials and Generation Z may have good reasons for feeling angry, anxious or melancholy about the state of human–environment conditions today (see Chapter 2 for more).[3]

Yes, histories of culture, politics and climate change at the human–environment interface clearly matter (Howe, 2017; Sutter, 2017). Paul Sutter (2017) has noted,

> History's uses are many. We need careful historical explanations of how and why humans turned to fossil fuels, unleashing huge stores of fossil carbon and other greenhouse gases into the atmosphere and pushing the global system into unprecedented territory. More than that, the discipline of history is essential to understanding the forces of economic, social, political, diplomatic, and cultural power that have driven us into a world of human-induced climate change, and to appreciating how effective solutions must contend with those power dynamics and the inequalities and injustices they have spawned. (pp. xi–xii)

Moreover, Josh Howe (2017) warned against the "presentist paradox" as a form of confirmation bias. This is the problem where history is read as a set of things interpreted through present concerns (social, political, environmental) rather than dynamic influences shaping pathways to the present set of conditions on their own terms through the past. As such, this presentist perspective papers over the complexities, contingencies and contradictions of the human–environment experience, distorts the past and then simply "histories become narrow 'just-so' stories that mostly tell us things we already know and reveal very little about the human experience" (p. 9).

This push from Howe is a push to accept and evaluate histories on their own terms, meeting history – and the people and environment in it – where they are. In terms of creative climate communications and engagements, this pivot can be an empowering one, recognizing that Millennials are working to reconstitute relationships with climate change through a variety of interdisciplinary and research–practice activities. As such, struggles of today – by Millennials, Generation Z and others – are not judged or constrained by markers of the

---

[3] Newly elected New York Congressperson (and Millennial) Alexandria Ocasio-Cortez embodies a spirit of action and engagement on climate change with her embrace of a "New Green Deal" for the USA in order to catalyze a path for deep decarbonization of US industry and society (Anapol, 2018b).

past. In fact, numerous polls show that younger generations tend to be more accepting of climate science in general than older generations are (Nuccitelli, 2016). In particular, a 2018 Pew survey found that US Republicans ages 22 to 37 support coal, oil and natural gas sources for energy in much lower numbers than older generations. The Pew survey also found that fewer than half of Republican Millennials surveyed supported offshore oil and gas drilling compared to three-quarters of older Republicans (Neuhauser, 2018). Therefore, creative climate communications can find traction and resonance in unprecedented and not previously conceived of ways. Creative climate communications are not what they used to be.

Maria Ojala and Yuliya Lakew (2017) outlined six reasons why young people are important to climate change discussions (pp. 609–610):

1. The young of today are the future leaders, decision makers and researchers of tomorrow.
2. From an ethical perspective, it is vital to listen to and learn from this group, as young people will most probably bear a larger burden of the negative consequences of climate change than older people.
3. Young people are consumers and citizens today.
4. Different groups of young people could have unique knowledge about local places.
5. Young people, especially children, are perhaps more vulnerable to experiencing negative affect and low well-being in relations to climate change.
6. Many are involved in the formal education system and thus are relatively easy to reach with information about climate change . . . and not yet set in their ways.

Through creative communications, younger generations are expressing their ambitions and concerns about climate change more than ever before. Whether by choice or by necessity, many younger people – Millennials and Generation Z in particular – are both "talking the talk" and "walking the walk." Rob Nixon has commented, "There are a lot of creative strategies that we are seeing emerge . . . I find a lot of hope in my students and the activists of their generation, for whom the future stands at the center of their agendas" (Christensen, 2018, pp. 7–8). As examples, these hopes and worries are emergent through creative nonfiction and fiction literature (e.g., Lloyd, 2011; Dunlap and Cohen, 2016; Dembicki, 2017). Also, young people are stepping forward through heartfelt exhortations for change, and as young filmmakers and musicians.

To illustrate, Slater Jewell-Kemker is a Canadian filmmaker who has produced and directed *An Inconvenient Youth*. This film is about the power of youth voices in addressing climate change in contemporary times (Restauri, 2014). While Slater has been working for many years in creative arts and environmental issues, her voice was amplified through a powerful speech on Parliament Hill in Ottawa, Canada at a rally immediately following the UN climate talks in Copenhagen, Denmark (COP15).[4] She exclaimed, "We will not be silent as our environmental future is thrown out the window. In 1992, the year I was born, our world leaders gathered at the first climate conference .... and told us they would take action on climate change. Seventeen years have passed. Seventeen years, and still our leaders are not leading. We've given our governments enough time, and now there's no time left to give. If they aren't willing to lead us into a future that we would be proud to leave our children, then they should get out of the way. We can't afford to wait any longer."

As another example of youth leadership and creative action, Xiuhtezcatl Martinez is a musician, an indigenous environmental leader, and youth director in an environmental organization called Earth Guardians. He has spoken about climate change and environmental stewardship for many years of his young life. In 2015 he took an opportunity to speak at the United Nations High-Level Event on Climate Change in New York. In his prepared remarks, Martinez said, "Climate change is a human rights issue . . . we have to realize that what is at stake is no longer just the planet, is no longer just the environment. But what is at stake right now is the existence of my generation. What is at stake right now . . . what is in our hands today is the survival of this generation and the continuation of the human race."[5] Also in 2015, Martinez joined twenty-one other youth plaintiffs in a US federal lawsuit against the government and the fossil fuel industry, claiming that inaction on climate change is jeopardizing their constitutional right to future life, liberty and prosperity. This claim is lodged as a case of violation of a public trust doctrine. Reflecting on his work and that of young people stepping forward as leaders, Xiuhtezcatl has saga-ciously commented, "Kids shouldn't have to be on the frontline of climate change. It's the job of our leaders to take action for us" (Cumming, 2015). Nonetheless, in the absence of leadership by leaders, Xiuhtezcatl has worked to mobilize a youth movement to take action on climate change for their genera-tions and those to come.

---

[4] This was an anti-prorogation rally contesting then Prime Minister Steven Harper's decision to end the 2009 parliamentary session in Canada. Protests against this prorogation included the assertion that this move was a way to avoid addressing international climate change policy action that was raised in Copenhagen at the UN climate talks (COP15).

[5] His brief speech at the High-level Event on Climate Change at New York UN Headquarters in June 2015 can be viewed here: www.youtube.com/watch?time_continue=2&v=27gtZ1oV4kw

Should kids be allowed to be kids while they are kids, and rely on adults and leaders to take the lead on their behalf? Ideally, yes. Does the youth activism emanating from the Parkland Florida shootings, from the Occupy movement and from Xiuhtezcatl and Slater, among many others, mark a sub-optimal reality, and a failing of the Baby Boomer generation and of Generation X? Sadly, yes.

In spaces of scholarship, Elliott Honeybun-Arnolda and Noam Obermeister (2018) have argued that young people must take "a frontline position in the constant struggle to overcome longstanding antagonisms between the scholarship of fact-finding and that of meaning-making" (p. 1). They concluded, "We are confident that old dichotomies can be broken down, from which a new role for science in society can emerge. In what seems to be the age of political disenfranchisement and post-truth politics, let us confront rhetoric with rhetoric, emotion with emotion, and together explore the new realms of thought where previous antagonisms used to be" (p. 6) (see Chapter 1 for more). Together, many research surveys and assessments have identified the power of the next generation or wave of "boundary spanning" interdisciplinary researchers (Meyer et al., 2016) and bridging leaders (Gosselin et al., 2016).

About four years after John Schellnhuber made the comment to *Der Spiegel* that began Chapter 1, he offered some updated considerations for Kevin Kaners of the *Elephant Podcast*. Schellnhuber spoke about the current generation, "People who are 50, 60, 70 run the world … elderly men. We benefit from exploiting the past, fossil fuels: resources that were created by nature over millions of years, destroyed in these generations. We choose to withhold from future generations. We withhold resources and leave environmental waste." He continued, "Younger generations should be very angry about this" (Kaners, 2015).

However, instead of choosing anger and resentment, many kids are choosing compassion. For example, kids are often informing their parents and grandparents about climate change. Amanda Stone's "Climate in the Classroom" program in New Hampshire has featured middle school–aged kids as speakers about dimensions of climate change. Amanda Stone said, "It was a really effective way to get this new audience out and listening to new information about climate change" (Harrington, 2018). Meanwhile university programs and curricula are engaging in real-world applications. For example, the College of the Atlantic has joined other academic institutions in project-based learning as it relates to climate change issues (explicitly or implicitly) (Cardwell, 2015). There are now many new resources available to K–12 teachers as well that help them to engage students on climate change (e.g., Beach et al., 2017).

But new offerings, perspectives and resources can come from various places, from a "97% consensus" or from outlier perspectives (see Chapter 3 for more).

For example, in 2017 the right-wing Heartland Institute think tank sent out 350,000 copies of a booklet and DVD, "Why Scientists Disagree about Global Warming" to middle and high school teachers as well as university instructors across the USA. These materials challenged consensus that humans contribute to climate change, and sought essentially to influence teachers and their students with outlier perspectives. While many teachers potentially incorporated these materials into their curriculum, some teachers were outspoken about these mailings as "misinformation" (Banerjee and Lee, 2017). This is one among a number of factors, like insufficient grasp of the details of climate science, perceived as potential pushback from skeptical parents and wariness of entering the politically infused climate science–policy arena, that hinder effective teaching (Schwartz, 2016). In fact, Eric Plutzer, Mark McCaffrey, A. Lee Hannah, Joshua Rosenau, Minda Berbeco and Ann Reid have pointed to a national survey that found that US teachers managed to spend just one to two hours per academic year on the topic of climate change (Plutzer et al., 2016).

Time horizons regarding communications of the urgency of the present, mixed with the urgency of the future, permeate creative and effective (climate) considerations in our present-day communications environment. However, Sabine Pahl, Stephen Sheppard, Christine Boomsma and Christopher Groves have posited that there are fundamental mismatches between the temporal unfolding of climate change and the human mind's functioning (Pahl et al., 2014). Nonetheless, Adam Corner, Olga Roberts and Agathe Pellisier have examined these considerations in the UK context as they led focus group research with 18- to 25-year-olds.[6] Among provocative insights gained, the participants in this UK study recommended that effective climate communication should not talk about how climate change will impact future generations. Corner et al. (2014) reported, "Young people see this as a problem for the *here and now* and will respond positively to messages that frame climate change as a contemporary concern that requires an immediate response" (p. 4) (see Chapter 1 for more). In the US context, Ezra Markowitz and Meaghan Guckian (2018) have called for work to "either 'bring the future into the present' or else to extend the present into the future, thus decreasing people's tendency to discount the future costs and benefits of today's (in)action on climate change" (p. 51). These findings and recommendations prove to be helpful insights to avoid the pitfalls of diminishing responsibility in the present by emphasis on the future.

---

[6] In corollary work, Rob White (2011) has examined how future burdens of climate change shapes perceptions of Australian youth.

An example of temporal awareness prompting present urgency can be found in the book by long-time climate scientist James Hansen. Motivated by concern for the future of his grandchildren, in 2009 he wrote the book *Storms of My Grandchildren*. This is the same James Hansen who (as a NASA scientist at the time) forcefully warned Congress in 1988 that global warming was a reality. He said on the Senate floor that he was "99 percent certain" that warmer temperatures were caused by the burning of fossil fuels and that they were not solely a result of natural variation (Weisskopf, 1988). He also asserted, "It is time to stop waffling so much and say that the evidence is pretty strong that the greenhouse effect is here" (Shabecoff, 1988, A1). In *Storms of My Grandchildren*, Hansen drew on motivations for intergenerational well-being in calling for immediate collective action. He concluded, "It may be necessary to take to the streets to draw attention to injustice . . . civil resistance may be our best hope" (Hansen, 2009, p. 277). As another example, in 2015 well-known UK scientist Chris Rapley worked with Duncan Macmillan to write and produce a play called *2071: The World We'll Leave Our Grandchildren*. The play proved to be a successful show at the Royal Court in London. Narrated from the perspective of renowned climate scientist Chris Rapley, the play finished with Chris stating, "I look at my eldest grandchild who is now the age I was during that world-changing year (1957, the International Geophysical Year). I tell her I think she should become an engineer. She will reach the age I am now in 2071" (Rapley and Macmillan, 2015, p. 197).

Research mentioned in previous chapters also points to effective ways to address time scales to bring climate change "home" (Slocum, 2004). As examples, research by Lisa Zaval, Ezra Markowitz and Elke Weber (2015) regarding motivation to leave behind positive legacies, along with the "Dear Tomorrow" project work of Trisha Shrum and Jill Kubit, point to the resonance along these temporal dimensions of engagement.

As a reminder, however, biophysical agency of the climate remains a real challenge in these considerations. For example, in 2018 the World Health Organization (WHO) released a report on ambient outdoor air quality, climate change and public health. Among their findings, they reported that 93% of the world's children are exposed to fine particulate matter levels above WHO air quality guidelines, and ambient and household air pollution caused 7 million deaths (600,000 children) globally in 2016. Journalist Matthew Taylor (2018) from the *Guardian* wrote, "The study found that more than . . . 1.8 billion children are breathing toxic air, storing up a public health time bomb for the next generation. The WHO said medical experts in almost every field of children's health are uncovering new evidence of the scale of the crisis in both rich and poor countries – from low birth weight to poor neurodevelopment, asthma to heart disease." *CNN* journalist Mary McDougall (2018) reported, "Air pollution is one of the leading threats to health in children under 5, accounting for almost one in 10 deaths among this age group."

US comedian Stephen Colbert once joked, "I don't trust young people. They're here to replace us."[7] Nonetheless, going forward a decidedly optimistic and trust-laden approach is critical. Social science and humanities research has found that communications approaches that empower youth to make change can be effective, as is listening to their perspectives, minimizing their burdensome worries and amplifying their voices as new knowledge brokers in a contemporary communications environment (e.g., Ojala, 2012). Research has also found that trust in this next generation of leaders and trust in progress in creatively communicating about climate change are vital to effectively tackling this twenty-first-century challenge (e.g., Makri, 2017).

Decarbonization and Sustainability

While focused on creative climate communications, this book also intersects with pathways of communication about decarbonization and sustainability. Decarbonization is a process involving decreasing the carbon content of energy-generating fuels that contribute to climate change. Tools to ratchet up decarbonization efforts involve efficiency gains (e.g., switching from coal to natural gas) and mode-switching (from carbon-based sources to renewable energy sources). There have been market signals as well as regulatory interventions that have served to accelerate or slow these movements (Pielke, 2007).

Decarbonization is pursued through a combination of political economic measures and cultural as well as societal demands. As such, decarbonization involves many social, political and communication challenges (Luers, 2013; McNally, 2018). The 2018 IPCC Special Report on 1.5°C warming mentioned in Chapter 3 and elsewhere called on decarbonization by mid-century (Allen et al., 2018). This is a bold charge that impacts our everyday lives, lifestyles, relationships and livelihoods. This is a call to examination of the carbon-based ways in which we live, work, play and relax in contemporary society.

Heide Hackman, Susi Moser and Asuncion St. Clair (2014) have noted that "The interaction of climate change problems with social crises such as poverty, multidimensional inequalities and growing social discontent, and the inevitable trade-offs across communities, sectors, space and time all make climate change a wicked problem" (p. 653). These "wicked problems" – defined as something

---

[7] This can be viewed in the context of his 2007 White House correspondent's dinner speech here: www.c-span.org/video/?c4519958/stephen-colbert-white-house-correspondents-association-dinner

difficult to solve (Churchman, 1967) – therefore need to be confronted with creativity, boldness, authenticity, ambition and imagination (Hulme, 2009; Brown et al., 2010).

Sustainability and sustainable development thread through decarbonization. They do so through demands to amend and shape pursuits of human needs and desires in line with stewardship requirements of the environment, nature and ecosystem services (Matson et al., 2016). When put in terms of clean air, clean water, biodiversity and public health (or articulations regarding how climate change engagement ultimately improves our shared lives and livelihoods), communications present appealing propositions (see Chapter 7 for more). However, when confronting resistance from vested interests, along with infra-structural limitations in the built environments and social and cultural barriers, these aspirations are more complex and challenging.

The best known definition of sustainable development emanated from a 1987 UN report called *Our Common Future*. This defined sustainable development as "development that meets the needs of present generations without compro-mising the ability of future generations to meet their needs" (Brundtland, 1987, p. 1). Building from this framework since the 1990s, the UN has more recently devised a set of Sustainable Development Goals (SDGs). In 2015, the UN General Assembly meeting convened in New York City announced this set of 17 global goals (with 169 targets within them) to succeed the Millennium Development Goals, and to place a focus on a "2030 Agenda for Sustainable Development" (see Figure 8.1).[8]

In addition to the Brundtland Report and Millennium Development Goals, the SDGs also built upon the 1992 UN "Earth Summit" in Rio de Janeiro, Brazil as well as its successors RIO+10 and RIO+20 (ten and twenty years following that initial meeting). This process produced a nonbinding resolution entitled "The Future We Want" that focused on poverty eradication, energy, water and sanitation and public health. Climate challenges have been woven through each of these SDGs, and it is made explicit in goal thirteen on climate action. This goal has been articulated as efforts to "take urgent action to combat climate change and its impacts by regulating emissions and promoting devel-opments in renewable energy."[9]

These broad articulations have been "large tent" statements of principles. They have been subject to praise as well as critique. Regarding the former, there has been praise for the committed work to explicitly move forward with pursuits of

[8] The UN resolution that led to the adoption of these SDGs can be viewed here www.un.org/ga/search/view_doc.asp?symbol=A/RES/70/1&Lang=E

[9] An elaboration of this goal can be found here: www.undp.org/content/undp/en/home/sustainable-development-goals/goal-13-climate-action.html

Figure 8.1 The seventeen United Nations Sustainable Development Goals, established in September 2015.

meeting human needs as well as stewardship of the environment and ecosystem services at the international scale. Regarding the latter, critiques have included the perspective that these have placed an excessive emphasis on humans (anthropocentric), and have too narrowly focused on future generations (intergenerational) over present inequalities (intragenerational). Some criticisms also pointed out that this has been a technocentric or policycentric approach at the sacrifice of ethical and moral considerations (Orr, 2002). Other blends of praise and critique have included some concerns that this general definition made it susceptible to broad interpretation and manipulation for other goals (Redclift, 1987; Boehnert, 2018). For example, corporations, state and local governments and universities may brand themselves as "sustainable" or working toward "sustainability" in laudable attempts to build sustainability goals, measures and metrics into their business plans and supply chains. But they may also do so just to sell more products and services (thereby "greenwashing", a practice making something appear to be more environmentally friendly or environmentally beneficial than it really may be).

These pursuits and debates are not new. They can be traced through associated human–environment challenges over time, seen through writings of

Thoreau, Leopold, Steinbeck and Carson where stewardship is considered in the context of human development. Sustainable development has commonly been seen as the convergence of the three components of economic growth, social justice and healthy environments (Giddings et al., 2002). More recent conceptualizations have emerged from the sustainability sciences. This has been an interdisciplinary set of pursuits that has focused on creating and harnessing many different kinds of knowledge and assets to address social and environmental problems and enhance "well-being" through use-inspired research (Matson et al., 2016). For example, some scholars have conceptualized sustainability and sustainable development as the assemblages and connections among social, manufactured, knowledge-based, natural and human assets. Social assets include the economic, political and social arrangements. Manufactured assets involve things people make (e.g., infrastructure, technologies, art). Human assets refer to demographics of the human population. Natural assets encompass stocks and flows of ecosystem services, natural resources and biodiversity. Knowledge assets incorporate processes of learning along with artifacts of information (Hinckley and Matson, in press). Here, notions of "well-being" emerge through social interactions (where well-being emerges as more than just the sum of individuals' well-being) and through equity considerations (both within and between generations) in order to navigate dynamic, multiscale and complex social-biophysical systems.

Efforts in these areas have sought to increase our capacities to manage interactions at the human–environment interface where sustainable development pursuits play out. As an offshoot to the UN SDGs, a nongovernmental organization Futerra teamed up with the UN Sustainable Lifestyles and Education and the World Business Council for Sustainable Development to launch the Good Life Goals at the 2018 UN General Assembly in New York City (see Figure 8.2). This set of Good Life Goals was designed as a way to empower individuals to find their place in these large sustainable development goals and challenges, "to be relevant, easily understood and accessible." Upon its launch, Futerra co-founder Solitaire Townsend remarked, "I believe that individuals are just as important as institutions when it comes to changing the world. As citizens, parents, neighbors and consumers, our decisions matter. By sharing a set of clear Good Life Goals and meaningful actions for each Sustainable Development Goal, we can build confidence that everyone is important, needed, and can contribute to achieving the SDGs. Changing the world has never just been about policies or products, it always comes down to people."[10]

---

[10] An elaboration of these Good Life Goals can be found here: www.wbcsd.org/Programs/People/ Sustainable-Lifestyles/News/Personal-actions-that-everyone-can-take-to-support-the-SDGs

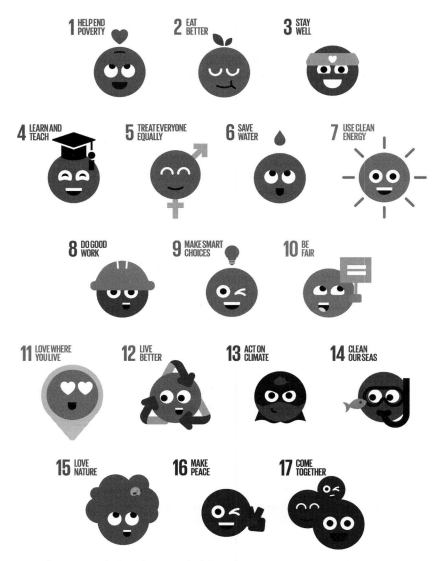

Figure 8.2 United Nations Sustainable Development Goals reconceptualized and reframed in 2018 as "Good Life Goals."

So goal thirteen, for example, makes a subtle shift in framing as it moves from "climate action" to "act on climate." These reframings of what might be considered imposing SDGs, and these reconsiderations of past conditions and capacities, can help individuals then view themselves as more capable of confronting challenges to climate change, decarbonization and sustainability issues. This process of reframing is one strategy wisely considered in a spatial and temporal context. Moreover, this is a way to discuss sustainability alongside the notion that these are pathways to make our lives better (see the Chapter 7, "Features of a Road Map," for more). Heide Hackman et al. (2014) have noted that processes of framing and reframing are "one key strategy alongside a broader set of changes required to integrate the social perspective more fully into research and practice and to give the sustainability transformation a better start and a better chance of long-term success in environmental and social terms" (p. 4). Framing – in context and with a mindfulness of surrounding cultural, social, political, environmental, psychological factors – helps to create conditions for more effective engagements with climate change.

Conclusion = Get (It) Together

In the 1924 book *Tract on Money Reform*, John Maynard Keynes wrote about long-run consequences in the context of economically infused decision-making. He commented, "In the long run we are all dead" (Keynes, 1924). Keynes has been credited with the foundational thinking and writing that paved a way for macroeconomics. Trained in mathematics, Keynes sought to more capably explain causes, consequences and remedies for larger-scale behaviors associated with the Great Depression that could not be explained by supply and demand. To do this, he looked to flows of income at the national level, and laid the groundwork for what became the Gross Domestic Product (or GDP) as an indicator of economic well-being.

This work and associated influences earned him great notoriety that continues through this day. Not as well known, however, have been his ongoing analyses about the long term beyond mere large-scale economics. In 1930, he wrote a book entitled *Economic Possibilities for Our Grandchildren* (2010), despite not personally having grandchildren. In this book, Keynes explored a 100-year time horizon, and pondered what mix of spending, savings and government support would be needed to achieve some semblance of "sustainability" by the year 2030. Embedded in this writing was also an acknowledgment that free markets and economic competition tended to suffer from short time horizons, neglectful of longer-term futures (Heilbroner, 2011). While Keynes did not use the terms sustainability, sustainable development or environment explicitly, the tenets of sustainability were present in his deliberations regarding how equity, power, access and values comprised key facets of sustainable development at the human–environment interface.

Noam Chomsky meditated on these challenges in an interview with Kevin Kaners in 2016. When asked by Kaners for advice for the future, Chomsky commented, "First, try to understand the world as it is, not as you wish it were; second, decide what outcomes you think are desirable, then ask yourself 'am I willing to commit my energy, effort, dedication to try to achieve the outcomes that I think are beneficial?" Chomsky then advised to "commit to efforts to increase the likelihood that my [or your] grandchildren have a decent world to live in." He then noted, "simple choices, choices for people like us are vast."

Tying these Keynesian and Chomskian insights together, we can contemplate the existential worry that "We're all going to die." Well, sooner or later, yes we are all going to die. However, in the meantime, we all need to get (it) together and to work things out. With this work comes a need for leadership, boldness, creativity and ambition (see Chapter 7). And, our opportunities for engagement are as great as they have ever been.

Going forward, to address the scale of the challenge with a collective, aggregated and sufficient scale of responses, social science and humanities research has found great value in coordinated work through community, combating fragmentation and isolation. Amy Poteete, Marco Janssen and Elinor Ostrom (2010) have noted, "A crucial challenge ... is the need to move toward theories of collective action and common pool resources that acknowledge complexity and multiple levels of analysis, yet offer meaningful leverage, can be tested and can be improved over time" (pp. 232–233). Philip Smith and Nicolas Howe (2015) have also argued a shift to civil solidarity, among other changes needed, "not the accurate communication of yet more scientific facts, is the biggest question facing climate activism today" (p. 200). Moreover, Gwendolyn Blue (2017) has pointed out the value of collective participation in the public sphere. She commented that participatory engagements "hold much promise for opening public policy to diverse perspectives and values" (p. 200). While participation is not in and of itself a panacea, it contributes meaningfully to citizen empowerment, social learning and institutional reflexivity (Arnstein, 1969).

In part, movements to collective cooperation and coordinated action can be considered through notions of mobilizations of social capital. Social capital was first defined through social science research by Joseph Coleman (1988) as "the set of resources that are inherent in family relations and in community social organizations and that are useful for the cognitive or social development of a child or young person" (p. S95). This original conception was then further developed by Robert Putnam (2001), through analyses of social relations in Northern and Southern Italy, where social capital was "embodied in norms and

networks of civic engagement," economic development and effective governance (p. 14). Following these foundational examinations, Thomas Macias and
Kristin Williams explored elements of social capital through General Social
Survey data. They found that time spent with neighbors correlated positively
with pro-environmental lifestyles (Macias and Williams, 2016).[11]

Happiness is another important measure of well-being that has been
investigated by social scientists and humanities researchers and practitioners
as it relates to climate change. Their research has found that the doom and
gloom of these stark climate challenges, mixed with political and social
partisanship and polarization, can lead to pervasive feelings of suspicion,
distrust, cynicism and pessimism. Leading scholar Anthony Leiserowitz has
described a "hope gap" that often sits between increased awareness of the
challenges and engagement in the face of understanding, from the individual to the collective levels (Upton, 2015). However, to productively
respond to the scale of these challenges requires positive and cooperative
movements toward creativity, capacity-building and hope. As an example of
practitioner engagement, former New York City mayor Michael Bloomberg
teamed up with former Sierra Club director Carl Pope to write a book,
Climate of Hope: How Cities, Business and Citizens Can Save the Planet
(2017). Their perspectives earned traction in creative ways to address
climate change through this collaborative assessment and guide. Similarly,
former US secretary of the treasury Henry Paulson (a Republican) and
philanthropist Tom Steyer (a Democrat) have teamed up with Michael
Bloomberg (a Republican-turned-Democrat) on a project called "Risky
Business" that examined the economic risks posed by climate change as
well as economic opportunities for engagement.

In *The Happy Hero*, Solitaire Townsend (2017) argued that positive
approaches along with happy heroism are keys to success at the human–
environment interface. While we must acknowledge loss often involved in
the causes and consequences of climate change (e.g., Rush, 2018), to generate
collective action we must also bring in the gains (and fun) of engagement.
Instead of thinking (or saying), "What have future generations done for me
lately?" a collective and forward-looking commitment over long time horizons
is required to match the scale of these challenges.

We must be nimble as we integrate lessons from previous creative – and
hence effective – communications forays (see Chapter 7 for more). These
lessons include common refrains that getting together in active dialogue and

---

[11] However, they found a negative correlation between time spent with family and proenvironmental behaviors.

exchange – through participation in open and respectful conversations that also include active listening – provide elements of communication success.

Furthermore, we must encourage "new knowledge brokers" as creative climate communicators (Priest, 2016; Lam, 2017; Sieber, 2017) (see Chapter 1 for more). We must also help academic researchers feel confident and competent to speak out in the public sphere (see Chapter 6 for more). Nancy Baron (2016) has pointed out, "Today, most consider such communication crucial: they recognize that the publication of research ... is not an end in itself, but a launch pad to further things" (p. 518).

Participation by diverse perspectives in the public sphere offers both opportunities and challenges in combating political polarization, and building credibility among segments of the public that may be critical scientific ways of knowing about climate change (Wynne, 2006). It can also help to recapture a "missing middle ground" (Kirk, 2018). While climate change discourse is not specifically in the province of science (see Chapter 3 for more), it also is not merely an issue subject to "belief" or "opinion." Facts are not optional. In the advent of "post truth" and "fake news," there is a need for steadfast commitments to facts, accuracy and "meeting people where they are" on climate change (see Chapter 1 for more).

We must be agile in a changing communications ecosystem. Joe Smith, George Revill and Kim Hammond (2018) have commented that "the fast-evolving media ecology calls for a commitment to ongoing attention and innovation from scholars if they are to interpret and critique, and increasingly take opportunities to participate in, the generation of content. The softening of boundaries between production and consumption can be matched by more active commentary and participation by the research community in meaning-making and the voicing of constituencies" (p. 611).

Through social science research investigations, Terry Burns, John O'Connor and Susan Stocklmayer (2003) have offered five purposes for communication of science: to raise awareness, to enhance enjoyment, to garner interest, to help form opinions, and to build understanding. Molly Simis, Haley Madden, Michael Cacciatore and Sara Yeo (2016) have added "political persuasion, education and civic duty" to this list (p. 411). In a similar vein, after the release of the IPCC Special Report on 1.5°C warming (see Chapter 3 for more), Matt Nisbet (2018a) provided this set of insights at *Medium*:

> For too long, as scholars, we have focused narrowly on general public attitudes, evaluating by way of opinion polls or experiments strategies to more effectively communicate climate change risks. Like many environmentalists and philanthropists, our research has been motivated by a desire to create a sense of public urgency, believing that intensifying voter pressure on elected officials was the

key to policy change. But in doing so, we have never adequately articulated the conditions by which public mobilization might translate into effective public policy action. Nor have we examined closely the process by which political elites, those in the best position to make decisions about our collective future, might come to agree on the same effective policy approaches *but for different reasons*.

This call captures the spirit of the book's objectives, to broaden the scope of considerations that can then help creatively and effectively link climate science and policy action, and connect these deliberations to everyday engagements in the public sphere.

As I have called for throughout this volume, we must be smarter. While there is a time and place for nearly everything, social science and humanities research has shown that finding common ground on climate change involves creative and systematic work to effectively make connections through issues, people and things that everyday citizens care about. In this way, creative (climate) communications can help to bridge divides that impede progress on climate science and policy. In so doing, we need to think openly about how climate communications-as-interventions have contributed to (mis)perceptions and (mis)understandings. We must carefully and honestly assess how creative (climate) communications can enhance possibilities for meaningful responses to contemporary climate challenges. Amid the fast-paced swirl of communications in contemporary society along with an inundation of information in the public sphere, Philip Smith and Nicolas Howe (2015) have commented that "paradoxically, there is an urgent need to slow down – and think" (p. 207).

This book has sought to provoke, or in the words of Amartya Sen (2013), to "inform agitation." This book has incorporated elements of intersubjective meaning systems and the value of nuanced difference, as it situates cultural processes in political economic spaces. It has examined research into what works, how, when, why and under what conditions. I have therefore deployed both "the hatchet" as critique, and "the seed" as emancipatory work (Robbins, 2011) to trigger innovation and improvement in creative climate communications. My efforts here work through various communication strategies, approaches, experiments and struggles to better understand interactions between changing environments and human societies, along with relations of resource use and the societal, economic and political sets of relations that shape those behaviors (Forsyth, 2004). These explorations in this long-form text can then increase our ability to smarten up our communications creatively and encourage authentic participatory engagement in the public sphere.

Going forward, there remain many challenges involving open questions linking climate change and population, poverty, consumption, water quality, local air quality, public health and so on. These threats connect to other challenges such as cancer rates, infectious diseases and movements like human rights (for better wages, immigrant and indigenous rights). Through these many considerations, one thing is clear: to address these connected challenges we must be prepared to take communication risks, to potentially make mistakes and apply imagination in order to come up with new and innovative ways of retelling climate stories in the twenty-first century.

Going forward, there needs to be ongoing social science and humanities research – with support for systematic analyses therein – regarding how creative (climate) communications elicit varying levels of awareness and engagement. Creative and effective (climate) communications shape perspectives, attitudes, intentions, beliefs and behaviors among public citizens around the world. Put simply, they make climate change meaningful for everyday people.

While there is no shortage of challenges, there is no shortage of hope either, grounded in commitment, creativity and endeavor. And there is no shortage in the renewable power of our collective enterprise. I hope this book helps us to rethink and revamp the templates we use to communicate about and engage with climate change.

This is an important moment in history. Collective imagination is on the rise. Creativity (as applied imagination) is expanding rather than retracting from this core challenge of meeting people where they are on climate change in the twenty-first century.

Many who are reading this book are likely privileged to get to academically wrestle with research and practice challenges at the human–environment interface. I am too. With that privilege comes responsibility and purpose to find common ground on climate change, and to search for meaning amid multiple culturally, politically, socially, ideologically, and economically infused perspectives and aspirations. If we squander these chances we are fools. Instead, we must recognize that we have many wonderful opportunities to integrate lessons learned through past experimentation and analyses, to then get more organized and steer our discussions and actions into new, collective and positive futures. With sustained funding commitments and sustained willingness to innovate, the future of creative (climate) communications looks bright. So, let's get (it) together' and let's get on with it.

Ambition, leadership, innovation, boldness, audacity, courage, humility and consistent determination are all needed across all overlapping sectors comprising a global public citizenry. It is important to lean into the formidable challenges of decarbonization and urgent policy and everyday engagement in the face of climate change. Nelson Mandela has said, "It always seems impossible until it's done." Moreover, Frederick Douglass (1857) said, "If there is not struggle, there is no progress."

There is much more work to be done, so **we must get stuck in rather than getting stuck**. We must struggle, together.

Up with hope.

References

Abbasi, D. R. (2006). *Closing the Gap between Science and Action: A Synthesis of Insights and Recommendations from the 2005 Yale F and ES Conference on Climate Change*. New Haven, CT: Yale School of Forestry and Environmental Studies.

ActionMedia (2005). *Naming Global Warming*. Minneapolis, MN: ActionMedia.

Adams, P. C. and Gynnild, A. (2013). Environmental messages in online media: The role of place. *Environmental Communication*, 7(1): 113–130.

Adler, B. (2015). The climate movement doesn't need conservatives. *Grist*, May 28. Available at: https://grist.org/climate-energy/the-climate-movement-doesnt-need-conservatives/

Adorno, T. (1991). *The Culture Industry*. London: Routledge.

Agyeman, J. (2008). Toward a "just" sustainability? *Continuum*, 22(6): 751–756.

Agyeman, J., Doppelt, B., Lynn, K. and Hatic, H. (2007). The climate–justice link: Communicating risk with low-income and minority audiences. In S. C. Moser and L. Dilling (eds.), *Creating a Climate for Change: Communicating Climate Change and Facilitating Social Change*, pp. 119–138. Cambridge: Cambridge University Press.

Akter, S., Bennett, J. and Ward, M. B. (2012). Climate change scepticism and public support for mitigation: Evidence from an Australian choice experiment. *Global Environmental Change*, 22(3): 736–745.

Albeck-Ripka, L. (2018). How they changed their minds about global warming. *New York Times*, February 22, A11.

Alexander, C. S. and Sysko, J. M. (2012). A study of the cognitive determinants of generation Y's entitlement mentality. *Academy of Educational Leadership Journal*, 16(2): 63.

Allan, M. R., Stott, P. A., Mitchell, J. F. B., et al. (2000). Quantifying the uncertainty in forecasts of anthropogenic climate change. *Nature*, 407: 617.

Alperin, J. P., Muñoz Nieves, C., Schimanski, L., et al. (2019). Meta-research: How significant are the public dimensions of faculty work in review, promotion, and tenure documents? *eLife*, e42254.

Amel, E. L., Manning, C. M. and Scott, B. A. (2009). Mindfulness and sustainable behavior: Pondering attention and awareness as means for increasing green behavior. *Ecopsychology*, 1: 14–25. DOI:10.1089/eco.2008.0005

Anapol, A. (2017). Centers for Disease Control banned for using "evidence-based" and "science-based" in official documents: Report. *The Hill*, December 15. Available at: http://thehill.com/news-by-subject/healthcare/365204-trump-admin-bans-cdc-from-using-evidence-based-and-science-based

Anapol, A. (2018a). Photographer cropped inauguration photos to make crowd look larger after Trump intervention: Report. *The Hill*, September 6. Available at: https://thehill.com/homenews/administration/405304-nps-photographer-cropped-inauguration-photos-to-make-crowd-appear

Anapol, A. (2018b). Ocasio-Cortez to join Bernie Sanders for climate change town hall. *The Hill*, November 29. https://thehill.com/policy/energy-environment/418980-ocasio-cortez-to-join-bernie-sanders-for-climate-change-town-hall?amp

Anderegg, W. R., Prall, J. W., Harold, J. and Schneider, S. H. (2010). Expert credibility in climate change. *Proceedings of the National Academy of Sciences of the USA*, 107 (27): 12107–12109.

Anderson, A. (1997). *Media, Culture and the Environment*. London: Routledge.

Anderson, A. (2009). Media, politics and climate change: Towards a new research agenda. *Sociology Compass*, 3(2): 166–182.

Anderson, A. (2015). Reflections on environmental communication and the challenges of a new research agenda. *Environmental Communication*, 9(3): 379–383.

Anderson, A. and Becker, A. B. (2018). Not just funny: Sarcasm as a catalyst for public engagement with climate change. *Science Communication*, 40(4): 524–540.

Anderson, A. A. (2017). Effects of social media use on climate change opinion. In M. Nisbet (ed.), *Oxford Research Encyclopedia of Climate Science*, Vol. 2, pp. 486–500. Oxford: Oxford University Press.

Anderson, A. A. and Huntington, H. E. (2017). Social media, science, and attack discourse: How Twitter discussions of climate change use sarcasm and incivility. *Science Communication*, 39(5): 598–620.

Anderson, B. (2012). Affect and biopower: Towards a politics of life. *Transactions of the Institute of British Geographers*, 37(1): 28–43.

Anderson, B. (2017). *Encountering Affect: Capacities, Apparatuses, Conditions*. London: Routledge.

Anderson, K. and Kuhn, K. (2014). *Cowspiracy: The Sustainability Secret*. San Francisco: AUM Films and First Spark Media.

Anderson, L. W., Krathwohl, D. R. and Bloom, B. S. (2001). *A Taxonomy for Learning, Teaching, and Assessing: A Revision of Bloom's Taxonomy of Educational Objectives*. New York, NY: Longman.

Anshelm, J. and Hultman, M. (2014). A green fatwā? Climate change as a threat to the masculinity of industrial modernity. *NORMA: International Journal for Masculinity Studies*, 9(2): 84–96.

Appiah, K. A. (2016). There is no such thing as Western civilization. *The Guardian*, November 9. Available at: www.theguardian.com/world/2016/nov/09/western-civilisation-appiah-reith-lecture

Argrawala, S. (1998). Structural and process history of the Intergovernmental Panel on Climate Change. *Climatic Change*, 39: 621–642.

Aristotle. (1996). *Poetics*. Translated by Malcolm Heath. London: Penguin.

Arnold, A. (2018). *Climate Change and Storytelling: Narratives and Cultural Meaning in Environmental Communication*. New York, NY: Springer Nature.

Arnstein, S. R. (1969). A ladder of citizen participation. *Journal of the American Institute of Planners*, 35(4): 216–224.

Aronoff, K. (2018). "Hothouse Earth" co-author: The problem is neoliberal economics. *The Intercept*, August 14. Available at: https://theintercept.com/2018/08/14/hot house-earth-climate-change-neoliberal-economics/

Asher, K. and Wainwright, J. (2018). After post-development: On capitalism, difference, and representation. *Antipode*. Available at: https://doi.org/10.1111/anti.12430

Asimov, I. (1980). A cult of ignorance. *Newsweek*, January 24, p. 19.

Atkins, G. (1994). *Improv! A Handbook for the Actor*. Portsmouth, NH: Heineman.

Atkinson, M. D. and DeWitt, D. (2018). Does celebrity issue advocacy mobilize issue publics? *Political Studies*, 0032321717751294.

Attari, S. Z. (2018). No one's perfect: How to advocate for climate conservation anyway. *Behavioral Scientist*, April 16. Available at: http://behavioralscientist.org/no-ones-perfect-how-to-advocate-for-climate-conservation-anyway/

Attari, S. Z., Krantz, D. H. and Weber, E. U. (2016). Statements about climate researchers' carbon footprints affect their credibility and the impact of their advice. *Climatic Change*, 138(1–2): 325–338.

Aufrecht, M. (2017). Leave only footprints? Reframing climate change, environmental stewardship, and human impact. *Ethics, Policy and Environment*, 20(1): 84–102.

Bain, P. G., Hornsey, M. J., Bongiorno, R., et al. (2012). Promoting pro-environmental action in climate change deniers. *Nature Climate Change*, 2(8): 600.

Bain, P. G., Milfont, T. L., Kashima, Y., et al. (2016). Co-benefits of addressing climate change can motivate action around the world. *Nature Climate Change*, 6(2): 154.

Bakhtin, M. M. (1984). *Rabelais and His World*. Translation by Helene Iswolsky. Bloomington, IN: Indiana University Press.

Balaraman, K. (2016). Climate change and the "fake news" problem. *Energy & Environment News*, December 5. Available at: www.eenews.net/stories/1060046634

Bandhuim, E. S. and Davis, M. K. (1972). Scales of the measurement of ethos: Another attempt. *Speech Monographs*, 39: 296–301

Banerjee, N. and Lee, J. (2017). Science teachers respond to climate materials sent by Heartland Institute. *Inside Climate News*, December 22. Available at: https://inside climatenews.org/news/22122017/science-teachers-heartland-institute-anti-climate-booklet-survey

Baptiste, N. (2017). He was a professional climate skeptic. Then he switched sides. *Mother Jones*, November 6. Available at: www.motherjones.com/environment/2017/11/i-was-a-professional-climate-denier-i-was-wrong/

Barbaro, N. and Pickett, S. M. (2015). Mindfully green: Examining the effect of connectedness to nature on the relationship between mindfulness and engagement in pro-environmental behavior. *Personality and Individual Differences*, 93: 137–142. DOI:10.1016/j.paid.2015.05.026

Barkham, P. (2018). We're doomed: Mayer Hillman on the climate reality no one else will dare mention. *The Guardian*, April 26. www.theguardian.com/environment/2018/apr/26/were-doomed-mayer-hillman-on-the-climate-reality-no-one-else-will-dare-mention

Barnes, T. J. (1991). Metaphors and conversations in economic geography: Richard Rorty and the gravity model. *Geografiska Annaler, Human Geography, Series B*, 73(2): 111–120.

Barnes, T. J. (2001). Retheorizing economic geography: From the quantitative revolution to the "cultural turn." *Journal of Economic Geography*, 91(3): 546–565.

Baron, N. (2016). So you want to change the world? *Nature*, 540: 517–519.

Barringer, F. (2012). From "Frontline", a look at the skeptics' advance. *New York Times Green Blog*, October 24. Available at: http://green.blogs.nytimes.com/2012/10/24/from-frontline-a-look-at-the-skeptics-advance/

Bauer, M. W. (2016). Results of the essay competition on the "deficit concept." *Public Understanding of Science*, 25(4):389–399.

Baum, M. A. and Groeling, T. (2008). New media and the polarization of American political discourse. *Political Communication*, 25(1): 345–365.

BBC. (2017). *"We're not lazy, we're innovative" – Generation Z hits back in live debate*. September 26. Available at: www.bbc.co.uk/newsbeat/article/41348207/were-not-lazy-were-innovative-generation-z-hits-back-in-live-debate

Beach, R., Share, J. and Webb, A. (2017). *Teaching Climate Change to Adolescents: Reading, Writing, and Making a Difference*. London: Routledge.

Beck, M. (2017). Telling stories with models and making policy with stories: An exploration. *Climate Policy*, 7(5): 1–14.

Beck, S. and Mahony, M. (2017). The IPCC and the politics of anticipation. *Nature Climate Change*, 7(5): 311.

Beck, U. (1992). *Risk Society: Towards a New Modernity*. London: SAGE.

Becker, K. (2013). Magic bullet game analysis. Available at: http://minkhollow.ca/becker/doku.php?id=pf:game-reviews:magic-bullet

Bedard, P. (2018). Heritage Foundation: 64% of Trump's agenda already done, faster than Reagan. *Washington Examiner*, February 27.

Beebe, J. R., Baghramian, M., Drury, L. O. C., et al. (2018). Divergent perspectives on expert disagreement: Preliminary evidence from climate science, climate policy, astrophysics, and public opinion. *Environmental Communication*, 13(1): 35–50.

Beene, R. (2018). EPA drops mention of "climate change" in auto-emissions reversal, *Bloomberg, Manhattan*, April 3. Available at: www.bloomberg.com/news/articles/2018-04-03/epa-drops-mention-of-climate-change-in-auto-emissions-reversal

Béland, D. and Howlett, M. (2016). How solutions chase problems: Instrument constituencies in the policy process. *Governance*, 29(3): 393–409.

Benford, R. D. and Snow, D. A. (2000). Framing processes and social movements: An overview and assessment. *Annual Review of Sociology*, 26(1): 611–639.

Bennett, D. J. and Jennings, R. C. (2011). *Successful Science Communication*. Cambridge: Cambridge University Press.

Berger, A. A. (2018). The mind and psyche of millennials. In *Cultural Perspectives on Millennials*, pp. 11–27. Cham, Switzerland: Palgrave Macmillan.

Bergman, S. M., Fearrington, M. E., Davenport, S. W. and Bergman, J. Z. (2011). Millennials, narcissism, and social networking: What narcissists do on social networking sites and why. *Personality and Individual Differences*, 50(5): 706–711.

Berlant, L. and Ngai, S. (2017). Comedy has issues. *Critical Inquiry*, 43(2): 233–249.

Bernauer, T. and McGrath, L. F. (2016). Simple reframing unlikely to boost public support for climate policy. *Nature Climate Change*, 6(7): 680.

Besley, J. and Dudo, A. (2017). Scientists' view about public engagement. In M. Nisbet (ed.), *Oxford Research Encyclopedia of Climate Science*, Vol. 3, pp. 399–415. Oxford: Oxford University Press.

Besley, J. C. (2018). Audiences for science communication in the United States. *Environmental Communication*, 1–18.

Besley, J. C. and Nisbet, M. (2013). How scientists view the public, the media and the political process. *Public Understanding of Science*, 22(6): 644–659.

Besley, J. C., Dudo, A., Yuan, S. and Lawrence, F. (2018). Understanding scientists' willingness to engage. *Science Communication*, 40(5): 559–590.

Besley, J. C., Oh, S. H. and Nisbet, M. (2013). Predicting scientists' participation in public life. *Public Understanding of Science*, 22(8): 971–987.

Bevis, M. (2013). *Comedy: A Very Short Introduction*. Oxford: Oxford University Press.

Bissell, D. (2008). Comfortable bodies, sedentary affects. *Environment and Planning*, 40: 1697–1712.

Black, R. (2015). No more summaries for wonks. *Nature Climate Change*, 5(4): 282.

Blanding, M. (2017). Covering climate change, with urgency and creativity. Nieman Reports, Cambridge, MA, August 28. Available at: http://niemanreports.org/articles/covering-climate-change-with-urgency-and-creativity/

Bleys, B., Defloor, B., Van Ootegem, L. and Verhofstadt, E. (2018). The environmental impact of individual behavior: Self-assessment versus the ecological footprint. *Environment and Behavior*, 50(2): 187–212.

Bloom, B., ed. (1959). *Taxonomy of Educational Objectives: Handbook I: Cognitive Domain*. New York, NY: Longman.

Bloomberg, M. and Pope, C. (2017). *Climate of Hope: How Cities, Business and Citizens Can Save the Planet*. New York, NY: St. Martin's Press

Bloomfield, E. F. and Tillery, D. (2018). The circulation of climate change denial online: Rhetorical and networking strategies on Facebook. *Environmental Communication*, 1–17. DOI:doi.org/10.1080/17524032.2018.1527378

Blue, G. (2017). Participatory and deliberative approaches to climate change. In M. Nisbet (ed.), *Oxford Research Encyclopedia of Climate Science*, Vol. 3, pp. 191–203. Oxford: Oxford University Press.

Boehnert, J. (2018). *Design, Ecology, Politics: Towards the Ecocene*. London: Bloomsbury Press.

Böhm, G., Doran, R., Rødeseike, A. and Pfister, H. R. (2018). Laypeople's affective images of energy transition pathways. *Frontiers in Psychology*, 9: 1904.

Bohr, J. (2016). The "climatism" cartel: Why climate change deniers oppose market-based mitigation policy. *Environmental Politics*, 25(5): 812–830. DOI:10.1080/09644016.2016.1156106

Bok, R. (2018). "By our metaphors you shall know us." The "fix" of geographical political economy. *Progress in Human Geography*, 18(7): 928–941.

Bolderdijk, J. W., Gorsira, M., Keizer, K., et al. (2013). Values determine the (in)effectiveness of informational interventions in promoting pro-environmental behavior. *PloS ONE*, 8(12): e83911.

Bolsen, T. and Shapiro, M. A. (2017). *Strategic framing and persuasive messaging to influence climate change perceptions and decisions*. In M. Nisbet (ed.), *Oxford Research Encyclopedia of Climate Science*, Vol. 3, pp. 491–508. Oxford: Oxford University Press.

Bolsen, T. and Shapiro, M. A. (2018). The US news media, polarization on climate change, and pathways to effective communication. *Environmental Communication*, 12(2): 149–163.

Bolstad, E. (2016). Scientists' struggle: Talking climate "in plain English." *Energy and Environmental Daily*, March 28. Available at: www.eenews.net/stories/1060034657

Borel, B. (2017). Fact-checking won't save us from fake news. *FiveThirtyEight* blog, January 4. Available at: https://fivethirtyeight.com/features/fact-checking-wont-save-us-from-fake-news/

Bosetti, V., Weber, E., Berger, L., et al. (2017). COP21 climate negotiators' responses to climate model forecasts. *Nature Climate Change*, 7(3): 185.

Bostrom, A. (2017). Mental models and risk perceptions related to climate change. In M. Nisbet (ed.), *Oxford Research Encyclopedia of Climate Science*, Vol. 3, pp. 55–74. Oxford: Oxford University Press.

Bottrill, C., Liverman, D. and Boykoff, M. (2010). Carbon soundings: GHG emissions in the UK music industry. *Environmental Research Letters*, 5(1): 1–8.

Boulianne, S. and Theocharis, Y. (2018). Young people, digital media, and engagement: A meta-analysis of research. *Social Science Computer Review*, 1–17. DOI:doi.org/10.1177/0894439318814190

Bower, G. H., Black, J. B. and Turner, T. J. (1979). Scripts in memory for text. *Cognitive Psychology*, 11(2): 177–220.

Bowling, A. (1997). *Research Methods in Health*. Buckingham, UK: Open University Press.

Boyce, T. and Lewis, J., eds. (2009). *Climate Change and the Media*, Vol. 5. New York, NY: Peter Lang.

Boykoff, M. T. (2011). *Who Speaks for the Climate? Making Sense of Media Coverage of Climate Change*. Cambridge: Cambridge University Press:

Boykoff, M. (2013). Public Enemy no.1? Understanding media representations of outlier views on climate change. *American Behavioral Scientist*, 57(6): 796–817.

Boykoff, M. and Boykoff, J. (2007). Climate change and journalistic norms: A case study of U.S. mass-media coverage. *Geoforum*, 38(6): 1190–1204.

Boykoff, M. and Goodman, M. K. (2009). Conspicuous redemption? Reflections on the promises and perils of the "celebritization" of climate change. *Geoforum*, 40: 395–406.

Boykoff, M. and Olson, S. (2013). "Wise contrarians" in contemporary climate science-policy-public interactions. *Celebrity Studies Journal*, 4(3): 276–291.

Boykoff, M. and Oonk, D. (2019). Evaluating the perils and promises of academic climate advocacy. *Climate Change*, DOI:doi.org/10.1007/s10584-018-2339-3

Boykoff, M. and Osnes, B. (2019). A laughing matter? Confronting climate change through humor. *Political Geography*. Available at: https://doi.org/10.1016/j.polgeo.2018.09.006

Boykoff, M. and Rajan, S. R. (2007). Signals and noise: Mass media coverage of climate change in the USA and UK. *European Molecular Biology Organization Reports*, 8(3): 207–211.

Boykoff, M. and Yulsman, T. (2013). Political economy, media and climate change – The sinews of modern life. *Wiley Interdisciplinary Reviews: Climate Change*, 4(5): 359–371.

Boykoff, M., Goodman, M. and Curtis, I. (2010). The cultural politics of climate change: interactions in everyday spaces. In *The Politics of Climate Change*, pp. 136–154. London: Routledge.

Boykoff, M., Osnes, B. and Safran, R. (2018). "Contando estorias de la ciencia del cambio de clima "Dentro del Invernadero."" In *"Comunicación Audiovisual de la*

Ciencia" [co-editors G. O. Gómez, B. León and M. Francés i Domènec]. Special Issue, TV Morfosis CNTD.

Bravender, R. (2018). "Can we say … climate?" Agency grapples with Trump's view. *Energy & Environment Daily*, April 2. Available at: www.eenews.net/stories/1060077853

Brenan, M. and Saad, L. (2018). Global warming concern steady despite some partisan shifts. Gallup, Washington, DC, March 28. Available at: https://news.gallup.com/poll/231530/global-warming-concern-steady-despite-partisan-shifts.aspx

Brevini, B. and Lewis, J. (2018) Introduction. In B. Brevini, and J. Lewis (eds.), *Climate Change and the Media*, pp. 1–8. London: Routledge.

Brewer, P. R. and McKnight, J. (2015). Climate as comedy: The effects of satirical television news on climate change perceptions. *Science Communication*, 37(5): 635–657.

Brewer, P. R. and McKnight, J. (2017). A statistically representative climate change debate: Satirical television news, scientific consensus, and public perceptions of global warming. *Atlantic Journal of Communication*, 25(3): 166–180.

Breyer, M. (2018). Radiohead's Thom Yorke releases song to help Antarctica. *Treehugger*. October 18. Available at: www.treehugger.com/conservation/radio heads-thom-yorke-greenpeace-release-song-help-antarctica.html

Brieger, S. A. (2018). Social identity and environmental concern: The importance of contextual effects. *Environment and Behavior*, 1–28.

Brown, D. (2017). Murray suit against John Oliver "unconstitutional" – HBO *Energy & Environment News*, July 25. Available at: www.eenews.net/greenwire/2017/07/25/stories/1060057868

Brown, V. A., Harris, J. A. and Russell, J. Y., eds. (2010). *Tackling Wicked Problems through the Transdisciplinary Imagination*. London: Earthscan.

Brugger, K. (2018a). Skepticism drops when people are told of scientific consensus – study. *Energy & Environment Daily*, April24. Available at: www.eenews.net/climate wire/2018/04/24/stories/1060079871

Brugger, K. (2018b). Climate activists get tips for talking to conservatives. *Energy & Environment Daily*, June 12. Available at: www.eenews.net/climatewire/stories/1060084157/feed≥

Brulle, R. J. (2010). From environmental campaigns to advancing a public dialogue: Environmental communication for civic engagement. *Environmental Communication*, 4(1): 82–98.

Brulle, R. J. (2014). Institutionalizing delay: Foundation funding and the creation of U. S. climate change counter-movement organizations. *Climatic Change*, 122: 681–694.

Brulle, R. J. (2018a). March: Critical reflections on the march for science. *Sociological Forum*, 33(1): 255–258.

Brulle, R. J. (2018b). The climate lobby: A sectoral analysis of lobbying spending on climate change in the USA, 2000 to 2016. *Climatic Change*, 149(3): 1–15.

Brulle, R. J. and Antonio, R. J. (2015). The Pope's fateful vision of hope for society and the planet. *Nature Climate Change*, 5(10): 900.

Brulle, R. J., Carmichael, J. and Jenkins, J. C. (2012). Shifting public opinion on climate change: An empirical assessment of factors influencing concern over climate change in the US, 2002–2010. *Climatic Change*, 114(2): 169–188.

Brundtland, G. H. (1987). *Our common future*. Report for the World commission on Environment and Development, United Nations.

Brysse, K., Oreskes, N., O'Reilly, J. and Oppenheimer, M. (2013). Climate change prediction: Erring on the side of least drama? *Global Environmental Change*, 23(1): 327–337.

Bucchi, M. (2015). Norms, competition and visibility in contemporary science: The legacy of Robert K. Merton. *Journal of Classical Sociology*, 15(3): 233–252.

Bucchi, M. and Trench, B., eds. (2008). *Handbook of Public Communication of Science and Technology*. New York, NY: Routledge.

Bulfin, A. (2017). Popular culture and the "new human condition": Catastrophe narratives and climate change. *Global and Planetary Change*. DOI:10.1016/j.gloplacha.2017.03.002

Burns, N. and Grove, S. K. (1997). *The Practice of Nursing Research Conduct, Critique, and Utilization*. Philadelphia, PA: W. B. Saunders.

Burns, T. W., O'Connor, D. J. and Stocklmayer, S. M. (2003). Science communication: A contemporary definition. *Public Understanding of Science*, 12(2): 183–202.

Bushell, S., Colley, T. and Workman, M. (2015). A unified narrative for climate change. *Nature Climate Change*, 5(11): 971.

Cacciatore, M. A., Anderson, A. A., Choi, D., et al. (2012). Coverage of emerging technologies: A comparison between print and online media. *New Media and Society*, 14(6): 1039–1059.

Callison, C. (2014). *How Climate Change Comes to Matter: The Communal Life of Facts*. Durham, NC: Duke University Press.

Callison, C. (2017). Climate change and indigenous publics. In M. Nisbet (ed.), *Oxford Research Encyclopedia of Climate Science*, Vol. 1, pp. 112–132. Oxford: Oxford University Press.

Cama, T. (2017). EPA's Pruitt: Bring back "true environmentalism." *The Hill*, December 27. Available at: https://thehill.com/policy/energy-environment/366478-epas-pruitt-bring-back-true-environmentalism

Cambridge Dictionary Online. (2008). *Cambridge online dictionary*, Open Educational Resources (OER) Portal.

Campbell, B. and Manning, J. (2016). Campus culture wars and the sociology of morality. *Comparative Sociology*, 15(2): 147–178.

Cann, H. W. and Raymond, L. (2018). Does climate denialism still matter? The prevalence of alternative frames in opposition to climate policy. *Environmental Politics*, 27(3): 433–454.

Capstick, S. B. and Pidgeon, N. F. (2014). What is climate change scepticism? Examination of the concept using a mixed methods study of the UK public. *Global Environmental Change*, 24: 389–401.

Cardwell, D. (2015). Tackling climate change, one class at a time. *New York Times*, June 30, B1.

Carmichael, J. T. and Brulle, R. J. (2017). Elite cues, media coverage, and public concern: An integrated path analysis of public opinion on climate change, 2001–2013. *Environmental Politics*, 26(2): 232–252.

Carmichael, J. T. and Brulle, R. J. (2018). Media use and climate change concern. *International Journal of Media & Cultural Politics*, 14(2): 243–253.

Carmichael, J. T., Brulle, R. J. and Huxster, J. K. (2017). The great divide: Understanding the role of media and other drivers of the partisan divide in public concern over climate change in the USA, 2001–2014. *Climatic Change*, 141(4): 599–612.

Carter, B. E. and Wiles, J. R. (2016). Comparing effects of comedic and authoritative video presentations on student knowledge and attitudes about climate change. *Bioscene*, 42(1): 16–24.

Carvalho, A. (2007). Ideological cultures and media discourses on scientific knowledge: Re-reading news on climate change. *Public Understanding of Science*, 16: 223–243.

Carvalho, A. and Burgess, J. (2005). Cultural circuits of climate change in UK broadsheet newspapers, 1985–2003. *Risk Analysis*, 25(6): 1457–1469.

Cash, D. W., Borck, J. C. and Patt, A. G. (2006). Countering the loading-dock approach to linking science and decision making: Comparative analysis of El Niño/Southern Oscillation (ENSO) forecasting systems. *Science, Technology, and Human Values*, 31(4): 465–494.

Cassady, J. C., ed. (2010). Test anxiety: Contemporary theories and implications for learning. In *Anxiety in Schools: The Causes, Consequences, and Solutions for Academic Anxieties*, pp. 7–26. New York, NY: Peter Lang.

Castree, N. (2010). Neoliberalism and the biophysical environment 2: Theorising the neoliberalisation of nature. *Geography Compass*, 4(12): 1734–1746.

Castree, N. (2016). Geography and the new social contract for global change research. *Transactions of the Institute of British Geographers*, 41(3): 328–347.

Chang, Clio. (2017). Minutes: News and notes. *New Republic*, October 19. Available at: https://newrepublic.com/minutes/140069/trump-calls-environmentalist-pledging-cut-auto-regulations-approving-oil-pipelines

Chapman, D. A., Lickel, B. and Markowitz, E. M. (2017). Reassessing emotion in climate change communication. *Nature Climate Change*, 7(12): 850.

Chappell, A. (2006). Using the "grieving" process and learning journals to evaluate students' responses to problem-based learning in an undergraduate geography curriculum. *Journal of Geography in Higher Education*, 30: 15–31.

Chattoo, C. B. (2017). The laughter effect: The serious role of comedy in social change, p. 53. Center for Media and Social Impact, School of Communication, American University.

Chen, B. X. (2018). The internet trolls have won. Get used to it. *New York Times*, August 8, B7.

Chiao, J. Y., Mathur, V. A., Harada, T. and Lipke, T. (2009). Neural basis of preference for human social hierarchy versus egalitarianism. *Annals of the New York Academy of Sciences*, 1167: 174–181. DOI:10.1111/j.1749-6632.2009.04508.x

Chinn, S., Lane, D. S. and Hart, P. S. (2018). In consensus we trust? Persuasive effects of scientific consensus communication. *Public Understanding of Science*, 27(7): 807–823.

Cho, R. (2017). What changes minds about climate change? Earth Institute blogs, August 9. Available at: https://blogs.ei.columbia.edu/2017/08/09/what-changes-minds-about-climate-change/

Christensen, M. (2018). Slow violence in the Anthropocene: An interview with Rob Nixon on communication, media, and the environmental humanities. *Environmental Communication*, 12(1): 7–14.

Churchman, C. W. (1967). Wicked problems. *Management Science*, 14(4): 141–146.

Clark, B. (2018). 400 consecutive hotter-than-average months and we're still pretending climate change isn't real. *TNW Science*, May 18. Available at: https://

thenextweb.com/science/2018/05/18/400-consecutive-hotter-than-average-months-and-were-still-pretending-climate-change-isnt-real/

Clayton, S. (2018). Mental health risk and resilience among climate scientists. *Nature Climate Change*, 8(4): 260.

Clayton, S., Devine-Wright, P., Stern, P. C., et al. (2015). Psychological research and global climate change. *Nature Climate Change*, 5(7): 640–646.

Clery, D. (2003). Bringing science to the cafes. *Science*, 300(5628):2026.

Clifford, K. R. and Travis, W. R. (2018). Knowing climate as a social-ecological-atmospheric construct. *Global Environmental Change*, 49: 1–9.

Cohen, J. (2017). CDC word ban? The fight over seven health-related words in the President's next budget. *Science*, 18(December). DOI:10.1126/science.aar7959

Coleman, J. S. (1988). Social capital in the creation of human capital. *American Journal of Sociology*, 94: S95–S120.

Collins, H. M. and Evans, R. (2002). The third wave of science studies: Studies of expertise and experience. *Social Studies of Science*, 32: 235–296.

Colman, Z. (2018). Mag prints 70-page climate story, leaves some unsatisfied. *Energy & Environment News*, August 2. Available at: www.eenews.net/stories/1060091933

Comstock, M. and Hocks, M. E. (2016). The sounds of climate change: Sonic rhetoric in the Anthropocene, the age of human impact. *Rhetoris Review*, 35(2): 165–175.

Concha, J. (2018). Schwarzenegger to CNN's Axelrod: Environmentalists doing a terrible job selling climate change concerns. *The Hill*, November 4. Available at: https://thehill.com/homenews/media/414752-schwarzenegger-to-cnns-axelrod-envir onmentalists-doing-a-terrible-job-selling?amp

Condon, P., Desbordes, G., Miller, W. B. and DeSteno, D. (2013). Meditation increases compassionate responses to suffering. *Psychological Science*, 24: 2125–2127. doi:10.1177/0956797613485603

Converse, P. E. (1964). The nature of belief systems in mass publics. *Critical Review*, 18 (1–3): 1–74.

Cook, J. (2017a). Countering climate science denial. In M. Nisbet (ed.), *Oxford Research Encyclopedia of Climate Science*, Vol. 2, pp. 347–367. Oxford: Oxford University Press.

Cook, J. (2017b). Response by Cook to "beyond counting climate consensus." *Environmental Communication*, 11(6): 733–735.

Cook, J. and Lewandowsky, S. (2016). Rational irrationality: Modeling climate change belief polarization using Bayesian networks. *Topics in Cognitive Science*, 8(1): 160–179.

Cook, J., Lewandowsky, S. and Ecker, U. K. (2017). Neutralizing misinformation through inoculation: Exposing misleading argumentation techniques reduces their influence. *PloS ONE*, 12(5): e0175799.

Cook, J., Nuccitelli, D., Green, S. A., et al. (2013). Quantifying the consensus on anthropogenic global warming in the scientific literature. *Environmental Research Letters*, 8(2): 024024.

Cooper, K. E. and Nisbet, E. C. (2017). Documentary and edutainment portrayals of climate change and their societal impacts. In M. Nisbet (ed.), *Oxford Research Encyclopedia of Climate Science*. Available at: http://oxfordre.com/climatescience/view/10.1093/acrefore/9780190228620.001.0001/acrefore-9780190228620-e-373

Corbet, J. and Clark, B. (2017). The arts and humanities in climate change engagement. In M. Nisbet (ed.), *Oxford Research Encyclopedia of Climate Science*, Vol. 1, pp. 48–66. Oxford: Oxford University Press.

Cormick, C., Nielssen, O., Ashworth, P., et al. (2014). What do science communicators talk about when they talk about science communications? Engaging with the engagers. *Science Communication*, 37(2): 274–282.

Corner, A. (2018). Britain, can we – really – talk about this weather we're having? *New York Times*, July 27. Available at: www.nytimes.com/2018/07/27/opinion/britain-heatwave-climate-change.html

Corner, A. and Clarke, J. (2017). *Talking Climate: From Research to Practice in Public Engagementung*. Cham, Switzerland: Springer International.

Corner, A., Roberts, O. and Pellisier, A. (2014). *Young Voices: How Do 18–25 Year Olds Engage with Climate Change?* London: Climate Outreach and Information Network.

Corner, A., Shaw, C. and Clarke, J. (2018). *Principles for Effective Comunication and Public Engagement on Climate Change: A Handbook for IPCC Authors*. Oxford: Climate Outreach.

Corner, J. (2017). Fake news, post-truth and media–political change. *Media, Culture and Society*, 39(7): 1100–1107.

Cortassa, C. (2016). In science communication, why does the idea of a public deficit always return? The eternal recurrence of the public deficit. *Public Understanding of Science*, 25(4): 447–459.

Cosgrove, D. (1994). Contested global visions: One world, whole Earth and the Apollo space photographs. *Annals of the Association of American Geographers*, 84: 270–294.

Crutzen, P. J. (2002). The anthropocene. *Journal de Physique*, 10: 1–5.

Cumming, ed. (2015). Xiuhtezcatl Roske-Martinez: "Our greed is destroying the planet." *The Guardian*, October 9. Available at: www.theguardian.com/environment/2015/oct/09/xiuhtezcatl-roske-martinez-earth-guardians

Cunningham, B. (2003). Re-thinking objectivity. *Columbia Journalism Review*, 42: 24–32.

Cunsolo, A. and Ellis, N. R. (2018). Ecological grief as a mental health response to climate change-related loss. *Nature Climate Change*, 8(4): 275.

Curtis, Lisa. (2012). Why I'm not an environmentalist. *Huffington Post*, April 25. Available at: www.huffingtonpost.com/lisa-curtis/environmentalism_b_1443311.html

Cusick, D. (2018). That image of a cow farting methane? It's a "Climoji". *Energy & Environment News*, January 17. Available at: www.eenews.net/climatewire/2018/01/17/stories/1060071111

Dalby, S. (2007). Anthropocene geopolitics: Globalisation, empire, environment and critique. *Geography Compass*, 1: 1–16.

Darling, S. B. and Sisterson, D. L. (2014). *How to Change Minds about Our Changing Climate Grenwhich* New York, NY: Workman.

Davenport, C. (2017). Climate change references are purged from the White House website. *New York Times*, January, 21, A20.

Davenport, C. (2018). How much as "climate change" been scrubbed from federal website? A lot. *New York Times*, January 10. Available at: www.nytimes.com/2018/01/10/climate/climate-change-trump.html

Davidson, J., Smith, M. M. and Bondi, L., eds. (2012). *Emotional Geographies.* Farnham, UK: Ashgate.

Davies, S. R. (2008). Constructing communication: Talking to scientists about talking to the public. *Science Communication*, 29(4): 413–434.

Davis, F. (1979). *Yearning for Yesterday: A Sociology of Nostalgia.* New York, NY: Free Press.

Davis, J. L., Love, T. P. and Killen, G. (2018). Seriously funny: The political work of humor on social media. *New Media & Society*, 20(10): 3898–3916.

Davis, L., Fähnrich, B., Nepote, A. C., et al. (2018). Environmental communication and science communication: Conversations, connections and collaborations. *Environmental Communication*, 12(4): 431–437.

Dearing, J. W., Meyer, G. and Rogers, E. M. (1994). *Diffusion Theory and HIV Risk Behavior Change: Preventing AIDS*, pp. 79–93. New York, NY: Springer.

Debord, G. (1983). *Society of the Spectacle.* New York, NY: Zone Books.

de Certeau, M. (1984). *The Practice of Everyday Life.* Translated by S. Rendall, Berkeley, CA: University of California Press.

Dekeyser, M., Raes, F., Leijssen, M., et al. (2008). Mindfulness skills and interpersonal behaviour. *Personality and Individual Differences*, 44: 1235–1245. DOI:10.1016/j.paid.2007.11.018

DeMarco, P. M. (2017). Moving from awareness to action: Acceptance speech for the 2017 William Freudenburg award. *Journal of Environmental Studies and Sciences*, 7(4): 469–472.

Dembicki, G. (2017). *Are We Screwed? How a New Generation Is Fighting to Survive Climate Change.* New York, NY: Bloomsbury Press.

Demeritt, D. (2000). The new social contract for science: Accountability, relevance, and value in US and UK science and research policy. *Antipode*, 32(3): 308–329.

Demeritt, D. (2001). The construction of global warming and the politics of science. *Annals of the Association of American Geographers*, 912: 307–337.

Deryugina, T. and Shurchkov, O. (2016). The effect of information provision on public consensus about climate change. *PloS ONE*, 11(4): e0151469.

DiFrancesco, A. D. and Young, N. (2011). Seeing climate change: The visual construction of global warming in Canadian national print media. *Cultural Geographies*, 18(4): 517–536.

Dilling, L. and Lemos, M. C. (2011). Creating usable science: Opportunities and constraints for climate knowledge use and their implications for science policy. *Global Environmental Change*, 21(2): 680–689.

Diprose, K., Fern, R., Vanderbeck, R. M., et al. (2018). Corporations, consumerism and culpability: sustainability in the British press. *Environmental Communication*, 12(5): 672–685.

Dixon, G., Hmielowski, J. and Ma, Y. (2017). Improving climate change acceptance among US conservatives through value-based message targeting. *Science Communication*, 39(4): 520–534.

Dobson, A. (2003). *Citizenship and the Environment.* Oxford: Oxford University Press.

Dolšak, N. and Prakash, A. (2018). The climate change hypocrisy of jet-setting academics. *Huffington Post*, March 31. Available at: www.huffingtonpost.com/entry/opinion-dolsak-prakash-carbon-tax_us_5abe746ae4b055e50acd5c80

Donner, S. D. (2014). Finding your place on the science-advocacy continuum: an editorial essay. *Climatic Change*, 124: 1–8.

Donner, S. D. (2017). Risk and responsibility in public engagement by climate scientists: Reconsidering advocacy during the Trump era. *Environmental Communication*, 11(3): 430–433.

Douglas, M. (1975). *Implicit Meanings: Essays in Anthropology*. London: Routledge.

Douglass, F. (1950). West India Emancipation. Speech delivered at Canandaigua, New York, August 4, 1857. In Philip S. Foner (ed.), *The Life and Writings of Frederick Douglass*, Vol. 2, p. 437. New York, NY: International Publishers.

Downie, J. (2012). It's global warming, stupid. *Washington Post*, November 1. Available at: www.washingtonpost.com/blogs/post-partisan/post/its-global-warming-stupid/2012/11/01/b49c4914-2435-11e2-ac85-e669876c6a24_blog.html

Downs, A. (1957). An economic theory of political action in a democracy. *Journal of Political Economy*, 65(2): 135–150.

Doyle, J. (2007). Picturing the clima(c)tic: Greenpeace and the representational politics of climate change communication. *Science as Culture*, 16(2): 129–150.

Doyle, J. (2016). *Mediating Climate Change*. New York, NY: Routledge.

Doyle, J., Farrell, N. and Goodman, M. K. (2017). Celebrities and climate change. In M. Nisbet (ed.), *Oxford Research Encyclopedia of Climate Science*, Vol. 1, pp. 95–112. Oxford: Oxford University Press.

Dreifus, C. (2016). There's a science to his art. *New York Times*, April, 19, D3.

Druckman, J. N. (2001). The implications of framing effects for citizen competence. *Political Behavior*, 23(3): 225–256.

Drury, I. (2008). Boris orders arts chiefs to stop "dumbing down" culture for young people. *Mail on Sunday Online*, November 24. Available at: www.mailonsunday.co.uk/news/article-1088956/Boris-orders-arts-chiefs-stop-dumbing-culture-young-people.html

Dryzek, J. S. and Lo, A. Y. (2015). Reason and rhetoric in climate communication. *Environmental Politics*, 24(1): 1–16.

Dudo, A., Copple, J. and Atkinson, A. (2017). Entertainment film and TV portrayals of climate change. In M. Nisbet (ed.), *Oxford Research Encyclopedia of Climate Science*, Vol. 2, pp. 557–577. Oxford: Oxford University Press.

Duncombe, S. (2012). *Dream: Reimagining Progressive Politics in an Age of Fantasy*. New York: The New Press.

Dunlap, J. and Cohen, S. A., eds. (2016). *Coming of Age at the End of Nature: A Generation Faces Living on a Changed Planet*. San Antonio, TX: Trinity University Press.

Dunlap, R. E. (2013). Climate change skepticism and denial: An introduction. *American Behavioral Scientist*, 57(6): 691–698.

Dunlap, R. E. and McCright, A. M. (2008). A widening gap: Republican and Democratic views on climate change. *Environment: Science and Policy for Sustainable Development*, 50(5): 26–35.

Dunlap, R. E. and McCright, A. M. (2011). Organized climate change denial. In J. Dryzek, R. Norgaard and D. Schlosberg (eds.), *The Oxford Handbook of Climate Change and Society*. Oxford: Oxford University Press.

Dunlap, R. E., McCright, A. M. and Yarosh, J. H. (2016). The political divide on climate change: Political polarization widens in the U.S. *Environment: Science and Policy for Sustainable Development*, 58(5): 4–23.

Dunwoody, S. and Kohl, P. A. (2017). Using weight-of-experts messaging to communicate accurately about contested science. *Science Communication*, 39(3): 338–357.

Easterby, S. (2018). Climate activists are lousy salesmen. *Wall Street Journal*, April 25. Available at: www.wsj.com/articles/climate-activists-are-lousy-salesmen-1524695895

Ehrhardt-Martinez, K., Rudel, T. K, Norgaard, K. M. and Broadbent, J. (2015). Mitigating climate change. In R. Dunlap and R. Brulle (eds.), *Climate Change and Society*, pp. 199–234. Oxford: Oxford University Press.

Elger, K. and Schwägerl, C. (2011). We are looting the past and future to feed the present. *Der Spiegel, Hamburg, HR*, March 23. Available at: www.spiegel.de/international/germany/leading-climatologist-on-fukushima-we-are-looting-the-past-and-future-to-feed-the-present-a-752474.html

Elias, N. and Parvulescu, A. (2017). Essay on laughter. *Critical Inquiry*, 43(2): 281–304.

Ellison, C. D. (2017). The climate change debate: Black people are being left out and that can be deadly. *The Root*, January 3. Available at: www.theroot.com/the-climate-change-debate-black-people-are-being-left-1791134127

Emmerson, P. (2017). Thinking laughter beyond humour: Atmospheric refrains and ethical indeterminacies in spaces of care. *Environment and Planning A*, 1–17.

Engdahl, E. and Lidskog, R. (2014). Risk, communication and trust: Towards an emotional understanding of trust. *Public Understanding of Science*, 23(6): 703–717.

Entman, R. M. (1993). Framing: Toward clarification of a fractured paradigm. *Journal of Communication*, 43(4): 51–58.

Ereaut, G. and Segnit, N. (2006). Warm words: How we are telling the climate story and can we tell it better. Institute for Public Policy Research, August. Available at: www.ippr.org/publications/warm-wordshow-are-we-telling-the-climate-story-and-can-we-tell-it-better

Evans Comfort Suzannah and Park, Young, Eun. (2018). On the field of environmental communication: A systematic review of the peer-reviewed literature. *Environmental Communication*, DOI:10.1080/17524032.2018.1514315

Fähnrich, B. (2017). Science diplomacy: Investigating the perspective of scholars on politics–science collaboration in international affairs. *Public Understanding of Science*, 26(6): 688–703.

Fahy, D. (2015). *The New Celebrity Scientists: Out of the Lab and into the Limelight.* Lanham, MD: Rowman & Littlefield.

Fahy, D. (2018). Objectivity as trained judgment: How environmental reporters pioneered journalism for a "post-truth" era. *Environmental Communication*, 12(7): 855–861.

Fahy, D. and Nisbet, M. C. (2011). The science journalist online: Shifting roles and emerging practices. *Journalism*, 12(7): 778–793.

Fairclough, N. (1995). *Media Discourse*. London: Edward Arnold.

Faris, S., Lipscombe, S., Whitehead, S. and Wilson, D. (2014). *From the Ground Up: Changing the Conversation about Climate Change*. London: BBC Media Action.

Farrell, J. (2016a). Network structure and influence of climate change countermovement. *Nature Climate Change*, 6(4): 370–374.

Farrell, J. (2016b). Corporate funding and ideological polarization about climate change. *Proceedings of the National Academy of Sciences of the USA*, 113(1): 92–97.

Feinberg, M. and Willer, R. (2013). The moral roots of environmental attitudes. *Psychological Science*, 24(1): 56–62.

Feldman, L. (2013). Cloudy with a chance of heat balls: The portrayal of global warming on The Daily Show and The Colbert Report. *International Journal of Communication*, 7: 430–451.

Feldman, L. (2017). Assumptions about science in satirical news and late-night comedy. In K. Hall Jamieson, D. Kahan and D. A. Scheufele (eds.), *The Oxford Handbook of the Science of Science Communication*, p. 321. Oxford: Oxford University Press.

Feldman, L. and Hart, P. S. (2018). Climate change as a polarizing cue: Framing effects on public support for low-carbon energy policies. *Global Environmental Change*, 51: 54–66.

Festinger, L. (1957). *Theory of Cognitive Dissonance*, Vol. 2. Stanford, CA: Stanford University Press.

Fielding, K. S., Head, B. W., Lafan, W., et al. (2012). Australian politicians' beliefs about climate change: Political partisanship and political ideology. *Environmental Politics*, 21: 712–733.

Fifeld, A. (2009). US rightwing activists curb efforts to cut CO_2 emissions. *Financial Times*, November 3.

Figueres, C., Schellnhuber, H. J., Whiteman, G., et al. (2017). Three years to safeguard our climate. *Nature*, 546: 593–595.

Fisher, D. (2013). Understanding the relationship between subnational and national climate change politics in the United States: Toward a theory of boomerang Federalism. *Environment and Planning C: Government and Policy*, 31(5): 769–784.

Fitzgerald, E. (2013). Greenland melt music. *Living on Earth*, April 5. Available at: www.loe.org/shows/segments.html?programID=13-P13-00014&segmentID=8

Flam, F. (2017). Why some scientists won't march for science. *Bloomberg*, March 7. Available at: www.bloomberg.com/view/articles/2017–03-07/why-some-scientists-won-t-march-for-science

Fleming, A., Vanclay, F., Hiller, C. and Wilson, S. (2014). Challenging dominant discourses of climate change. *Climatic Change*, 127(3–4): 407–418.

Fleming, J. R. (2014). Picturing climate control: Visualizing the unimaginable. *Image Politics of Climate Change: Visualizations, Imaginations, Documentations*, 55: 345.

Flora, J. A., Saphir, M., Lappé, M., Roser-Renouf, C., et al. (2014). Evaluation of a national high school entertainment education program: The Alliance for Climate Education. *Climatic Change*, 127(3–4): 419–434.

Flusberg, S. J., Matlock, T. and Thibodeau, P. H. (2017). Metaphors for the war (or race) against climate change. *Environmental Communication*, 11(6): 769–783.

Foley, J. (2017). The war on facts is a war on democracy. *Scientific American* blog, January 25. Available at: https://blogs.scientificamerican.com/guest-blog/the-war-on-facts-is-a-war-on-democracy/

Forchtner, B. and Kølvraa, C. (2015). The nature of nationalism: Populist radical right parties on countryside and climate. *Nature and Culture*, 10(2): 199–224. DOI:10.3167/nc.2015.100204

Forchtner, B., Kroneder, A. and Wetzel, D. (2018). Being skeptical? Exploring far-right climate-change communication in Germany. *Environmental Communication*, 12(5): 589–604, DOI:10.1080/17524032.2018.1470546

Ford, J. M. (2015). *Climate Change as Metaphor & Catalyst: The Deeper Meaning & Potential of an Environmental Crisis*. PhD thesis, Vrije Universiteit Amsterdam.

Forgas, J. P. (2008). Affect and cognition. *Perspectives on Psychological Science*, 3(2): 94–101.

Forsyth, T. (2004). *Critical Political Ecology: The Politics of Environmental Science*. London: Routledge.

Foucault, M. (1975). *Discipline and Punish*. Translated by A. Sheridan. New York, NY: Pantheon.

Foucault, M. (1978). *The History of Sexuality: An Introduction*, Vol. 1. London: Penguin.

Foucault, M. (1980). *Power/Knowledge*. New York, NY: Pantheon.

Foucault, M. (1984). Space, knowledge and power. In Paul Rabinow (ed.), *The Foucault reader*, pp. 239–256. New York, NY: Pantheon Books.

Fownes, J. R., Yu, C. and Margolin, D. B. (2018). Twitter and climate change. *Sociology Compass*, 12(6): 1–12.

Fraser, N. (2007). Transnationalizing the public sphere: On the legitimacy and efficacy of public opinion in a post-Westphalian world. *Theory, Culture & Society*, 24(4): 7–30.

Frey, A. (2018). Three steps to better climate conversations. Sightline Institute, July 5. Available at: www.sightline.org/2018/07/05/three-steps-to-better-climate-conversations/

Friedman, L. (2017). Climate report full of warnings awaits president. *New York Times*, August, 7, A1.

Friedman, T. (2010). Global weirding is here. *New York Times*, February 17, A23.

Fritze, J. G., Blashki, G. A., Burke, S. and Wiseman, J. (2008). Hope, despair and transformation: Climate change and the promotion of mental health and wellbeing. *International Journal of Mental Health Systems*, 2(1): 13.

Funk, C., Gottfried, J. and Mitchell, A. (2017). Most Americans see science-related entertainment shows and movies in either a neutral or positive light. Pew Research Center, September 20. Available at: www.journalism.org/2017/09/20/most-ameri cans-see-science-related-entertainment-shows-and-movies-in-either-a-neutral-or-positive-light/

Gabehart, K. (2015). Reframing global climate change in the US: Coupling pro-environmentalism with entrenched economic and patriotic values. *Praxis: Politics in Action*, 2(1): 43–56.

Gaffney, O. and Steffen, W. (2017). The Anthropocene equation. *The Anthropocene Review*, 4(1): 53–61.

Gaffney, O. (2017). Simple equation shows how human activity is trashing the planet. *New Scientist*, February 10. Available at: www.newscientist.com/article/2120951-simple-equation-shows-how-human-activity-is-trashing-the-planet/

Gal, N. (2018). Ironic humor on social media as participatory boundary work. *New Media & Society*, 1–21.

Galindo, G. R. (2016). Could films help to save the world from climate change? A discourse exploration of two climate change documentary films and an analysis of their impact on the UK printed media. In P. Almlund, P. Homann and S. Riis (eds.),

Rethinking Climate Change Research: Clean Technology, Culture and Communication, pp. 243–260. Farnham, UK: Ashgate.

Garrett, R. K. (2017). Strategies for countering false information and beliefs about climate change. In M. Nisbet (ed.), *Oxford Research Encyclopedia of Climate Science*, Vol. 3, pp. 508–526. Oxford: Oxford University Press.

Gauchat, G. (2018). Trust in climate scientists. *Nature Climate Change*, 8(6): 458–459.

Gauchat, G., O'Brien, T. and Mirosa, O. (2017). The legitimacy of environmental scientists in the public sphere. *Climatic Change*, 143(3–4): 297–306.

Gavin, N. T. (2018). Media definitely do matter: Brexit, immigration, climate change and beyond. *The British Journal of Politics and International Relations*, 1–19.

Geertz, C. (1973). *The Interpretation of Cultures: Selected Essays*. New York, NY: Basic Books.

Geiger, N. and Swim, J. K. (2016). Climate of silence: Pluralistic ignorance as a barrier to climate change discussion. *Journal of Environmental Psychology*, 47: 79–90.

Geiger, N., Middlewood, B. and Swim, J. (2017). Psychological, social and cultural barriers to communicating about climate change. In M. Nisbet (ed.), *Oxford Research Encyclopedia of Climate Science*, Vol. 3, pp. 312–328. Oxford: Oxford University Press.

Ghosh, A. (2018). *The Great Derangement: Climate Change and the Unthinkable*. London: Penguin.

Gibson, R. and Zillmann, D. (1994). Exaggerated versus representative exemplification in news reports: Perceptions of issues and personal consequences. *Communication Research*, 21: 603–624.

Giddings, B., Hopwood, B. and O'Brien, G. (2002). Environment, economy and society: Fitting them together into sustainable development. *Sustainable Development*, 10(4): 187–196.

Gieryn, T. (1999). *Cultural Boundaries of Science: Credibility on the Line*. Chicago, IL: University of Chicago Press.

Gifford, R. (2014). Environmental psychology matters. *Annual Review of Psychology*, 65.

Gifford, R. and Comeau, L. A. (2011). Message framing influences perceived climate change competence, engagement, and behavioral intentions. *Global Environmental Change*, 21(4): 1301–1307.

Gillis, J. (2012). Clouds' effect on climate change is last bastion for dissenters. *New York Times*, May 1. Available at: www.nytimes.com/2012/05/01/science/earth/clouds-effect-on-climate-change-is-last-bastion-for-dissenters.html?_r=0

Gold, A. U., Oonk, D. J., Smith, L. K., et al. (2015). Lens on climate change: Making climate meaningful through student-produced videos. *Journal of Geography*. DOI:10.1080/00221341.2015.1013974

Goldman, G. T., Berman, E., Halpern, M., et al. (2017). Ensuring scientific integrity in the age of Trump. *Science*, 355(6326): 696–698.

Goldstein, B. E., Wessells, A. T., Lejano, R. and Butler, W. (2015). Narrating resilience: Transforming urban systems through collaborative storytelling. *Urban Studies*, 52 (7): 1285–1303.

Goldstein, J. (2017). Parks and recreation: The sudden, widespread resistance of alter-native National Parks Twitter. *Think Progress*, January 27. Available at: https://thinkprogress.org/parks-and-recreation-the-sudden-widespread-resistance-of-alternative-national-parks-twitter-f76b35aa67ea/

Goldstein, J. and Keohane, R. O., eds. (1993). *Ideas and Foreign Policy: Beliefs, Institutions, and Political Change*. Ithaca, NY: Cornell University Press.

Goode, E. (2018). Climate skeptics say polar bears are fine. Scientists beg to differ. *New York Times*, April 10, A10.

Goodman, M. K. (2010). The mirror of consumption: Celebritization, developmental consumption and the shifting cultural politics of fair trade.*Geoforum*, 41(1): 104–116.

Goolsbee, A. (2018). Environmentalism is a long-term investment. *New York Times*, November 4, SB 4.

Gordon, E. and Schirra, S. (2011). Playing with empathy: Digital role-playing games in public meetings. In *Proceedings of ACM Conference Communities and Technologies 2011*, Brisbane, Australia.

Gordon, E., Haas, J. and Michelson, B. (2017). Civic creativity: Role-playing games in deliberative process. *International Journal of Communication*, 11: 19.

Gosselin, D., Cooper, S., Lawton, S., et al. (2016). Lowering the walls and crossing boundaries: Applications of experiential learning to teaching collaboration. *Journal of Environmental Studies and Sciences*, 6(2): 324–335.

Gosselin, D., Vincent, S., Boone, C., et al. (2016). Negotiating boundaries: Effective leadership of interdisciplinary environmental and sustainability programs. *Journals of Environmental Studies and Sciences*, 6: 268–274.

Gottlieb, R. (1993). *Forcing the Spring: The Transformation of the American Environmental Movement*. Washington, DC: Island Press.

Gottlieb, R. (2002). *Environmentalism Unbound: Exploring New Pathways for Change*. Cambridge, MA: MIT Press.

Graham, M., Schroeder, R. and Taylor, G. (2013). Re: Search. *New Media and Society*, 12(3): 1–8.

Graves, L. (2018). Which works better: Climate fear, or climate hope? Well it's complicated. *The Guardian*, January 4. Available at: www.theguardian.com/commen tisfree/2018/jan/04/climate-fear-or-hope-change-debate

Gravey, V., Hargreaves, T., Lorenzoni, I. and Seyfang, G. (2017). Theoretical theatre: Harnessing the power of comedy to teach social science theory. 3S Working Paper, pp. 2017–2030. Norwich: Science, Society and Sustainability Research Group.

Green, D. (2017). If academics are serious about research impact, they need to learn from advocates. *Oxfam* blogs, July 4. Available at: https://oxfamblogs.org/fp2p/if-academics-are-serious-about-research-impact-they-need-to-learn-from-advocates/

Green, J. L. (2018). Why scream about sound in space? The functions of audience discourse about unrealistic science in narrative fiction. *Public Understanding of Science*. DOI:doi.org/10.1177/0963662518808729

Greenblatt, S. (1988). *Shakespearean Negotiations: The Circulation of Social Energy in Renaissance England*. Berkeley, CA: University of California Press.

Guber, D. L. (2017). Partisan cueing and polarization in public opinion about climate change. In M. Nisbet (ed.), *Oxford Research Encyclopedia of Climate Science*, Vol. 3, pp. 203–230. Oxford: Oxford University Press.

Guilford, J. P. (1950). Creativity. *American Psychologist*, 5(9): 444–454.

Gunster, S. (2017). Engaging climate communication: Audiences, frames, values and norms 1. In *Journalism and Climate Crisis*, (pp. 49–76). Routledge.

Gunther, A. C. and Thorson, E. (1992). Perceived persuasive effects of product commercials and public service announcements: Third-person effects in new
domains. *Journalism and Climate Crisis*, 19(5): 574–596.

Gupta, J. (2001). *Our Simmering Planet: What to Do about Global Warming?* New
York, NY: Zed Books.

Gustafson, A. and Goldberg, M. (2018). Even Americans highly concerned about
climate change dramatically underestimate the scientific consensus. October 18.
Available at: http://climatecommunication.yale.edu/publications/even-americans-
highly-concerned-about-climate-change-dramatically-underestimate-the-scientific-
consensus/

Hackmann, H., Moser, S. C. and St. Clair, A. L. (2014). The social heart of global
environmental change. *Nature Climate Change*, 4(8): 653.

Hakim, D. and Lipton, E. (2018). Once-trusted studies are scorned by Trump's EPA.
New York Times, August 26, B1, B4.

Hall, C. (2013). What will it mean to be green? Envisioning positive possibilities
without dismissing loss. *Ethics, Policy and Environment*, 16(2): 125–141.

Hall, S. (1988). *The Hard Road to Renewal: Thatcherism and the Crisis of the Left.*
London: Verso.

Hall, S. (1997). *Representation: Cultural Representation and Signifying Practices.*
Thousand Oaks, CA: SAGE.

Haluza-Delay, R. (2017). Communicating about climate change with religious groups
and leaders. In M. Nisbet (ed.), *Oxford Research Encyclopedia of Climate Science*,
Vol. 2, pp. 85–94. Oxford: Oxford University Press.

Hamilton, L. C. (2011). Education, politics and opinions about climate change evidence
for interaction effects. *Climatic Change*, 104(2): 231–242.

Hammond, P. (2017). *Climate Change and Post-Political Communication: Media,
Emotion and Environmental Advocacy.* . London: Routledge.

Han, H. and Stenhouse, N. (2014). Bridging the research-practice gap in climate
communication: Lessons from one academic-practitioner collaboration. *Science
Communication*, 37(3): 396–404.

Han, H., Sparks, A. C., and Towery, N. D. (2017). Opening up the black box: Citizen
group strategies for engaging grassroots activism in the twenty-first century. *Interest
Groups and Advocacy*, 6: 22–43. DOI:10.1057/s41309-017-0010-4

Hannigan, J. (2014). *Environmental Sociology: A Social Constructionist Perspective.*
London: Routledge.

Hansen, A. (2010). *Environment, Media and Communication*. London: Routledge.

Hansen, A. (2015). Promising directions for environmental communication research.
Environmental Communication, 9(3): 384–391.

Hansen, A. and Machin, D. (2008). Visually branding the environment: Climate change
as a marketing opportunity. *Discourse Studies*, 10: 777–794.

Hansen, Anders and Machin, D. (2008). Visually branding the environment: Climate
change as a marketing opportunity. *Discourse Studies*, 10: 777–794.

Hansen, J. (2009). *Storms of My Grandchildren: The Truth about the Coming
Climate Catastrophe and Our Last Chance to Save Humanity.* New York, NY:
Bloomsbury.

Hansen, J., Holm, L., Frewer, L., et al. (2003). Beyond the knowledge deficit: recent
research into lay and expert attitudes to food risks. *Appetite*, 41(2): 111–121.

Hansman, H. (2015). This song is composed from 133 years of climate change data. *Smithsonian Magazine*, September 21. Available at: www.smithsonianmag.com/science-nature/this-song-composed-from-133-years-climate-change-data-180956225/#1xDmlCkEKuR4heFX.99

Hansson, S. O. (2018). Dealing with climate science denialism: Experiences from confrontations with other forms of pseudoscience. *Climate Policy*, 18(9): 1094–1102.

Harish, A. (2012). New law in North Carolina bans latest scientific predictions of sea-level rise. *ABC News*, August 2. Available at: https://abcnews.go.com/US/north-carolina-bans-latest-science-rising-sea-level/story?id=16913782

Harmon, A. and Fountain, H. (2017). For scientists, a political test. *New York Times*, February 7, D1, D3.

Harold, J., Lorenzoni, I., Shipley, T. F. and Coventry, K. R. (2016). Cognitive and psychological science insights to improve climate change data visualization. *Nature Climate Change*, 6(12): 1080.

Harrington, S. (2017). Artist turns climate science into a coloring book. *Yale Climate Connections*, August 29. Available at: www.yaleclimateconnections.org/2017/08/the-climate-science-coloring-book/

Harrington, S. (2018). These kids are teaching their own parents about climate change. *Yale Climate Connections*, August 28. Available at: www.yaleclimateconnections.org/2018/08/kids-teach-parents-about-climate-change/

Hart, D. M. and Victor, D. G. (1993). Scientific elites and the making of US policy for climate change research, 1957–74. *Social Studies of Science*, 23(4): 643–680.

Hart, P. S. and Nisbet, E. C. (2012). Boomerang effects in science communication: How motivated reasoning and identity cues amplify opinion polarization about climate mitigation policies. *Communication Research*, 39(6): 701–723.

Harvey, D. (1973). *Social Justice and the City*. Athens, GA: University of Georgia.

Hassol, S. J., Torok, S., Lewis, S. and Luganda, P. (2016). (Un)natural disasters: communicating linkages between extreme events and climate change. *World Meteorological Organization Bulletin*, 65(2). Available at: https://public.wmo.int/en/resources/bulletin/unnatural-disasters-communicating-linkages-between-extreme-events-and-climate

Hawkins, H. and Kanngieser, A. (2017). Artful climate change communication: overcoming abstractions, insensibilities, and distances. *Wiley Interdisciplinary Reviews: Climate Change*, 8(5): 472.

Hayhoe, K. (2018). When facts are not enough. *Science*, 360(6392): 943.

Heilbroner, R. L. (2011). *The Worldly Philosophers: The Lives, Times and Ideas of the Great Economic Thinkers*. New York, NY: Simon and Schuster.

Helmuth, B., Gouhier, T. C., Scyphers, S. and Mocarski, J. (2016). Trust, tribalism and tweets: Has political polarization made science a "wedge issue"? *Climate Change Responses*, 3(1): 3.

Heras, M. and Tàbara, J. D. (2016). Conservation theatre: Mirroring experiences and performing stories in community management of natural resources. *Society and Natural Resources*, 29(8): 948–964.

Hersher, R. (2017). Climate scientists watch their words, hoping to stave off funding cuts. *National Public Radio*, November 29. Available at: www.npr.org/sections/thetwo-way/2017/11/29/564043596/climate-scientists-watch-their-words-hoping-to-stave-off-funding-cuts

Hertsgaard, M. (1998). *Earth Odyssey: Around the World in Search of Our Environmental Future*. New York, NY: Broadway Books.

Hestres, L. E. (2018). Take action now: Motivational framing and action requests in climate advocacy. *Environmental Communication*, 12(4): 462–479 DOI:10.1080/17524032.2018.1424010

Hetland, P. (2014). Models in science communication: Formatting public engagement and expertise. *Nordic Journal of Science and Technology Studies*, 2(2): 5–17.

Hickman, L. (2015). The IPCC in an age of social media. *Nature Climate Change*, 5(4): 284.

Hijri, Z. (2016). Do IPCC reports communicate effectively? *Inside Climate News*, August 5. Available at: https://insideclimatenews.org/news/05082016/ipcc-reports-communicate-effectively-climate-change-consensus-un

Hinckley, E. S. and Matson, P. A. (in press). Reversing the trend from domination of the Earth System to sustainability: How can science help? *Proceedings of the National Academy of Science of the USA*.

Hmielowski, J. D., Feldman, L., Myers, T. A., et al. (2014). An attack on science? Media use, trust in scientists, and perceptions of global warming. *Public Understanding of Science*, 23(7): 866–883.

Hobson, K. (2008). Reasons to be cheerful: Thinking sustainably in a (climate) changing world. *Geography Compass*, 2: 1–16.

Hobson, K. and Niemeyer, S. (2014). What sceptics believe: The effects of information and deliberation on climate change scepticism. *Public Understanding of Science*, 22(4): 396–412.

Hoewe, J. and Ahern, L. (2017). First-person effects of emotional and informational messages in strategic environmental communications campaigns. *Environmental Communication*, 11(6): 810–820.

Hoffman, A. J. (2015). *How Culture Shapes the Climate Change Debate*. Palo Alto, CA: Stanford University Press.

Hoffman, A. J. (2016). Reflections: Academia's emerging crisis of relevance and the consequent role of the engaged scholar. *Journal of Change Management*, 16(2): 77–96.

Hoffman, A. (2018). Rising insurance costs may convince Americans that climate change risks are real. *The Conversation*, October 23. Available at: https://theconversation.com/rising-insurance-costs-may-convince-americans-that-climate-change-risks-are-real-105192

Honeybun-Arnolda, E. and Obermeister, N. (2018). A climate for change: Millennials, science and the humanities. *Environmental Communication*, 13(1): 1–8.

Horan, T. J. (2013). "Soft" versus "hard" news on microblogging networks. *Information, Communication and Society*, 16(1): 43–60.

Horkheimer, M. and Adorno, T. (1947). *Dialectic of Enlightenment: Philosophical Fragments*. Translated by Edmund Jephcott, 2002. Palo Alto, CA: Stanford University Press.

Hornsey, M. J., Harris, E. A. and Fielding, K. S. (2018). Relationships among conspiratorial beliefs, conservatism and climate scepticism across nations. *Nature Climate Change*, 8(7): 614.

Houghton, J. T., Ding, Y., Griggs, D., et al. (2001). *Climate change 2001: the scientific basis*. Geneva, Switzerland: Intergovernmental Panel on Climate Change.

Houghton, J. T., Meiro Filho, L. G., Callander, B. A., et al. (1995). *Climate Change 1995: The Science of Climate Change: Contribution of Working Group I to the Second Assessment Report of the Intergovernmental Panel on Climate Change.* Geneva, Switzerland: Intergovernmental Panel on Climate Change.

Howarth, C. and Black, R. (2015). Local science and media engagement on climate change. *Nature Climate Change*, 5(6): 506–508.

Howarth, C. and Sharman, A. (2017). Influence of labeling and incivility on climate change communication. In M. Nisbet (ed.), *Oxford Research Encyclopedia of Climate Science*, Vol. 2, pp. 772–784. Oxford: Oxford University Press.

Howe, J. P. (2017). *Making Climate Change History: Documents from Global Warming's Past.* Seattle, WA: University of Washington Press.

Hsu, A., Weinfurter, A. J. and Xu, K. (2017). Aligning subnational climate actions for the new post-Paris climate regime. *Climatic Change*, 142(3–4): 419–432.

Hsu, A., Weinfurter, A. J. and Xu, K. (2017). Aligning subnational climate actions for the new post-Paris climate regime. *Climatic Change*, 142(3–4): 419–432.

Huertas, A. (2016). Despite fact-checking, zombie myths about climate change persist. *Poynter Institute*, December 22. Available at: www.poynter.org/news/despite-fact-checking-zombie-myths-about-climate-change-persist

Huiberts, E. and Joye, S. (2018). Close, but not close enough? Audience's reactions to domesticated distant suffering in international news coverage. *Media, Culture & Society*, 40(3): 333–347.

Hulme, M. (2009). *Why We Disagree about Climate Change: Understanding Controversy, Inaction and Opportunity.* Cambridge: Cambridge University Press.

Hunt, E. (2017). Trump's inauguration crowd: Sean Spicer's claim versus the evidence. *The Guardian*, January 22. Available at: www.theguardian.com/us-news/2017/jan/22/trump-inauguration-crowd-sean-spicers-claims-versus-the-evidence

Hurlstone, M. J., Lewandowsky, S., Newell, B. R. and Sewell, B. (2014). The effect of framing and normative messages in building support for climate policies. *PloS ONE*, 9(12): e114335.

Huxster, J. K., Slater, M. H., Leddington, J., et al. (2017). Understanding "understanding" in public understanding of science. *Public Understanding of Science.* DOI:0963662517735429

Inglis, F. (2005). *Culture and sentiment: Principle and practice in development.* Atlantic Coast Opportunities Agency, Government of Canada Conference, Halifax, Nova Scotia, p. 16. Available at: www.fredinglis.org.uk/papers/culture_and_sentiment.pdf

IPCC (Intergovernmental Panel on Climate Change). (2010). Statement on IPCC principles and procedures. Available at: www.ipcc.ch/pdf/press/ipcc-statement-prin ciples-procedures-02–2010.pdf

Jacobson, S. (2012). Transcoding the news: An investigation into multimedia journalism. *New Media and Society*, 14(5): 867–885.

Jacques, P. J. (2012). A general theory of climate denial. *Global Environmental Politics*, 12(2): 9–17.

Jacques, P. J., Dunlap, R. E. and Freeman, M. (2008). The organization of denial. *Environmental Politics*, 17: 349–385.

Jasanoff, S. (1996). Beyond epistemology: Relativism and engagement in the politics of science. *Social Studies of Science*, 26(2): 393–418.

Jasanoff, S. (2014). A mirror for science. *Public Understanding of Science*, 23(1): 21–26.

Jasanoff, S., ed. (2004). *States of Knowledge: The Co-production of Science and the Social Order*. London: Routledge.

Jasanoff, S. and Kim, S. H. (2009). Containing the atom: Sociotechnical imaginaries and nuclear power in the United States and South Korea. *Minerva*, 47(2): 119.

Jaspal, R., Nerlich, B. and Cinnirella, M. (2014). Human responses to climate change: Social representation, identity and socio-psychological action. *Environmental Communication*, 8(1): 110–130.

Johnson, B. (2014). *Carbon Nation: Fossil Fuels in the Making of American Culture*. Lawrence, KS: University Press of Kansas.

Johnson, B. B. and Dieckmann, N. F. (2017). Lay Americans' views of why scientists disagree with each other. *Public Understanding of Science*, 0963662517738408.

Johnson, E. J. and Tversky, A. (1983). Affect, generalization, and the perception of risk. *Journal of Personality and Social Psychology*, 45: 20–31.

Johns-Putra, A. (2016). Climate change in literature and literary studies: From cli-fi, climate change theater and ecopoetry to ecocriticism and climate change criticism. *Wiley Interdisciplinary Reviews: Climate Change*, 7(2): 266–282.

Jones, M. D. (2010). *Heroes and Villains: Cultural Narratives, Mass Opinions, and Climate Change*. Norman, OK: The University of Oklahoma.

Jones, M. D. and Peterson, H. (2017). Narrative persuasion and storytelling as climate communication strategies. In M. Nisbet (ed.), *Oxford Research Encyclopedia of Climate Science*, Vol. 3, pp. 127–140. Oxford: Oxford University Press.

Jutkowitz, A. (2017). *The Strategic Storyteller: Content Marketing in the Age of the Educated Consumer*. Hoboken, NJ: John Wiley & Sons.

Jylhä, K. M., Cantal, C., Akrami, N. and Milfont, T. L. (2016). Denial of anthropogenic climate change: Social dominance orientation helps explain the conservative male effect in Brazil and Sweden. *Personality and Individual Differences*, 98: 184–187.

Kagubare, I. (2018). Artists strive to make climate impacts "visceral," November 21. Available at: www.eenews.net/climatewire/stories/1060107179/

Kahan, D. (2014) Making climate-science communication evidence-based: All the way down. In D. Crow and M. Boykoff (eds.), *Culture, Politics and Climate Change: How Information Shapes Our Common Future*, pp. 203–220. New York, NY: Earthscan from Routledge.

Kahan, D. (2015a). What is the "science of science communication"? *Journal of Science Communication*, 14(3): 1–12.

Kahan, D. M. (2015b). Climate-science communication and the measurement problem. *Political Psychology*, 36: 1–43.

Kahan, D., Jenkins-Smith, H. and Braman, D. (2011). Cultural cognition of scientific consensus. *Journal of Risk Research*, 14: 147–174.

Kahan, D. M. (2013). Ideology, motivated reasoning, and cognitive reflection. *Judgment and Decisionmaking*, 8(4): 407–424.

Kahan, D. M. (2017). The "gateway belief" illusion: Reanalyzing the results of a scientific-consensus messaging study. *Journal of Science Communication*, 16(5): A03.

Kahan, D. M., Braman, D., Cohen, G. L., et al. (2010). Who fears the HPV vaccine, who doesn't, and why? An experimental study of the mechanisms of cultural cognition. *Law and Human Behavior*, 34(6): 501–516.

Kahan, D. M. and Carpenter, K. (2017). Out of the lab and into the field. *Nature Climate Change*, 7(5): 309.

Kahan, D. M., Peters, E., Wittlin, M., et al. (2012). The polarizing impact of science literacy and numeracy on perceived climate change risks. *Nature Climate Change*, 2(10): 732.

Kahneman, D. (2011). *Thinking Fast and Slow*. New York, NY: Farrar, Straus and Giroux.

Kalmus, P. (2017). *Being the Change: Live Well and Spark a Climate Revolution*. Gabriola Island, BC, Canada: New Society Publishers.

Kaners, K. (2015). Meet the climate scientist behind the pope: John Schellnhuber. *The Elephant* podcast, October 25. Available at: www.elephantpodcast.org/episodes/meet-the-climate-scientist-behind-the-pope-john-schellnhuber

Kaners, K. (2016). The human species has never faced a question like This. *The Elephant* podcast, May 23. Available at: www.elephantpodcast.org/episodes/noam-chomsky-the-human-species-has-never-faced-a-question-like-this

Kaplan, S. (2017a). This group wants to fight "anti-science" rhetoric by getting scientists to run for office. *Washington Post*, January 17. Available at: www.washingtonpost.com/news/speaking-of-science/wp/2017/01/17/this-group-wants-to-fight-anti-science-rhetoric-by-getting-scientists-to-run-for-office/?utm_term=.36c3179eeff2

Kaplan, S. (2017b). Activism is a hot topic at the world's biggest Earth and planetary science conference. *Washington Post*, December 15. Available at: www.washingtonpost.com/news/speaking-of-science/wp/2017/12/15/activism-is-a-hot-topic-at-the-worlds-biggest-earth-and-planetary-science-conference/?utm_term=.f804d72dc1e6

Kaplan, S. and McNeil, D. G., Jr. (2017). Uproar over purported ban at C.D.C. of words like "fetus." *New York Times*, December 17, A20.

Karp, D. G. (1996). Values and their effect on pro-environmental behavior. *Environment and Behavior*, 28(1): 111–133.

Kasperson, R. E., Renn, O., Slovic, P., et al. (1988). The social amplification of risk: A conceptual framework. *Risk Analysis*, 8(2): 177–187.

Katz-Kimchi, M. and Atkinson, L. (2014). Popular climate science and painless consumer choices: Communicating climate change in the hot pink flamingos exhibit, Monterey Bay Aquarium, California. *Science Communication*, 36(6): 754–777.

Keene, D. K. and Handrich, R. R. (2010). Between coddling and contempt: Managing and mentoring Millennials. *Jury Expert*, 22:1.

Kelly, S. (2018). Congressman asks gas industry to join him as "3 percenters," fails to mention movement's Ties to white supremacists. DeSmogBlog, October 18. Available at: www.desmogblog.com/2018/10/18/congressman-clay-higgens-gas-industry-three-percenters?amp

Keltner, D. and Bonanno, G. A. (1997). A study of laughter and dissociation: Distinct correlates of laughter and smiling during bereavement. *Journal of Personality and Social Psychology*, 73(4): 687–702.

Kemeny, M. E., Foltz, C., Cavanagh, J. F., et al. (2012). Contemplative/emotion training reduces negative emotional behavior and promotes prosocial responses. *Emotion*, 12(2): 338.

Kemmis, S. and McTaggart, R., eds. (2000). *Participatory Action Research. Handbook of Qualitative Research*. London: SAGE.

Kennedy, E. B., Jensen, E. A. and Verbeke, M. (2018). Preaching to the scientifically converted: Evaluating inclusivity in science festival audiences. *International Journal of Science EducationB*, 8(1): 14–21.

Keynes, J. M. (1924). *A Tract on Money Reform*. Published for the Royal Economic Society by Cambridge University Press.

Keynes, J. M. (2010). Economic Possibilities for Our Grandchildren. In *Essays in Persuasion*, pp. 321–332. London: Palgrave Macmillan.

Kinsley, M. (2008). To swift-boat or not. *Time Magazine*, June 12.

Kirk, K. (2017). Changing minds on a changing climate. *Yale Climate Connections*, April 4. Available at: www.yaleclimateconnections.org/2017/04/changing-minds-on-a-changing-climate/

Kirk, K. (2018). Middle ground: Fertile for climate change dialogue. *Yale Climate Connections*, January 8. Available at: www.yaleclimateconnections.org/2018/01/middle-ground-fertile-for-climate-change-dialogue/

Kneas, D. (2017). Chest hair and climate change: Harrison Ford and the making of "Lost There, Felt Here." *Environmental History*, 22(3): 516–526.

Knox, H. (2018). Not flying: Steps toward a post-carbon anthropology. Personal blog, March 24. Available at: https://hannahknox.wordpress.com/2018/03/24/not-flying-steps-towards-a-post-carbon-anthropology/

Ko, H. (2016). In science communication, why does the idea of a public deficit always return? How do the shifting information flows in healthcare affect the deficit model of science communication? *Public Understanding of Science*, 25(4): 427–432.

Kobayashi, K. (2018). The impact of perceived scientific and social consensus on scientific beliefs. *Science Communication*, 40(1): 63–88.

Koelle, B. and Annecke, W. (2014). Community based climate change adaptation (CBA). *Adaptation and Beyond*. Available at: www.indigo-dc.org

Koelle, B. and Oettle, N. (2009). *Adapting with Enthusiasm: Climate Change Adaptation in the Context of Participatory Action Research. Sustainable Land Use Conference*. Windhoek, Namibia: Ministry of Environment and Tourism, Namibia.

Kolb, D. A. (1984). *Experiential Learning: Experience as the Source of Learning and Development*. Upper Saddle River, NJ: Prentice Hall.

Kotcher, J. E., Myers, T. A., Vraga, E. K., et al. (2017). Does engagement in advocacy hurt the credibility of scientists? Results from a randomized national survey experiment. *Environmental Communication*, 11(3): 415–429.

Krange, O. Kaltenborn, B. and Hultman, M. (2018). Cool dudes in Norway: Climate change denial among conservative Norwegian Men. *Environmental Sociology*. DOI:10.1080/23251042.2018.1488516

Kristof, N. (2018). The "greatest hoax" strikes Florida. *New York Times*, 10 October 10, A25.

Krosnick, J. A., Holbrook, A. L., Lowe, L. and Visser, P. S. (2006). The origins and consequences of democratic citizens' policy agendas: A study of popular concern about global warming. *Climatic Change*, 77(1): 7–43.

Krugman, P. (2018). The G.O.P.'s climate of paranoia. *New York Times*, August 28, A20.

Kukkonen, A., Ylä-Anttila, T., Swarnakar, P., et al. (2018). International organizations, advocacy coalitions, and domestication of global norms: Debates on climate change in Canada, the US, Brazil, and India. *Environmental Science & Policy*, 81: 54–62.

Kunda, Z. (1990). The case for motivated reasoning. *Psychological Bulletin*, 108(3): 480.

Kurz, T., Augoustinos, M. and Crabb, S. (2010). Contesting the "national interest" and maintaining "our lifestyle": A discursive analysis of political rhetoric around climate change. *British Journal of Social Psychology*, 49(3): 601–625.

Lachapelle, E., Borick, C. P. and Rabe, B. (2012). Public attitudes toward climate science and climate policy in federal systems: Canada and the United States compared. *Review of Policy Research*, 29(3): 334–357.

Lachman, R. (2018). STEAM not STEM: Why scientists need arts training. *The Conversation*, January 17. Available at: https://theconversation.com/steam-not-stem-why-scientists-need-arts-training-89788

Laclau, E. and Mouffe, C. (2001). *Hegmony and Radical Democracy in Hegemony and Socialist Strategy*. London: Verso.

Lahsen, M. (2013). Anatomy of dissent: A cultural analysis of climate skepticism. *American Behavioral Scientist*. DOI:10.1177/0002764212469799

Lakoff, G. (2014). *The All New Don't Think of an Elephant. Know Your Values and Frame the Debate*. White River Junction, VT: White River Junction (Chelsea Green) Publishing.

Lam, A. (2017). Boundary-crossing careers and the "third space of hybridity": Career actors as knowledge brokers between creative arts and academia. *Environment and Planning A*, 1–26.

Landsburg, S. E. (2012). Why I am not an environmentalist: The science of economics versus the religion of ecology. In *The Armchair Economist: Economics & Everyday Life*, 223–231. New York, NY: Free Press.

Larson, B. (2011). *Metaphors for Environmental Sustainability: Redefining Our Relationship with Nature*. New Haven, CT: Yale University Press.

Latour, B. (2004). *Politics of Nature*. Cambridge, MA: Harvard University Press.

Leal Filho, W. (2019). An overview of the challenges in climate change communication across various audiences. In W. Leal Filho, B. Lackner, H. McGhie (eds.), *Addressing the Challenges in Communicating Climate Change across Various Audiences*, pp. 1–11. Climate Change Management. Cham, Switzerland: Springer.

Leavenworth, S. (2017). Real estate industry blocks sea-level warnings that could crimp profits on coastal properties. *News & Observer*, September 13. Available at: www.newsobserver.com/news/business/article173114701.html

Leber, R. (2018a). The *New York Times* fails to name and shame climate villains. *Mother Jones*. August 1. Available at: www.motherjones.com/politics/2018/08/the-times-fails-to-name-and-shame-climate-villains/

Leber, R. (2018b). Tom Steyer and the link between hate groups and climate denial. *Grist*, November 3. Available at: https://grist.org/article/tom-steyer-trump-and-the-link-between-hate-groups-and-climate-denial/

Lee, A. P. (2017). The climate fight isn't just about facts. *High Country News*, February 3. Available at: www.hcn.org/articles/the-trouble-with-facts-we-dont-like

Lehtonen, J. and Välimäki, J. (2012). Discussion: The difficult problem of anxiety in thinking about climate change. In S. Weintrobe (ed.), *Engaging with Climate Change: Psychoanalytic and Interdisciplinary Perspectives*, pp. 48–52. London: Routledge.

Leiserowitz, A. A. (2006). Climate change risk perception and policy preferences: The role of affect, imagery, and values. *Climatic Change*, 77(1): 45–72.

Leiserowitz, A. and Smith, N. (2017). Affective imagery, risk perceptions, and climate change communication. In M. Nisbet (ed.), *Oxford Research Encyclopedia of Climate Science*, Vol. 1, pp. 1–21. Oxford: Oxford University Press.

Leiserowitz, A., Maibach, E. and Roser-Renouf, C. (2015). *Climate Change in the American Mind*. New Haven, CT: Yale Program on Climate Change Communication.

Leiserowitz, A., Maibach, E., Roser-Renouf, C., et al. (2017). *Climate Change in the American Mind: May* 2017. New Haven, CT: Yale Program on Climate Change Communication.

Leiserowitz, A. A. (2006). Climate change risk perception and policy preferences: The role of affect, imagery, and values. *Climatic Change*, 77(1): 45–72.

Lemke, T. (2002). Foucault, governmentality, and critique. *Rethinking Marxism*, 14(3): 49–64.

Leombruni, L. V. (2015). How you talk about climate change matters: A communication network perspective on epistemic skepticism and belief strength. *Global Environmental Change*, 35: 148–161.

León, B. and Erviti, M. C. (2015). Science in pictures: Visual representation of climate change in Spain's television news. *Public Understanding of Science*, 24 (2): 183–199.

Leopold, T. (2015). John Stewart: Court jester with a knife. *CNN*, February 16. Available at: www.cnn.com/2015/02/11/entertainment/jon-stewart-jester-feat/index.html

Lerman, D., Morais, R. J. and Luna, D. (2018). *The Language of Branding: Theory, Strategies, and Tactics*. London: Routledge.

Lertzman, R. (2015). *Environmental Melancholia: Psychoanalytic Dimensions of Engagement*. London: Routledge.

Leshner, A., Scheufele, D. A., Bostrom, A., et al. (2017). *Communicating Science Effectively: A research agenda*. Washington, DC: National Academies of Sciences, Engineering, and Medicine, National Academies Press.

Lester, L. (2015). Three challenges for environmental communication research. *Environmental Communication*, 9(3): 392–397.

Levy, D. L. and Kolk, A. (2002). Strategic responses to global climate change: Conflicting pressures on multinationals in the oil industry. *Business and Politics*, 4(3): 275–300.

Lewandowsky, S. and Whitmarsh, L. (2018). Climate communication for biologists: When a picture can tell a thousand words. *PLoS Biology*, 16(10): e2006004. https://doi.org/10.1371/journal. pbio.2006004

Lewandowsky, S., Ecker, U. K. and Cook, J. (2017). Beyond misinformation: Understanding and coping with the "post-truth" era. *Journal of Applied Research in Memory and Cognition*, 6(4): 353–369.

Lewandowsky, S., Ecker, U. K. H., Seifert, C. M., Schwarz, N. and Cook, J. (2012). Misinformation and its correction: Continued influence and successful debiasing. *Psychological Science in the Public Interest*, 13(3): 106–131.

Lewandowsky, S., Oberauer, K. and Gignac, G. E. (2013). NASA faked the moon landing—therefore, (climate) science is a hoax: An anatomy of the motivated rejection of science. *Psychological Science*, 24(5): 622–633.

Lewis, G. B., Palm, R. and Feng, B. (2018). Cross-national variation in determinants of climate change concern. *Environmental Politics*, 1–29.

Ley, A. (2018). Mobilizing doubt: The legal mobilization of climate denialist groups. *Law and Policy*, 40(3): 221–242.

Li, Y., Johnson, E. J. and Zaval, L. (2011). Local warming: Daily temperature change influences belief in global warming. *Psychological Science*, 22(4): 454–459.

Lieberman, D. A. (2006). What Can We Learn from Playing Interactive Games? In P. Vorderer and J. Bryant (eds.), *Playing Video Games: Motives, Responses, and Consequences*, pp. 379–397. Mahwah, NJ: Lawrence Erlbaum Associates.

Lieberman, M. D. (2013). *Social: Why Our Brains Are Wired to Connect*. Oxford: Oxford University Press.

Lievens, M. and Kenis, A. (2018). Social constructivism and beyond: On the double bind between politics and science. *Ethics, Policy and Environment*, 21(1): 81–95.

Lifton, R. J. (2017a). *The Climate Swerve: Reflections on Mind, Hope, and Survival*. New York, NY: The New Press.

Lifton, R. J. (2017b). Our changing climate mindset. *New York Times*, October 7, SR10.

Light, A. and Bartlein, P. J. (2004). The end of the rainbow? Color schemes for improved data graphics. *Eos, Transactions American Geophysical Union*, 85(40): 385–391.

Likert, R. (1932). A Technique for the Measurement of Attitudes. *Archives of Psychology*, 140: 1–55.

Lim, D., Condon, P. and DeSteno, D. (2015). Mindfulness and compassion: An examination of mechanism and scalability. *PLoS ONE*, 10: e0118221. DOI:10.1371/journal.pone.0118221

Lin, S. J. (2013). Perceived impact of a documentary film: An investigation of the first-person effect and its implications for environmental issues. *Science Communication*, 35(6): 708–733.

Linville, P. W. and Fischer, G. W. (1991). Preferences for separating or combining events. *Journal of Personality and Social Psychology*, 60(1): 5.

Littler, J. (2009). *Radical Consumption: Shopping for Change in Contemporary Culture*. London: Open University Press.

Liu, J. C. E. (2015). Low carbon plot: Climate change scepticism with Chinese characteristics. *Environmental Sociology*, 1(4): 280–292. DOI:10.1080/23251042.2015.1049811

Lloyd, S. (2011). *The Carbon Diaries* 2017. London: Hachette UK.

Lockwood, A. (2016). Graphs of grief and other green feelings: The uses of affect in the study of environmental communication. *Environmental Communication*, 10(6): 734–748.

Lockwood, D. (2018). Public art with a goal: Spur climate conversations. *Yale Climate Connections*, October 4. Available at: www.yaleclimateconnections.org/2018/10/public-art-with-a-climate-change-goal/

Lokhorst, A. M., Werner, C., Staats, H., et al. (2013). Commitment and behavior change: A meta-analysis and critical review of commitment-making strategies in environmental research. *Environment and Behavior*, 45(1): 3–34.

Lord, C. G., Ross, L. and Lepper, M. R. (1979). Biased assimilation and attitude polarization: The effects of prior theories on subsequently considered evidence. *Journal of Personality and Social Psychology*, 37(11): 2098.

Lorenzoni, I., Nicholson-Cole, S. and Whitmarsh, L. (2007). Barriers perceived to engaging with climate change among the UK public and their policy implications. *Global Environmental Change*, 17(3–4): 445–459.

Lövbrand, E., Beck, S., Chilvers, J., et al. (2015). Who speaks for the future of Earth? How critical social science can extend the conversation on the Anthropocene. *Global Environmental Change*, 32: 211–218.

Lubchenco, J. (1998). Entering the century of the environment: A new social contract for science. *Science*, 279(5350): 491–497.

Lubchenco, J. (2017). Delivering on science's social contract. *Michigan Journal of Sustainability*, 5(1): 95–108.

Luers, A. (2013). Rethinking US climate advocacy. *Climatic Change*, 120: 13–19.

Lukacs, M. (2017). Neoliberalism has conned us into fighting climate change as individuals. *The Guardian*, July 17. Available at: www.theguardian.com/environment/true-north/2017/jul/17/neoliberalism-has-conned-us-into-fighting-climate-change-as-individuals

Luke, T. W. (1997). *Ecocritique: Contesting the Politics of Nature, Economy, and Culture*. Minneapolis, MN: Minnesota University Press.

Luke, T. W. (1999). Environmentality as green governmentality. In Éric Darier (ed.), *Discourses of the Environment*, pp. 121–151. Oxford: Blackwell.

Luke, T. W. (2008). The politics of true convenience or inconvenient truth: Struggles over how to sustain capitalism, democracy, and ecology in the 21st century. *Environment and Planning A*, 40(8): 1811–1824.

Luntz, F. (2003). *The Environment: A Cleaner, Safer, Healthier America*, pp. 131–146. Washington, DC: The Luntz Research Companies – Straight Talk.

Lyotard, J. F. (1984). *The Postmodern Condition: A Report on Knowledge*. Minneapolis, MN: University of Minnesota Press.

Macias, T. and Williams, K. (2016). Know your neighbors, save the planet: Social capital and the widening wedge of pro-environmental outcomes. *Environment and Behavior*, 48(3): 391–420.

Maes, A. (2017). The visual divide. *Nature Climate Change*, 7(4): 231.

Maeseele, P. (2015). Beyond the post-political zeitgeist. In A. Hansen and R. Cox (eds.), *Routledge Handbook of Environment and Communication*, pp. 429–443. London: Routledge.

Maeseele, P. and Peppermans, Y. (2017). Ideology in climate change communication. In M. Nisbet (ed.), *Oxford Research Encyclopedia of Climate Science*, Vol. 2, pp. 741–754. Oxford: Oxford University Press.

Maibach, E., Leiserowitz, A., Rosenthal, S., et al. (2016). *Is There a Climate "Spiral of Silence" in America? March,* 2016. New Haven, CT: Yale Program on Climate Change Communication.

Maibach, E., Roser-Renouf, C. and Leiserowitz, A. (2009). Global warming's six Americas 2009: An audience segmentation analysis. Yale University/George Mason University. Available at: http://climatecommunication.yale.edu/publications/global-warmings-six-americas-2009/

Maibach, E. W. and van der Linden, S. L. (2016). The importance of assessing and communicating scientific consensus. *Environmental Research Letters*, 11(9): 091003.

Makarovs, K. and Achterberg, P. (2018). Science to the people: A 32-nation survey. *Public Understanding of Science*, 1–21. DOI:0963662517754047

Makri, A. (2017). Give the public the tools to trust scientists. *Nature*, 541(7637): 261.

Malka, A., Krosnick, J. A. and Langer, G. (2009). The association of knowledge with concern about global warming: Trusted information sources shape public thinking. *Risk Analysis: An International Journal*, 29(5): 633–647.

Manfredo, M. J., Teel, T. L. and Zinn, H. (2009). Understanding global values toward wildlife. In M. J. Manfredo, J. J., Vaske, P. J., Brown, et al. (eds.), *Wildlife and Society: The Science of Human Dimensions*, pp. 31–43. Washington, DC: Island Press.

Mann, C. C. (2014). How to talk about climate change so people will listen. *The Atlantic*, September. Available at: www.theatlantic.com/magazine/archive/2014/09/how-to-talk-about-climate-change-so-people-will-listen/375067/

Mann, M., Hassol, S. J. and Toles, T. (2017). Doomsday scenarios are as harmful as climate change denial. *Washington Post*, July 12. Available at: www.washingtonpost.com/opinions/doomsday-scenarios-are-as-harmful-as-climate-change-denial/2017/07/12/880ed002-6714-11e7-a1d7-9a32c91c6f40_story.html?utm_term=.1e5c200af006

Mann, M. E., Bradley, R. S. and Hughes, M. K. (1998). Global-scale temperature patterns and climate forcing over the past six centuries. *Nature*, 392(6678): 779.

Markowitz, E. M. and Guckian, M. L. (2018). Climate change communication: Challenges, insights, and opportunities. *Psychology and Climate Change*, 35–63.

Markowitz, E. M., Hodge, C. and Harp, G. (2014). Connecting on climate: A guide to effective climate change communication. Center for Research on Environmental Decisions.

Marshall, C. (2017). Climate researchers, professors join immigration protests. *Energy & Environment News*, January 30. www.eenews.net/greenwire/2017/01/30/stories/1060049201

Marshall, C. and Reilly, S. (2018). Scientists report political meddling, self-censorship. *Energy & Environment Daily*, August 14. Available at: www.eenews.net/stories/1060094079

Marshall, G. (2014). We need more than scientists to engage the public on climate change. *Climate Outreach and Information Network*, July 3. Available at: https://climateoutreach.org/we-need-more-than-scientists-to-engage-the-public-on-climate-change/

Martin, R. A. (2002). Is laughter the best medicine? Humor, laughter, and physical health. *Current Directions in Psychological Science*, 11(6): 216–220.

Mastrandrea, M. and Schneider, S. H. (2004). Probabilistic integrated assessment of dangerous climate change. *Science*, 304: 571–575.

Mateu, A. (2016). Communicating Climate Change: What can the IPCC learn from environmental communication? *Mètode*, April 12. Available at: https://metode.org/news/communicating-climate-change.html

Mathiesen, K. (2016). Polar opposites in US election. *Nature Climate Change*, 6: 979–981.

Matson, P., Clark, W. C. and Andersson, K. (2016). *Pursuing Sustainability: A Guide to the Science and Practice*. Princeton, NJ: Princeton University Press.

Matthews, S. (2017). Alarmism is the argument we need to fight climate change. *Slate Magazine*, July 10. Available at: www.slate.com/articles/health_and_science/ science/2017/07/we_are_not_alarmed_enough_about_climate_change.html

Mauss-Hanke, A. (2012). Discussion: The difficult problem of anxiety in thinking about climate change. In S. Weintrobe (ed.), *Engaging with Climate Change: Psychoanalytic and Interdisciplinary Perspectives*, pp. 52–55. London: Routledge.

Mayer, J. (2010). Covert operations: The billionaire brothers who are waging a war against Obama. *The New Yorker*, August 30.

Mayer, J. (2016). *Dark Money*. New York, NY: Doubleday.

Maynard, A. and Scheufele, D. A. (2016). What does research say about how to effectively communicate about science? *The Conversation*, December 13. Available at: https://theconversation.com/what-does-research-say-about-how-to-effectively-communicate-about-science-70244

Maza, C. (2016). One year later, how a pope's message on climate change has resonated. *Christian Science Monitor*, June 24. Available at: www.csmonitor.com/Environment/ 2016/0624/One-year-later-how-a-Pope-s-message-on-climate-has-resonated

McCarthy, J. (2002). First world political ecology: Lessons from the Wise Use movement. *Environment and Planning*, 34: 1281–1302.

McCarthy, J. J., Canziani, O. F., Leary, N. A., Dokken, D. J. and White, K. S., eds. (2001). *Climate Change 2001: Impacts, Adaptation, and Vulnerability: Contribution of Working Group II to the Third Assessment Report of the Intergovernmental Panel on Climate Change*. Cambridge: Cambridge University Press.

McCormack, D. (2003). An affect of geographical ethics in spaces of affect. *Transactions of the Institute of British Geographers*, 4: 488–507

McCright, A. and Dunlap, R. (2011). Cool dudes: The denial of climate change among conservative white males in the United States. *Global Environmental Change*, 21(4): 1163–1172.

McCright, A. M. (2007). Climate contrarians. In S. C. Moser and L. Dilling (eds.), *Creating a Climate for Change: Communicating Climate Change and Facilitating Social Change*, pp. 200–212. Cambridge: Cambridge University Press.

McCright, A. M. and Dunlap, R. E. (2000). Challenging global warming as a social problem: An analysis of the conservative movement's counter-claims. *Social Problems*, 47(4): 499–522.

McCroskey, J. C. and Teven, J. J. (1999). Goodwill: A re-examination of the construct and its measurement. *Communication Monographs*, 66(1): 90–103.

McDonald, R. I. (2017). PerGuardianceived temporal and geographic distance and public opinion about climate change. In M. Nisbet (ed.), *Oxford Research Encyclopedia of Climate Science*, Vol. 3, pp. 230–242. Oxford: Oxford University Press.

McDougall, M. (2018). More than 90% of world's children breathe toxic air, report says, as India prepares for most polluted season. *CNN*, October 30. Available at: www.cnn.com/2018/10/29/health/air-pollution-children-health-who-india-intl/ index.html

McGonigal, J. (2011). *Reality Is Broken: Why Games Make Us Better and How They Can Change the World*. New York, NY: Penguin Group.

McGraw, A. P. and Warren, C. (2010). Benign violations: Making immoral behavior funny. *Psychological Science*, 21(8): 1141–1149.

McGraw, A. P., Schiro, J. L. and Fernbach, P. M. (2015). Not a problem: A downside of humorous appeals. *Journal of Marketing Behavior*, 1: 187–208.

McGraw, A. P., Warren, C., Williams, L. E. and Leonard, B. (2012). Too close for comfort, or too far to care? Finding humor in distant tragedies and close mishaps. *Psychological Science*, 25(10): 1215–1223.

McGraw, A. P., Williams, L. E. and Warren, C. (2014). The rise and fall of humor: Psychological distance modulates humorous responses to tragedy. *Social Psychology and Personality Science*, 5: 566–572.

McGraw, P. and Warner, J. (2014). *The Humor Code: A Global Search for What Makes Things Funny*. New York, NY: Simon and Schuster.

McKeown, E. (2012). Talking points ammo: The use of neoliberal think tank fantasy themes to deligitimise scientific knowledge of climate change in Australian newspapers. *Journalism Studies*, 13: 277–297.

McKibben, B. (2018). Life on a shrinking planet. *The New Yorker*, November 26. Available at: www.newyorker.com/magazine/2018/11/26/how-extreme-weather-is-shrinking-the-planet?utm_medium=email&utm_source=actionkit

McKie, R. (2016). Nicholas Stern: Cost of global warming "is worse than I feared." *The Guardian*, November 5. Available at: www.theguardian.com/environment/2016/nov/06/nicholas-stern-climate-change-review-10-years-on-interview-decisive-years-humanity

McKinnon, C. (2014). Climate change: Against despair. *Ethics and the Environment*, 19(1): 31–48.

McLeod, K. (2017). Keep advocating for science, but don't forget to engage. *Compass* blogs, February 2. Available at: www.compassscicomm.org/single-post/2017/02/02/Keep-Advocating-For-Science-But-Dont-Forget-To-Engage

McNally, B. (2018). Mapping press narratives of decarbonisation: Insights on communication of climate responses. *The International Journal of Climate Change: Impacts and Responses*, 10(1): 39–57. DOI:10.18848/1835-7156/CGP/v10i01/39-57

McPhee, J. (1998). *Annals of the Former World*. London: Farrar, Straus and Giroux.

Meadows, D., Sweeney, L. B. and Mehers, G. M. (2016). *The Climate Change Playbook: 22 Systems Thinking Games for More Effective Communication about Climate Change*. Hartford, VT: Chelsea Green Publishing.

Meet the Press. (2018). *Meet The Press* interview with Rudy Giuliani. *NBC News*, August 19. Available at: www.nbcnews.com/meet-the-press/meet-press-august-19-2018-n901986

Mendler de Suarez, J., Suarez, P. and Bachofen, C. (2012). *Games for a New Climate: Experiencing the Complexity of Future Risks*. Boston, MA: Pardee Center Task Force Report.

Merton, R. K. (1942). A note on science and democracy. *Journal of Legal and Political Sociology*, 1: 115.

Merton, R. K. (1973). *The Sociology of Science: Theoretical and Empirical Investigations*. Chicago, IL: University of Chicago Press.

Mervis, J. (2017). Lamar Smith, unbound, lays out political strategy at climate doubters conference. *Science*, March 24. DOI:10.1126/science.aal0957

Metag, J. and Schäfer, M. (2018). Audience segments in environmental and science communication: Recent findings and future perspectives. *Environmental Communication*, 12(8): 995–1004.

Metag, J., Füchslin, T. and Schäfer, M. S. (2017). Global warming's five Germanys: A typology of Germans' views on climate change and patterns of media use and information. *Public Understanding of Science*, 26(4): 434–451.

Meyer, G. (2016). In science communication, why does the idea of a public deficit always return? *Public Understanding of Science*, 25(4): 433–446.

Meyer, J. M. (2015). *Engaging the Everyday: Environmental Social Criticism and the Resonance Dilemma*. Cambridge, MA: MIT Press.

Meyer, S. R., Levesque, V. R., Bieluch, K.H ., et al. (2016). Sustainability science graduate students as boundary spanners. *Journal of Environmental Studies and Sciences*, 6: 344–353.

Milfont, T. L. and Sibley, C. G. (2014). The hierarchy enforcement hypothesis of environmental exploitation: A social dominance perspective. *Journal of Experimental Social Psychology*, 55: 188–193. DOI:10.1016/j.jesp.2014.07.006

Milfont, T. L., Richter, I., Sibley, C. G., et al. (2013). Environmental consequences of the desire to dominate and be superior. *Personality and Social Psychology Bulletin*, 39: 1127–1138. DOI:10.1177/0146167213490805

Miller, R. W. (2018). March for Science 2018: Passionate advocates push the cause for research across the globe. *USA Today*, April 14. Available at: www.usatoday.com/story/news/nation/2018/04/14/march-science-2018/517294002/

Miller-McDonald, S. (2018). What must we do to live? *The Trouble*, October 14. Available at: www.the-trouble.com/content/2018/10/14/what-must-we-do-to-live

Milman, O. (2017). Climate scientists face harassment, threats and fears of "McCarthyist attacks." *The Guardian,* 22February 22. Available at: www.theguardian.com/environment/2017/feb/22/climate-change-science-attacks-threats-trump

Milman, O. (2018). Local climate efforts won't be enough to undo Trump's inaction, study says. *The Guardian*, August 30. Available at: www.theguardian.com/environment/2018/aug/29/local-climate-efforts-wont-undo-trump-inaction

Minsky, M. (2007). *The Emotion Machine: Commonsense Thinking, Artificial Intelligence, and the Future of the Human Mind*. New York, NY: Simon and Schuster.

Mintrom, M. and Luetjens, J. (2017). Creating public value: Tightening connections between policy design and public management. *Policy Studies Journal*, 45(1): 170–190.

Mintrom, M., Salisbury, C. and Luetjens, J. (2014). Policy entrepreneurs and promotion of Australian state knowledge economies. *Australian Journal of Political Science*, 49 (3): 423–438.

Miskimmon, A., O'Loughlin, B. and Roselle, L. (2014). *Strategic Narratives: Communication Power and the New World Order*. London: Routledge.

Mitgutsch, K. and Alvarado N. (2012). Purposeful by design? A serious game design assessment framework. In *Proceedings of the International Conference on the Foundations of Digital Games*, pp. 121–128. New York, NY: ACM.

Mock, B. (2014). Please scientists, tell us how you really feel about climate change. *Grist*, March 19. Available at: https://grist.org/climate-energy/please-scientists-tell-us-how-you-really-feel-about-climate-change/

Molina, M., McCarthy, J., Alley, R., et al. (2014). What we know: The reality, risks, and response to climate change. *American Academy for the Advancement of Science*. Available at: http://whatweknow.aaas.org/wp-content/uploads/2014/07/whatwe know_website.pdf

Monroe, M. C., Plate, R. R., Adams, D. C., et al. (2015). Harnessing homophily to improve climate change education. *Environmental Education Research*, 21(2): 221–238.

Mooney, C. (2016). 30 years ago scientists warned Congress on global warming. What they said sounds eerily familiar. *Washington Post*, June 11. Available at: www .washingtonpost.com/news/energy-environment/wp/2016/06/11/30-years-ago-scien tists-warned-congress-on-global-warming-what-they-said-sounds-eerily-familiar/? noredirect=on&utm_term=.7f61076a100b

Mooney, C. (2017). Scientists challenge magazine story about "uninhabitable Earth." *Washington Post*, July 12. Available at: www.washingtonpost.com/news/energy-environment/wp/2017/07/12/scientists-challenge-magazine-story-about-uninhabita ble-earth/

Mooney, C. and Dennis, B. (2018). Climate scientists are struggling to find the right words for very bad news. *Washington Post*, October 3. Available at: www.washington post.com/energy-environment/2018/10/03/climate-scientists-are-struggling-find-right-words-very-bad-news/

Mooney, C. and Kirshenbaum, S. (2010). *Unscientific America: How Scientific Illiteracy Threatens Our Future*. New York, NY: Basic Books.

Mooney, C. C. (2016). The vicious cycle that makes people afraid to talk about climate change. *Washington Post*, May 12. Available at: www.washingtonpost.com/news/ energy-environment/wp/2016/05/12/the-vicious-cycle-that-makes-people-afraid-to-talk-about-climate-change/?utm_term=.f384e0a95cee

Morano, M. (2018). UN issues yet another climate tipping point – Humans given only 12 more years to make "unprecedented changes in all aspects of society" *Climate Depot*, October 8. Available at: www.climatedepot.com/2018/10/08/un-issues-yet-another-climate-tipping-point-earth-given-only-12-more-years-to-make-unprece dented-changes-in-all-aspects-of-society/

Morgan, M. G., Fischhoff, B., Bostrom, A., et al. (2002). *Risk Communication: A Mental Models Approach*. Cambridge University Press.

Morreall, J. (1983). *Taking Laughter Seriously*. Albany, NY: SUNY Press.

Moser, S. (2009). Costly politics – unaffordable denial: The politics of public under-standing and engagement in climate change. In Maxwell Boykoff (ed.), *The Politics of Climate Change: A Survey*, pp. 155–182. London: Routledge/Europa.

Moser, S. C. (2007). *More Bad News: The Risk of Neglecting Emotional Responses to Climate Change Information*. New York, NY: Cambridge University Press.

Moser, S. C. (2010). Communicating climate change: History, challenges, process and future directions. *Wiley Interdisciplinary Reviews: Climate Change*, 1(1): 31–53.

Moser, S. C. (2017). Reflections on climate change communication research and practice in the second decade of the 21st century: What more is there to say? *Wiley Interdisciplinary Reviews: Climate Change*, 7(3): 345–369.

Moser, S. C. and Berzonsky, C. L. (2015). There must be more: Communication to close the cultural divide. In K. O'Brienand and E. Silbo (eds.), *The Adaptive Challenge of Climate Change*, pp. 287–310. Cambridge: Cambridge University Press.

Moser, S. C. and Dilling, L. (2011). Communicating climate change: Closing the science–action gap. In J. S. Dryzek, R. B. Norgaard and D. Schlosberg (eds.), *The Oxford Handbook of Climate Change and Society*, pp. 161–174. Oxford: Oxford University Press.

Motta, M. (2018). The enduring effect of scientific interest on trust in climate scientists in the United States. *Nature Climate Change*, 8(6): 485–490.

Mouffe, C. (2005). *The Return of the Political*, Vol. 8. London: Verso.

Moynihan, D. P. (2010). *Daniel Ptrick Moynihan: A Portrait in Letters of an American Visionary*. New York, NY: PublicAffairs.

Mukherjee, I. and Howlett, M. (2015). Who is a stream? Epistemic communities, instrument constituencies and advocacy coalitions in public policy-making. *Politics and Governance*, 3(2): 65–75.

Mukherjee, I. and Howlett, M. (2017). Communicating about climate change with policymakers. In M. Nisbet (ed.), *Oxford Research Encyclopedia of Climate Science*, Vol. 2, pp. 69–84. Oxford: Oxford University Press.

Myers, T., Kotcher, J., Cook, J., et al. (2018). *March for Science* 2017: *A Survey of Participants and Followers*. George Mason University, Fairfax, VA: Center for Climate Change Communication.

Myers, T. A., Nisbet, M. C., Maibach, E. W., and Leiserowitz, A. A. (2012). A public health frame arouses hopeful emotions about climate change. *Climatic Change*, 113 (3–4): 1105–1112.

Myles, A., Mustafa, B., Yang, C., et al. (2018). Global warming of 1.5 °C. Intergovernmental Panel on Climate Change. Available at: http://report.ipcc.ch/sr15/pdf/sr15_spm_final.pdf

Nabi, R. L., Gustafson, A., and Jensen, R. (2018). Framing climate change: Exploring the role of emotion in generating advocacy behavior. *Science Communication*, 40(4): 442–468.

Nair, S. K. (2016). Addressing environmental degradation and rural poverty through climate change adaptation: An evaluation of social learning in drought-affected districts of Southern India. CCAFS Working Paper no. 174. Copenhagen, Denmark: CGIAR Research Program on Climate Change, Agriculture and Food Security (CCAFS).

National Academies of Sciences, Engineering, and Medicine. (2017). *Communicating Science Effectively: A Research Agenda*. Washington, DC: National Academies Press.

Nature Climate Change. (2017). Connecting with climate science. *Nature Climate Change*, 7(159). DOI:10.1038/nclimate3246

Nature Climate Change Editors. (2017). Politics of climate change belief. *Nature Climate Change*, 7(1): 1.

Nature Geoscience Editors. (2018). Geoscientists online. *Nature Geoscience*, 11: 701.

Neate, R. (2016). ExxonMobil CEO: Ending oil production "not acceptable for humanity." *The Guardian*, May 25. Available at: www.theguardian.com/business/2016/may/25/exxonmobil-ceo-oil-climate-change-oil-production

Neffinger, J. and Kohut, M. (2013). *Compelling People: The Hidden Qualities That Make Us Influential*. London: Hachette UK.

Negri, A. (1999). Value and affect. *Boundary*, 2: 26–86.

Nelson, M. P. and Vucetich, J. A. (2009). On advocacy by environmental scientists: What, whether, why, and how. *Conservation Biology*, 23(5): 1090–1101.

Nerlich, B., Koteyko, N. and Brown, B. (2010). Theory and language of climate change communication. *Wiley Interdisciplinary Reviews: Climate Change*, 1(1): 97–110.

Neuhauser, A. (2018). On climate, sharp generational divide within GOP. *US News and World Report*, May 14. Available at: www.usnews.com/news/national-news/articles/2018–05-14/on-climate-sharp-generational-divide-within-gop

Newman, M. G. and Stone, A. A. (1996). Does humor moderate the effects of experimentally-induced stress? *Annals of Behavioral Medicine*, 18(2): 101–109.

Newman, T. P., Nisbet, E. C. and Nisbet, M. C. (2018). Climate change, cultural cognition, and media effects: Worldviews drive news selectivity, biased processing, and polarized attitudes. *Public Understanding of Science*, 27(8): 985–1002.

New York Times Editors. (2016). On climate change, look to the States. *New York Times*, December 26, A20.

Nichols, M. (2017). U.N. chief hopes storms will sway climate skeptics like Trump. *Reuters*, October 4. Available at: www.reuters.com/article/us-un-usa-climatechange/u-n-chief-hopes-storms-will-sway-climate-skeptics-like-trump-idUSKBN1C92HO

Nietzsche, F. (2019). *The Joyous Science*. New York, NY: Penguin Classics.

Nikunen, K. (2016). Media, passion and humanitarian reality television. *European Journal of Cultural Studies*, 19(3): 265–282.

Nisbet, E. C., Cooper, K. E. and Garrett, R. K. (2015). The partisan brain: How dissonant science messages lead conservatives and liberals to (dis) trust science. *The Annals of the American Academy of Political and Social Science*, 658(1): 36–66.

Nisbet, M. C. (2009). Communicating climate change: Why frames matter for public engagement. *Environment: Science and Policy for Sustainable Development*, 51(2): 12–23.

Nisbet, M. C. and Scheufele, D. A. (2009). What's next for science communication? Promising directions and lingering distractions. *American Journal of Botany*, 96(10): 1767–1778.

Nisbet, M. C., Hixon, M. A., Moore, K. D. and Nelson, M. (2010). Four cultures: New synergies for engaging society on climate change. *Frontiers in Ecology and the Environment*, 8(6): 329–331.

Nisbet, M. (2017). The March for Science: Partisan protests put public trust in scientists at risk. *Skeptical Inquirer*, 41(4): 18–19.

Nisbet, M. (2018a). The IPCC Report is a wake up call for communication scholars. *The Medium*, October 10. Available at: https://medium.com/wealth-of-ideas/the-ipcc-report-is-a-wake-up-call-for-scholars-advocates-and-philanthropists-36415d4882f

Nisbet, M. C. (2018b). *Scientists in Civic Life: Facilitated Dialogue-Based Communication*. Washington, DC: American Association for the Advancement of Science, Dialogue on Science, Ethics & Religion.

Nissan, H. and Conway, D. (2018). From advocacy to action: Projecting the health impacts of climate change. *PLoS Medicine*, 15(7): e1002624.

Norgaard, K. M. (2011). *Living in Denial: Climate Change, Emotions, and Everyday Life*. Cambridge, MA: MIT Press.

Norgaard, K. M. (2006). "People want to protect themselves a little bit": Emotions, denial, and social movement nonparticipation. *Sociological Inquiry*, 76(3): 372–396.

Nuccitelli, D. (2016). The climate change generation gap. *Bulletin of the Atomic Scientists*, April 21. Available at: https://thebulletin.org/2016/04/the-climate-change-generation-gap/

Nuccitelli, D., Cook, J., van der Linden, S., Leiserowitz, A. and Maibach, E. (2017). Why the 97% climate consensus is important. *The Guardian*, October 2. Available at: www.theguardian.com/environment/climate-consensus-97-per-cent/2017/oct/02/why-the-97-climate-consensus-is-important

Nyhan, B. and Reifler, J. (2018). The roles of information deficits and identity threat in the prevalence of misperceptions. *Journal of Elections, Public Opinion and Parties,* 1–23.

O'Connor, M. (1999). Dialogue and debate in a post-normal practice of science: a reflection. *Futures*, 31: 671–687.

O'Harrow, R., Jr. (2017). How charities' long fight fuelled climate-pact exit. *Washington Post*, September 6, A1.

Ojala, M. (2012). Regulating worry, promoting hope: How do children, adolescents, and young adults cope with climate change? *International Journal of Environmental and Science Education*, 7(4): 537–561.

Ojala, M. and Lakew, Y. (2017). Young people and climate change communication. In M. Nisbet (ed.), *Oxford Research Encyclopedia of Climate Science*, Vol. 3, pp. 609–628. Oxford: Oxford University Press.

Oliver, J. E. and Wood, T. J. (2014). Conspiracy theories and the paranoid style (s) of mass opinion. *American Journal of Political Science*, 58(4): 952–966.

Olofsson, K. L., Weible, C. M., Heikkila, T. and Martel, J. C. (2017). Using nonprofit narratives and news media framing to depict air pollution in Delhi, India. *Environmental Communication*, 12(7), 956–972.

Olson-Hazboun, S. K., Howe, P. D. and Leiserowitz, A. (2018). The influence of extractive activities on public support for renewable energy policy. *Energy Policy*, 123: 117–126.

O'Neill, S. (2017). Engaging with climate change imagery. In M. Nisbet (ed.), *Oxford Research Encyclopedia of Climate Science*, Vol. 2, pp. 541–557. Oxford: Oxford University Press.

O'Neill, S. and Nicholson-Cole, S. (2009). "Fear won't do it": Promoting positive engagement with climate change through visual and iconic representations. *Science Communication*, 30: 355–379.

O'Neill, S. J. and Boykoff, M. (2010a). The role of new media in engaging the public with climate change. In L. Whitmarsh, S. J. O'Neill and I. Lorenzoni (eds.), *Engaging the Public with Climate Change: Communication and Behaviour Change.* London: Earthscan.

O'Neill, S. J. and Boykoff, M. (2010b). Climate denier, skeptic, or contrarian? *Proceedings of the National Academy of Sciences of the USA*, DOI:doi/10.1073/pnas.1010507107

O'Neill, S. J. and Smith, N. (2014). Climate change and visual imagery. *Wiley Interdisciplinary Reviews: Climate Change*, 5(1): 73–87.

O'Neill, S. J., Boykoff, M., Niemeyer, S. and Day, S. A. (2013). On the use of imagery for climate change engagement. *Global Environmental Change*, 23(2): 413–421.

Oransky, I. and Marcus, A. (2017). Should scientists engage in activism? *Conversation*, February 6. Available at: http://theconversation.com/should-scientists-engage-in-activism-72234

Oreskes, N. (2004a). Science and public policy: What's proof got to do with it? *Environmental Science and Policy*, 7: 369–385.

Oreskes, N. (2004b). Beyond the ivory tower: The scientific consensus on climate change. *Science*, 306(5702): 1686.

Oreskes, N. (2017). Response by Oreskes to "Beyond counting climate consensus." *Environmental Communication*, 11(6): 731–732.

Oreskes, N. (2018). The scientific consensus on climate change: How do we know we're not wrong? *Climate Modelling*, 31–64.

Oreskes, N. and Conway, E. M. (2011). *Merchants of Doubt: How a Handful of Scientists Obscured the Truth on Issues from Tobacco Smoke to Global Warming.* New York, NY: Bloomsbury Press.

Oreskes, Naomi and Conway, Erik. (2018). Science isn't enough to save us. *New York Times*, October 17, A25.

Ormrod, J. E. (2017). *How We Think and Learn: Theoretical Perspectives and Practical Implications.* Cambridge: Cambridge University Press.

Orr, D. W. (2002). Four challenges of sustainability. *Conservation Biology*, 16(6): 1457–1460.

Osnes, B. (2008). Remembering the price and worth of freedom: The story of two Burmese comedians. *South Asian Popular Culture*, 6(1): 71–75.

Osnes, B. (2010). Empowering women's voices for energy justice. *Colorado Journal of International Environmental Law and Policy*, 21: 341–354.

Osnes, B. (2012). Voice strengthening and interactive theatre for women's productive income-generating activities in sustainable development. *Journal of Sustainable Development*, 5(6): 49–56.

Osnes, B. (2014). *Theatre for Women's Participation in Sustainable Development.* Routledge Studies in Sustainable Development. London: Routledge.

Osnes, B. (2017). *Performance for Resilience: Engaging Youth on Energy and Climate through Music, Movement, and Theatre.* London: Palgrave Pivot.

Osnes, B. and Gammon, M. (2013). Striking the Match: A web-based performance to illuminate issues of sustainability and ignite positive social change. *Sustainability: The Journal of Record*, 6(3): 167–170.

Osnes, B., Safran, R. and Boykoff, M. (2017). Student content production of climate communications. In *What Is Sustainable Journalism?* In P. Berglez, U. Olausson and M. Ots (eds.), pp. 93–111. New York, NY: Peter Lang.

Ostrom, E. (2010). Polycentric systems for coping with collective action and global environmental change. *Global Environmental Change*, 20(4): 550–557.

Oxford English Dictionary. (2016) Available at www.oed.com

Page, R., ed. (2009). *New Perspectives on Narrative and Multimodality.* New York, NY: Routledge.

Page, R. E. (2010). *New Perspectives on Narrative and Multimodality.* London: Routledge.

Pahl, S., Sheppard, S., Boomsma, C. and Groves, C. (2014). Perceptions of time in relation to climate change. *Wiley Interdisciplinary Reviews: Climate Change*, 5(3): 375–388.

Painter, J. (2013). *Climate Change in the Media: Reporting Risk and Uncertainty.* London: I. B. Tauris.

Panno, A., Giacomantonio, M., Carrus, G., et al. (2018). Mindfulness, pro-environmental behavior, and belief in climate change: The mediating role of social dominance. *Environment and Behavior*, 50(8): 864–888.

Parant, A., Pascual, A., Jugel, M., et al. (2017). Raising students awareness to climate change: An illustration with binding communication. *Environment and Behavior*, 49 (3): 339–353.

Pasek, J. (2017). It's not my consensus: Motivated reasoning and the sources of scientific illiteracy. *Public Understanding of Science*, 0963662517733681.

Patel, V. (2018). What Neil deGrasse Tyson thinks higher ed gets wrong. *Chronicle of Higher Education*, September 16. Available at: www.chronicle.com/article/What-Neil-deGrasse-Tyson/244522

Paulhus, D. L. (1984). Two-component models of socially desirable responding. *Journal of Personality and Social Psychology*, 46(3): 598.

Peach, S. (2016). New "global weirding" series informs, entertains. *Yale Climate Connections*, November 11. Available at: www.yaleclimateconnections.org/2016/11/new-global-weirding-series-informs-entertains/

Pearce, F. (1989). *Turning Up the Heat: Our Perilous Future in the Global Greenhouse*. London: Bodley Head.

Pearce, W., Grundmann, R., Hulme, M., et al. (2017a). Beyond counting climate consensus. *Environmental Communication*, 11(6): 723–730.

Pearce, W., Grundmann, R., Hulme, M., et al. (2017b). A reply to Cook and Oreskes on climate science consensus messaging. *Environmental Communication*, 11(6): 736–739.

Pedelty, M. (2011). *Ecomusicology: Rock, Folk, and the Environment*. Philadelphia, PA: Temple University Press.

Pennycook, G. and Rand, D. G. (2018). Lazy, not biased: Susceptibility to partisan fake news is better explained by lack of reasoning than by motivated reasoning. *Cognition*. 10.1016/j.cognition.2018.06.011

Peppermans, Y., & Maeseele, P. (2016). The politicization of climate change: Problem or solution? *Wiley Interdisciplinary Reviews: Climate Change*, 7(4): 478–485.

Peppermans, Y. and Maeseele, P. (2018). Manufacturing consent: Rereading news on four climate summits (2000–2012). *Science Communication*, 40(5): 621–649.

Pérez, R. and Simon, A. (2017). *Heartwired: Human Behaviour, Strategic Opinion Research and the Audacious Pursuit of Social Change*. Wonder Strategies for Good.

Perrault, S. (2013). *Communicating Popular Science: From Deficit to Democracy*. London: Palgrave Macmillan.

Persson, J., Sahlin, N. E. and Wallin, A. (2015). Climate change, values, and the cultural cognition thesis. *Environmental Science and Policy*, 52: 1–5.

Peters, J. W. (2016). Wielding claims of "fake news," conservatives take aim at mainstream media. *New York Times*, December 25, A11.

Pezzullo, P. C. (2009). *Toxic Tourism: Rhetorics of Pollution, Travel, and Environmental Justice*. Tuscaloosa, AL: University of Alabama Press.

Pezzullo, P. C. (2016). Hello from the other side: Popular culture, crisis, and climate activism. *Environmental Communication*, 10(6): 803–806.

Pezzullo, P. C. and Cox, R. (2017). *Environmental Communication and the Public Sphere*. Thousand Oaks, CA: SAGE.

Phillips, N. and Hardy, C. (2002). *Discourse Analysis: Investigating Processes of Social Construction*. Thousand Oaks, CA: SAGE.

Pielke, R. A., Jr. (2007). *The Honest Broker: Making Sense of Science in Policy and Politics*. Cambridge: Cambridge University Press.

Pielke, R. A., Jr. (2018). AAAS sends a message. *The Honest Broker* blog, 14 February 14. Available at: https://thehonestbroker.org/2018/02/14/aaas-sends-a-message/

Piirto, J. (2017). The five core attitudes and seven I's for enhancing creativity in the classroom. In R. A. Beghetto and J. C. Kaufman (eds.), *Nurturing Creativity in the Classroom*, pp. 142–171. New York, NY: Cambridge University Press.

Pintrich, P. R. (2003). A motivational science perspective on the role of student motivation in learning and teaching contexts. *Journal of Educational Psychology*, 95(4): 667.

Planck, M. (1936). *The Philosophy of Physics*. New York, NY: W. W. Norton.

Plutzer, E., McCaffrey, M., Hannah, A. L., et al. (2016). Climate confusion among US teachers. *Science*, 351(6274): 664–665.

Poortinga, W. and Pidgeon, N. F. (2003). Exploring the dimensionality of trust in risk regulation. *Risk Analysis*, 23(5): 961–972.

Porter, E. (2016). Climate change bias, but on both sides. *New York Times*, April 19, B1.

Poteete, A. R., Janssen, M. A. and Ostrom, E. (2010). *Working Together: Collective Action, the Commons, and Multiple Methods in Practice*. Princeton, NJ: Princeton University Press.

Pratto, F., Sidanius, J. and Levin, S. (2006). Social dominance theory and the dynamics of intergroup relations: Taking stock and looking forward. *European Review of Social Psychology*, 17(1): 271–320.

Pratto, F., Sidanius, J., Stallworth, L. M. and Malle, B. F. (1994). Social dominance orientation: A personality variable predicting social and political attitudes. *Journal of Personality and Social Psychology*, 67: 741–763. DOI:10.1037/0022-3514.67.4.741

Priest, S. (2016). *Communicating Climate Change: The Path Forward*. New York, NY: Springer Science+Business Media.

Priest, S. H. (2001). Misplaced faith: Communication variables as predictors of encouragement for biotechnology development. *Science Communication*, 23(2): 97–110.

Pulver, S. (2007). Making sense of corporate environmentalism: An environmental contestation approach to analyzing the causes and consequences of the climate change policy split in the oil industry. *Organization & Environment*, 20(1): 44–83.

Putnam, R. D. (2001). *Bowling Alone: The Collapse and Revival of American Community*. New York, NY: Simon and Schuster.

Rajan, S. R. (2006). *Modernizing Nature*. Oxford: Oxford University Press.

Randerson, J. (2007). Sir David Attenborough on global warming. *The Guardian*, June 25. Available at: www.theguardian.com/science/blog/2007/jun/25/sirdavidattenbor oughonglob

Rapley, C. and Macmillan, D. (2015). *2071: The World We'll Leave Our Grandchildren*. London: Royal Court Theatre.

Rayner, S. (2006). What drives environmental policy? *Global Environmental Change*, 16(1): 4–6.

Redclift, M. (1987). *Sustainable Development: Exploring the Contradictions*. London: Methuen.

Reddy, C. and Valentine, D. (2017). Scientists should be active but apolitical. *Huffington Post*, January 2. Available at: www.huffingtonpost.com/entry/scientists-should-be-active-but-apolitical-with-david_us_586ac375e4b068764965c3be

Reilly, A. (2016a). Top scientists find it hard to make public see risks. *Energy & Environment Daily*, March 25. Available at: www.eenews.net/stories/1060034628/.

Reilly, A. (2016b). Every American vulnerable to global warming – White House. *Energy & Environment Daily*, April 4. Available at: www.eenews.net/greenwire/2016/04/04/stories/1060035043

Reilly, A. (2018). Greens file suit about Heartland Institute influence. *Energy and Environment News*, March 15.

Reiners, D. S., Reiners, W. A. and Lockwood, J. A. (2013). The relationship between environmental advocacy, values and science: A survey of ecological scientists' attitudes. *Ecological Applications*, 23(5): 1226–1242.

Renn, O. (2011). The social amplification/attenuation of risk framework: Application to climate change. *Wiley Interdisciplinary Reviews: Climate Change*, 2(2): 154–169.

Restauri, D. (2014). *"An Inconvenient Truth"* tells the truth about climate change. *Forbes*, January 27. Available at: www.forbes.com/sites/deniserestauri/2014/01/27/4906/#32e3bb2323a5

Revkin, A. C. (2007). Climate change as news: Challenges in communicating environmental science. In Joseph F. C. DiMento (ed.), *Climate Change: What It Means for Us, Our Children, and Our Grandchildren*. Cambridge, MA: MIT Press.

Revkin, A. (2013). Fiddling while the world warms. *New York Times Dot Earth*, July 10. https://dotearth.blogs.nytimes.com/2013/07/10/fiddling-while-the-world-warms/?smid=tw-share

Ribot, J. (2014). Farce of the commons: Humor, irony, and subordination through a camera's lens. Swedish International Centre for Local Democracy. Research Report No. 2. Malmö, Sweden.

Rich, N. (2018). Losing Earth: The decade we almost stopped climate change. *The New York Times Magazine*, August 1. Available at: www.nytimes.com/interactive/2018/08/01/magazine/climate-change-losing-earth.html

Richardson, J. H. (2015). When the end of human civilization is your day job. *Esquire*, July. Available at: www.esquire.com/news-politics/a36228/ballad-of-the-sad-climatologists-0815/

Ring, W. (2016). Schools add "cli-fi" – climate fiction – to lit curriculums. *Associated Press*, March 6. Available at: https://apnews.com/71a6d9ca7c9644c9930e0df6a3a95bde

Risbey, J. S. (2008). The new climate discourse: Alarmist or alarming? *Global Environmental Change*, 18(1): 26–37.

Robbins, P. (2011). *Political Ecology: A Critical Introduction*, Vol. 16. Hoboken, NJ: John Wiley & Sons.

Roberts, D. (2017a). Did that *New York Magazine* article freak you out? Good. *Vox*, July 11. Available at: www.vox.com/energy-and-environment/2017/7/11/15950966/climate-change-doom-journalism

Roberts, D. (2017b). Does hope inspire more action on climate change than fear? We don't know. *Vox*, December 5. Available at: www.vox.com/energy-and-environment/2017/12/5/16732772/emotion-climate-change-communication

Roberts, J. (2017). Political scientist. *Science History Institute* magazine, Spring. Available at: www.sciencehistory.org/distillations/magazine/political-scientist

Rockström, J., Steffen, W., Noone, K., et al. (2009). Planetary boundaries: Exploring the safe operating space for humanity. *Ecology and Society*, 14(2): 32.

Rom, Zoë. (2017). Ryan Vachon: Top ice climber and climate scientist. *Rock: The Climber's Magazine*, April 20. Available at: https://rockandice.com/people/ryan-vachon-top-ice-climber-and-climate-scientist/

Romm, J. (2018). Here's what happens when you tell people the scientific consensus on climate change. *Think Progress*, April 18. Available at: https://thinkprogress.org/climate-message-turn-red-states-green-97-percent-7883780d457a/

Roos, J. M. (2014). Measuring science or religion? A measurement analysis of the National Science Foundation sponsored science literacy scale 2006–2010. *Public Understanding of Science*, 23(7): 797–813.

Rosenberg, E. (2017). "We stood there crying": Emaciated polar bear seen in "gut-wrenching" video and photos. *Washington Post*, December 9. Available at: www.washingtonpost.com/news/animalia/wp/2017/12/09/we-stood-there-crying-the-story-behind-the-emotional-video-of-a-starving-polar-bear/

Rosenthal, S. and Dahlstrom, M. F. (2017). Perceived influence of proenvironmental testimonials. *Environmental Communication*, 13(2): 222–238.

Roser-Renouf, C. and Maibach, E. W. (2018). Strategic communication research to illuminate and promote public engagement with climate change. In D. Hope and R. Bevins (eds.), *Change and Maintaining Change*, pp. 167–218. Cham, Switzerland: Springer.

Roser-Renouf, C., Maibach, E., Leiserowitz, A. and Rosenthal, S. (2016). Global warming's six Americas and the election, 2016. Yale Program on Climate Change Communication.

Roser-Renouf, C., Stenhouse, N., Rolfe-Redding, J., et al. (2015). Message strategies for global warming's six Americas. In *The Routledge Handbook of Environment and Communication*, pp. 368–386. London: Routledge.

Ruedy, N. E. and Schweitzer, M. E. (2010). In the moment: The effect of mindfulness on ethical decision making. *Journal of Business Ethics*, 95: 73–87. DOI:10.1007/s10551-011-0796-y

Ruiz, C., Marrero, R. and Hernández, B. (2018). Influence of emotions on the acceptance of an oil drilling project. *Environment and Behavior*, 50(3): 324–349.

Rumore, D., Schenk, T. and Susskind, L. (2016). Role-play simulations for climate change adaptation education and engagement. *Nature Climate Change*, 6(8): 745.

Rush, E. (2018). *Rising: Dispatches from the New American Shore*. Minneapolis, MN: Milkweed Editions.

Russell, C. (2008). Climate change: Now what? *Columbia Journalism Review*, July/August.

Rutherford, S. (2007). Green governmentality: Insights and opportunities in the study of nature's rule. *Progress in Human Geography*, 31: 291–307.

Ryzik, M. (2017). Climate-change film? Is Emma Stone in it? *New York Times*, 2 August 2, C1.

Sample, I. (2007). Scientists offered cash to dispute climate study. *The Guardian*, February 2, A1.

Sandell, C. and Blakemore, B. (2006). Making money by feeding confusion over global warming. *ABC News*, July 27.

Sanger-Katz, M. (2016). When was America great? It seems hard to decide. *New York Times*, April 26, A3.

Sawyer, R. K. (2017). Learning for creativity. In R. A. Beghetto and J. C. Kaufman (eds.), *Nurturing Creativity in the Classroom*. pp. 265–286. New York, NY: Cambridge University Press.

Schäfer, M. S. (2012). Online communication on climate change and climate politics: A literature review. *Wiley Interdisciplinary Reviews: Climate Change*, 3(6): 527–543.

Schäfer, M. S. and O'Neill, S. (2017). Frame analysis in climate change communication. In M. Nisbet (ed.), *Oxford Research Encyclopedia of Climate Science*, Vol. 2, pp. 646–668. Oxford: Oxford University Press.

Scheitle, C. P., Johnson, D. R. and Ecklund, E. H. (2018). Scientists and religious leaders compete for cultural authority of science. *Public Understanding of Science*, 27(1): 59–75.

Schell, J. (1989). Our fragile Earth. *Discover Magazine*, October: 47.

Schendler, A. and Jones, A. P. (2018). Stopping climate change is hopeless. Let's do it. *New York Times*, October 7: SR10.

Scheufele, D. A. (1999). Framing as a theory of media effects. *Journal of Communication*, 49(1): 103–122.

Scheufele, D. A. (2018). Beyond the choir? The need to understand multiple publics for science. *Environmental Communication*, 12(8): 1123–1126.

Schifeling, T. and Hoffman, A. J. (2017). Bill McKibben's influence on US climate change discourse: Shifting field-level debates through radical flank effects. *Organization and Environment*, 1–21.

Schlossberg, T. (2018). Trying to solve the problems that are affecting our world, and believing that they can make a difference. *New York Times*, May 27, F6.

Schmidt, G. and Wolfe, J. (2009). *Climate Change: Picturing the Science*. New York, NY: W. W. Norton.

Schneider, B. and Nocke, T., eds. (2014). *Image Politics of Climate Change: Visualizations, Imaginations, Documentations*. Bielefeld, Germany: Transcript-Verlag.

Schneider, S. H. (2001). A constructive deconstruction of deconstructionists: a response to Demeritt. *Annals of the Association of American Geographers*, 91(2): 338–344.

Schneider, S. H. (2009). *Science as a Contact Sport: Inside the Battle to Save Earth's Climate*. Washington, DC: National Geographic.

Schneider-Masterson, M. (2018). The influence of climate fiction. *Environmental Humanities*, 10(2): 473–500.

Schuldt, J. P. and Roh, S. (2014). Media frames and cognitive accessibility: What do "global warming" and "climate change" evoke in partisan minds? *Environmental Communication*, 8(4): 529–548.

Schuurman, N. (2013). Tweet me your talk: Geographical learning and knowledge production 2.0. *Professional Geographer*, 65(3): 369–377.

Schwartz, J. (2016). Science teachers lag on climate change. *New York Times*, 12 February 12, A18.

Schwartz, J. (2017). Climate scientists unite with lawyers to counter attacks on their efforts. *New York Times*, May 16, A16

Schwartz, J. (2018). Climate change is no joking matter. Except, this week, it was. *New York Times*, October 19. Available at: www.nytimes.com/2018/10/19/climate/snl-late-show-climate-change.html

Schwartz, S. H. (1992). Universals in the content and structure of values: Theoretical advances and empirical tests in 20 countries. *Advances in Experimental Social Psychology*, 25: 1–65.

Scientific American Editors. (2018a). Go public or perish. *Scientific American*, February.

Scientific American Editors. (2018b). Universities should encourage scientists to speak out about public issues. *Scientific American*, February

Searles, H. F. (1972). Unconscious processes in relation to the environmental crisis. *Psychoanalytic Review*, 59(3): 361.

See, Y. H. M., Valenti, G., Ho, A. Y. and Tan, M. S. (2013). When message tailoring backfires: The role of initial attitudes in affect–cognition matching. *European Journal of Social Psychology*, 43(6): 570–584.

Segerberg, A. (2017). Online and social media campaigns for climate change engagement. In M. Nisbet (ed.), *Oxford Research Encyclopedia of Climate Science*, Vol. 3, pp. 166–190. Oxford: Oxford University Press.

Sen, A. (2013). The ends and means of sustainability. *Journal of Human Development and Capabilities*, 14(1): 6–20.

Sengupta, S. (2018). U.N. Chief warns of a dangerous tipping point on climate change. *New York Times*, September 11, A6.

Shabecoff, P. (1988). Global warming has begun, expert tells Senate. *New York Times*, June 24, A1.

Shabecoff, P. (2003). *A Fierce Green Fire: The American Environmental Movement*. Washington, DC: Island Press.

Shanahan, J. (2017). Agenda building, narratives and attention cycles. In M. Nisbet (ed.), *Oxford Research Encyclopedia of Climate Science*, Vol. 1, pp. 21–32. Oxford: Oxford University Press.

Shank, M. (2017). 12 ways to improve climate change communications. *Huffington Post*, November 22. Available at: www.huffingtonpost.com/entry/12-ways-to-improve-climate-change-communications_us_5a15983fe4b0f401dfa7ec5f

Shapiro, S. L., Schwartz, G. E. and Bonner, G. (1998). Effects of mindfulness-based stress reduction on medical and premedical students. *Journal of Behavioral Medicine*, 21: 581–599. DOI:10.1023/A:1018700829825

Shellenberger, M. and Nordhaus, T. (2009). The death of environmentalism: Global warming politics in a post-environmental world. *Geopolitics, History, and International Relations*, 1(1): 121–164.

Shepherd, M. (2018). 3 reasons your uncle might think cold days disprove global warming. *Forbes*, November 23. Available at: www.forbes.com/sites/marshallshepherd/2018/11/23/3-reasons-your-uncle-might-think-cold-days-disprove-global-warming/#557409d31699

Sheppard, S. R. (2012). *Visualizing Climate Change: A Guide to Visual Communication of Climate Change and Developing Local Solutions*. London: Routledge.

Shi, J., Visschers, V. H., Siegrist, M. and Arvai, J. (2016). Knowledge as a driver of public perceptions about climate change reassessed. *Nature Climate Change*, 6(8): 759.

Shipley, T. F., Tikoff, B., Ormand, C. and Manduca, C. (2013). Structural geology practice and learning, from the perspective of cognitive science. *Journal of Structural Geology*, 54: 72–84.

Showstack, R. (2017). Researchers explore carbon footprints of superheroes. *Eos*, December 11. Available at: https://eos.org/articles/researchers-explore-carbon-foot prints-of-superheroes

Siciliano, J. (2017). Trump administration lining up climate change "red team." *Washington Examiner*, July 24.

Sidanius, J. and Pratto, F. (2001). *Social Dominance: An Intergroup Theory of Social Hierarchy and Oppression*. Cambridge: Cambridge University Press.

Sidanius, J., Pratto, F. and Mitchell, M. (1994). In-group identification, social dominance orientation, and differential intergroup social allocation. *The Journal of Social Psychology*, 134: 151–167. DOI:10.1080/00224545.1994.9711378

Sieber, A. (2017). Burst the bubbles: The future of climate campaigning in a post-factual age. *Climate Tracker*, September 11. Available at: http://climatetracker.org/burst-bubbles-future-climate-campaigning-post-factual-age/

Siegrist, M., Gutscher, H. and Earle, T. C. (2005). Perception of risk: The influence of general trust, and general confidence. *Journal of Risk Research*, 8(2): 145–156.

Siles, I. and Boczkowski, P. J. (2012). Making sense of the newspaper crisis: A critical assessment of existing research and an agenda for future work. *New Media and Society*, 14(8): 1375–1394.

Silvertown, J. (2009). A new dawn for citizen science. *Trends in Ecology & Evolution*, 24(9): 467–471.

Simis, M. J., Madden, H., Cacciatore, M. A. and Yeo, S. K. (2016). The lure of rationality: Why does the deficit model persist in science communication? *Public Understanding of Science*, 25(4): 400–414.

Simon, F. (2018). "Bad news" and "despair": Global carbon emissions to hit new record in 2018, IEA says. *Euractiv*. Available at: www.euractiv.com/section/climate-envir onment/news/bad-news-and-despair-global-carbon-emissions-to-hit-new-record-in-2018-iea-says/

Sims, J. (2018). Five questions for George R. R. Martin. *New York Times*, October 15, A2.

Skenazy, L. and Haidt, J. (2017). The fragile generation. *Reason Magazine*, December. Available at: https://reason.com/archives/2017/10/26/the-fragile-generation

Skibell, A. (2017a). This is our home. We don't want to live nowhere else. *Energy & Environment News*, May 24. Available at: www.eenews.net/stories/1060054972

Skibell, A. (2017b). Is "Game of Thrones" a parable about climate change? *Energy & Environmental News*, July 19. Available at: www.eenews.net/greenwire/2017/07/19/stories/1060057594

Skocpol, T. (2013). *Diminished Democracy: From Membership to Management in American Civic Life*. Norman, OK: University of Oklahoma Press.

Skuce, A. G., Cook, J., Richardson, M., et al. (2016). Does it matter if the consensus on anthropogenic global warming is 97% or 99.99%? *Bulletin of Science, Technology and Society*, 36(3): 150–156.

Skurka, C., Niederdeppe, J., Romero-Canyas, R. and Acup, D. (2018). Pathways of influence in emotional appeals: Benefits and tradeoffs of using fear or humor to promote climate change-related intentions and risk perceptions. *Journal of Communication*, 68(1): 169–193.

Slater, M. D. (1996). Theory and method in health audience segmentation. *Journal of Health Communication*, 1(3): 267–284.

Slocum, R. (2004). Polar bears and energy-efficient lightbulbs: Strategies to bring climate change home. *Environment and Planning*, 22: 413–438.

Slovic, P. (1987). Perception of risk. *Science*, 236(4799): 280–285.

Smit, D. (2018). Are millennials as bad as everyone makes out? *HR Future*, June 32–33.

Smith, J. (2017). Demanding stories: Television coverage of sustainability, climate change and material demand. *Philosophical Transactions of the Royal Society A*, DOI:10.1098/rsta.2016.0375

Smith, J., Revill, G. and Hammond, K. (2018). Voicing climate change? Television, public engagement and the politics of voice. *Transactions of the Institute of British Geographers*, 43(4): 601–614.

Smith, N. and Joffe, H. (2009). Climate change in the British press: The role of the visual. *Journal of Risk Research*, 12(5): 647–663.

Smith, N. and Leiserowitz, A. A. (2014). The role of emotion in global warming policy support and opposition. *Risk Analysis*, 34(5): 937–948.

Smith, P. and Howe, N. (2015). *Climate Change as Social Drama: Global Warming in the Public Sphere*. Cambridge: Cambridge University Press.

Smith, R. J. (2015). Katharine Hayhoe reframes Christian "belief" in climate change. *Blessed Tomorrow: Caring for Creation Today*, April 15. Available at: http://blessed tomorrow.org/blog/katharine-hayhoe-reframes-christian-belief-climate-change

Smith, W. R. (1956). Product differentiation and market segmentation as alternative marketing strategies. *Journal of Marketing*, 21(1): 3–8.

Sobczyk, N. (2018). UN report meets deaf ears but may aid push for carbon tax. *Energy and Environment Daily*, October 10. Available at: www.eenews.net/eedaily/2018/10/10/stories/1060102149

Sobel Fitts, A. (2014). Narrating climate change. *Columbia Journalism Review*, June 19.

Solomon, J. (1993). Reception and rejection of science knowledge: Choice, style and home culture. *Public Understanding of Science*, 2(2): 111–120.

Solomon, S., Qin, D., Manning, M., et al., eds. (2007). *Contribution of Working Group I to the Fourth Assessment Report of the Intergovernmental Panel on Climate Change*. Cambridge, UK and New York, NY: Cambridge University Press.

Song, H., McComas, K. A. and Schuler, K. L. (2018). Source effects on psychological reactance to regulatory policies: The role of trust and similarity. *Science Communication*, 40(5): 591–620.

Spatz, B. (2015). *What a Body Can Do: Technique as Knowledge, Practice as Research*. London: Routledge.

Starr, P. (2004) *The Creation of the Media: Political Origins of Modern Communications*. New York, NY: Basic Books.

Steffen, W., Grinevald, J., Crutzen, P. and McNeill, J. (2011). The Anthropocene: Conceptual and historical perspectives. *Philosophical Transactions of the Royal Society of London A: Mathematical, Physical and Engineering Sciences*, 369 (1938): 842–867.

Steffen, W., Rockström, J., Richardson, K., et al. (2018). Trajectories of the Earth System in the Anthropocene. *Proceedings of the National Academy of Sciences of the USA*. DOI:10.1073/pnas.1810141115

Stein, J. (2013). Millennials: The me generation. *Time Magazine*, May 20. Available at: http://time.com/247/millennials-the-me-me-me-generation/

Stenhouse, N., Myers, T., Vraga, E., et al. (2018). The potential role of actively open-minded thinking in preventing motivated reasoning about controversial science. *Journal of Environmental Psychology*, 57: 17–24.

Sterman, J. D. and Sweeney, L. B. (2007). Understanding public complacency about climate change: Adults' mental models of climate change violate conservation of matter. *Climatic Change*, 80(3–4): 213–238.

Stern, P. C., Perkins, J. H., Sparks, R. E. and Knox, R. A. (2016). The challenge of climate-change neoskepticism. *Science*, 353(6300): 653–654.

Sternberg, R. J. and Dess, N. K. (2001). Creativity for the new millennium. *American Psychologist*, 56(4): 332.

Stever, G. (2019). *The Psychology of Celebrity*. London: Routledge.

Stilgoe, J., Lock, S. J. and Wilsdon, J. (2014). Why should we promote public engagement with science? *Public Understanding of Science*, 23(1): 4–15.

Stoknes, P. E. (2015). *What We Think About When We Try Not to Think About Global Warming: Toward a New Psychology of Climate Action*. Chelsea, VT: Chelsea Green Publishing.

Stott, A. (2005). *Comedy*. London: Routledge.

Strauss, D. (2016). Clinton haunted by coal country comment. *Politico*, May 10. Available at: www.politico.com/story/2016/05/sanders-looking-to-rack-up-west-virginia-win-over-clinton-222952

Sturgis, P. and Allum, N. (2004). Science in society: Re-evaluating the deficit model of public attitudes. *Public Understanding of Science*, 13(1): 55–74.

Su, L. Y. F., Cacciatore, M. A., Scheufele, D. A., et al. (2014). Inequalities in scientific understanding: Differentiating between factual and perceived knowledge gaps. *Science Communication*, 36(3): 352–378.

Suarez, P. (2009). Linking Climate Knowledge and Decisions: Humanitarian Challenges. Boston University Frederick S. Pardee Center for the Study of the Longer-Range Future, Boston, MA.

Suarez, P. (2017). Virtual reality for a new climate: Red Cross innovations in risk management. *The Australian Journal of Emergency Management*, 32(2): 11.

Suarez, P., Mendler de Suarez, J., Koelle, B. and Boykoff, M. (2013). Serious fun: Scaling up community-based adaptation through experiential learning. In J. Ayers, L. Schipper, H. Reid, S. Huqand and A. Rahman (eds.), *Scaling up Community-based adaptation*, pp. 136–151. London: Earthscan.

Suarez, P., Otto, F., Kalra, N., et al. (2015). Loss and Damage in a Changing Climate: Games for Learning and Dialogue that Link HFA and UNFCCC. Red Cross/Red Crescent Climate Centre Working Paper Series No. 8.

Suldovsky, B. (2016). In science communication, why does the idea of the public deficit always return? Exploring key influences. *Public Understanding of Science*, 25(4): 415–426.

Suldovsky, B. (2017). The deficit model in climate change communication. In M. Nisbet (ed.), *Oxford Research Encyclopedia of Climate Science*, Vol. 2, pp. 784–804. Oxford: Oxford University Press.

Suldovsky, B., McGreavy, B. and Lindenfeld, L. (2017). Science communication and stakeholder expertise: Insights from sustainability science. *Environmental Communication*, 11(5): 587–592.

Sullivan, T. J., Driscoll, C. T., Beier, C. M., et al. (2018). Air pollution success stories in the United States: The value of long-term observations. *Environmental Science and Policy*, 84: 69–73.

Sun, L. H. and Eilperin, J. (2017). CDC gets list of forbidden words: Fetus, transgender, diversity. *Washington Post*, December 15. Available at: www.washingtonpost.com/national/health-science/cdc-gets-list-of-forbidden-words-fetus-transgender-diversity/2017/12/15/f503837a-e1cf-11e7-89e8-edec16379010_story.html?utm_term=.d4f713ad9995

Sun, Y., Shen, L. J. and Pan, Z. D. (2008). On the behavioral component of the third-person effect. *Communication Research*, 35: 257–278.

Sunstein, C. and Thaler, R. (2008). *Nudge: Improving Decisions about Health, Wealth, and Happiness*. New Haven, CT: Yale University Press.

Supran, G. and Oreskes, N. (2017). Assessing ExxonMobil's climate change communications (1977–2014). *Environmental Research Letters*, 12(8): 084019.

Sutter, P. S. (2017). *Making Climate Change History: Documents from Global Warming's Past*, pp. xi–xiv. Seattle, WA: University of Washington Press.

Svoboda, M. (2016). Cli-fi on the screen (s): Patterns in the representations of climate change in fictional films. *Wiley Interdisciplinary Reviews: Climate Change*, 7(1): 43–64.

Svrluga, S. (2018). Washington celebrates a day for marching and remembering. *Washington Post*, April 14. Available at: www.washingtonpost.com/local/march-for-science-returns-to-the-district-on-saturday-for-a-second-year/2018/04/13/40113f00-3f23-11e8-974f-aacd97698cef_story.html?utm_term=.9ede5ba31550

Swaine, J. (2018). Trump inauguration crowd photos were edited after he intervened. *The Guardian*, September 6. Available at: www.theguardian.com/world/2018/sep/06/donald-trump-inauguration-crowd-size-photos-edited

Swyngedouw, E. (2010). Apocalypse forever? *Theory, Culture & Society*, 27(2–3):213–232.

Tabary, Z. (2018). Scientists urged to "speak the same language" as public on climate. *Reuters*, June 5. Available at: www.reuters.com/article/us-global-science-climatechange/scientists-urged-to-speak-the-same-language-as-public-on-climate-idUSKCN1J12KD

Tabuchi, H. (2017a). Koch strategy mixes gospel and oil policy. *New York Times*, January 6, A1, A3.

Tabuchi, H. (2017b). Talk about the weather. *New York Times*, January 29, B1.

Tabuchi, H. (2017c). Spreading lies on climate science, and exploiting Google's algorithms to do it. *New York Times*, December 29, B3.

Tan, S., Zheng, K., Udal, P. and Jose, A. (2018). Bringing science to bars: A strategy for effective science communication. *Science Communication*, 40(6): 819–826.

Tandoc, E. C., Jr. and Eng, N. (2017). Climate change communication on Facebook, Twitter and Sina Weibo. In M. Nisbet (ed.), *Oxford Research Encyclopedia of Climate Science*, Vol. 1, pp. 602–615. Oxford: Oxford University Press.

Tandoc, E. C., Jr., Ling, R., Westlund, O., et al. (2017). Audiences' acts of authentication in the age of fake news: A conceptual framework. *New Media and Society*, 20(8): 2745–2763.

Tate, A. W. (2017). The storm-cloud of the twenty-first century: Biblical apocalypse, climate change and Ian McEwan's *Solar*. *Glass*, 29: 3–10.

Taylor, M. (2018). 90% of world's children are breathing toxic air, WHO study finds. *The Guardian*, October 29. Available at: https://amp.theguardian.com/environ ment/2018/oct/29/air-pollution-worlds-children-breathing-toxic-air-who-study-finds

Team GOP. (2018). The highly-anticipated 2017 Fake News Awards. January 17. Available at: www.gop.com/the-highly-anticipated-2017-fake-news-awards/

Tesch, D. and Kempton, W. (2004). Who is an environmentalist? The polysemy of environmentalist terms and correlated environmental actions. *Journal of Ecological Anthropology*, 8: 67–83.

Tesler, M. (2018). Elite domination of public doubts about climate change (not evolution). *Political Communication*, 35(2): 306–326.

Tett, Simon F. B., Stott, Peter A., Allen Myles R., et al. (1999). Causes of twentieth-century temperature change near the Earth's surface. *Nature*, 399: 569–572.

Thaker, J. and Leiserowitz, A. (2014). Shifting discourses of climate change in India. *Climatic Change*, 123(2): 107–119.

Thaler, A. D. (2017). When I talk about climate change, I don't talk about science. *Southern Fried Science*, January 3. Available at: www.southernfriedscience.com/ when-i-talk-about-climate-change-i-dont-talk-about-science/

Thierry, G., Athanasopoulos, P., Wiggett, A., et al. (2009). Unconscious effects of language-specific terminology on preattentive color perception. *Proceedings of the National Academy of Sciences of the USA*, 106(11): 4567–4570.

Thomas, E. F., McGarty, C. and Mavor, K. I. (2009). Aligning identities, emotions, and beliefs to create commitment to sustainable social and political action. *Personality and Social Psychology Review*, 13(3): 194–218.

Thrift, N. (2004). Intensities of feeling: Towards a spatial politics of affect. *Geografiska Annaler*, 86b: 57–78.

Thrush, G. (2018). In ruined fields, doubt about climate change. *New York Times*, October 20, A10.

Tjernström, E. and Tietenberg, T. (2008). Do differences in attitudes explain differences in national climate change policies? *Ecological Economics*, 65(2): 315–324.

Tøsse, S. E. (2013). Aiming for social or political robustness? Media strategies among climate scientists. *Science Communication*, 35(1): 32–55.

Townsend, S. (2017). *The Happy Hero: How to Change Your Life by Changing the World*. Unbound Digital.

Trench, B. (2008). Towards an analytical framework of science communication models. In *Communicating Science in Social Contexts*, pp. 119–135. Dordrecht: Springer.

Trolliet, M., Barbier, T. and Jacquet, J. (2019). From awareness to action: Taking into consideration the role of emotions and cognition for a stage toward a better communication of climate change. In W. Leal Filho, B. Lachner and H. McGhie (eds.), *Addressing the Challenges in Communicating Climate Change across Various Audiences*, pp. 47–64. Cham, Switzerland: Springer.

Trope, Y., Liberman, N. and Wakslak, C. (2007). Construal levels and psychological distance: Effects on representation, prediction, evaluation, and behavior. *Journal of Consumer Psychology*, 17(2): 83–95.

Tudge, C. (1997). The science gurus. *Independent*, January 5.

Turkle, S. (2016). *Reclaiming Conversation: The Power of Talk in a Digital Age*. New York, NY: Penguin Books.

Tversky, A. and Kahneman, D. (1973). Availability: A heuristic for judging frequency and probability. *Cognitive Psychology*, 5(2): 207–232.

Upton, J. (2015). Media are contributing to "hope gap" on climate change. *Climate Central*, March 28. Available at: www.climatecentral.org/news/media-hope-gap-on-climate-change-18822

Urry, A. (2016). Here's everything we know about how to talk about climate change. *Grist*, April 7. Available at: http://grist.org/climate-energy/heres-everything-we-know-about-how-to-talk-about-climate-change/

USGCRP (United States Global Change Research Program). (2018). Impacts, Risks, and Adaptation in the United States: Fourth National Climate Assessment, Volume II [Reidmiller, D. R., C. W. Avery, D. R. Easterling, K. E. Kunkel, K. L. M. Lewis, T. K. Maycock and B. C. Stewart (eds.)]. U.S. Global Change Research Program, Washington, DC, USA. DOI:10.7930/NCA4.2018

Valdesolo, P. and DeSteno, D. (2006). Manipulations of emotional context shape moral judgment. *Psychological Science*, 17(6): 476–477.

Van Assche, K., Beunen, R., Duineveld, M. and Gruezmacher, M. (2017). Power/ knowledge and natural resource management: Foucaultian foundations in the analysis of adaptive governance. *Journal of Environmental Policy and Planning*, 19(3): 308–322.

Van der Hel, S., Hellsten, I. and Steen, G. (2018). Tipping points and climate change: Metaphor between science and the media. *Environmental Communication*, 12(5): 605–620.

van der Linden, S., Leiserowitz, A., Rosenthal, S. and Maibach, E. (2017). Inoculating the public against misinformation about climate change. *Global Challenges*, 16(05): 1–14. DOI:10.1002/gch2.201600008

van der Linden, S., Maibach, E. and Leiserowitz, A. A. (2015). Improving public engagement with climate change: Five "best practice" insights from psychological science. *Perspectives on Psychological Science*, 10(6): 758–763.

van der Linden, S. L., Leiserowitz, A. A., Feinberg, G. D. and Maibach, E. W. (2014). How to communicate the scientific consensus on climate change: Plain facts, pie charts or metaphors?*Climatic Change*, 126 (1–2): 255–262.

van der Linden, S. L., Leiserowitz, A. and Maibach, E. (2017). Gateway illusion or cultural cognition confusion? *Journal of Science Communication*, 16(5): A04. Available at: https://ssrn.com/abstract=3094256 or http://dx.doi.org/10.2139/ssrn.3094256

van Dijk, J. (2006). *The Network Society*. London: SAGE.

van Renssen, S. (2017). The visceral climate experience. *Nature Climate Change*, 7(3): 168.

van Riper, Carena, Lum, Clinton, Kyle, et al. (2018). Values, motivations, and intentions to engage in proenvironmental behavior. *Environment & Behavior*, 1–26. DOI:doi.org/10.1177/0013916518807963

Vargo, C. J., Guo, L. and Amazeen, M. A. (2018). The agenda-setting power of fake news: A big data analysis of the online media landscape from 2014 to 2016. *New Media and Society*, 20(5): 2028–2049.

Verheggen, B., Strengers, B., Cook, J., et al. (2014). Scientists' views about attribution of global warming. *Environmental Science & Technology*, 48(16): 8963–8971.

Vezirgiannidou, S. E. (2013). Climate and energy policy in the United States: The battle of ideas. *Environmental Politics*, 22(4): 593–609.

Volcovici, V. (2017). EPA Chief wants scientists to debate climate change on TV. *Reuters*, July 11.

Wadsworth, Y., ed. (2001). *The Mirror, the Magnifying Glass, the Compass and the Map: Facilitating Participatory Action Research. Handbook of Action Research – Participative Inquiry and Practice*. London: SAGE.

Wagner, G. and Weitzman, M. (2015). *Climate Shock: The Economic Consequences of a Hotter Planet*. Princeton, NJ: Princeton University Press

Waldman, S. (2016). Scientists already worry about using the words "climate change." *Energy & Environment Daily*, November 23. Available at: www.eenews.net/stories/1060046191

Waldman, S. (2017). EPA asked Heartland for experts who question climate science. *Environment and Energy News*, September 21.

Waldman, S. (2018a). Talk of unicorns and "eco-terrorists" at alternative forum. *Energy & Environment News*, August 8. Available at: www.eenews.net/stories/1060092985

Waldman, S. (2018b). Will Trump listen to his science advisor? *Energy and Environment Daily*, August 23. Available at: www.eenews.net/climatewire/stories/1060095047/

Waldman, S. and Bravender, R. (2018). Pruitt is expected to restrict science: Here is what it means, *Environment and Energy News*, March 16.

Waldman, S. and Heikkinen, N. (2018). "Ugly fake scientist." Women say sexist attacks on the rise. *Energy & Environment Daily*, August 21. Available at: www.eenews.net/stories/1060094801

Waldman, S. (2018c). Did Trump create a new talking point for skeptics? *Energy and Environment Daily*, October 16. Available at: www.eenews.net/stories/1060102627

Waldman, S. (2018d). A look at the climate science sent to Trump. *Energy & Environment News*, October 25. Available at: www.eenews.net/climatewire/2018/10/25/stories/1060104341

Waldron, S. (2018). Michael E. Mann receives AAAS Public Engagement with Science Award. American Academy for the Advancement of Science, February 14. Available at: www.aaas.org/news/michael-e-mann-receives-aaas-public-engagement-science-award

Walker, B. J., Kurz, T. and Russel, D. (2017). Towards an understanding of when non-climate frames can generate public support for climate change policy. *Environment and Behavior*, 50(7): 781–806.

Wallach, M. A. (1988). Creativity and talent. In K. Gronhaug and G. Kaufmann (eds.), *Innovation: A Cross-Disciplinary Perspective*, pp. 13–27. Oslo: Norwegian University Press.

Walsh, L. (2015). The visual rhetoric of climate change. *Wiley Interdisciplinary Reviews: Climate Change*, 6(4): 361–368.

Walter, N., Ball-Rokeach, S. J., Xu, Y. and Broad, G. M. (2018). Communication ecologies: Analyzing adoption of false beliefs in an information-rich environment. *Science Communication*, 1–19.

Walter, S., Brüggemann, M. and Engesser, S. (2018). Echo chambers of denial: Explaining user comments on climate change. *Environmental Communication*, 12 (2): 204–217.

Wang, A. B. (2018). A coal exec sued John Oliver for calling him a "geriatric Dr. Evil." A judge tossed the case. *Washington Post*, February 26. Available at: www.washingtonpost.com/news/arts-and-entertainment/wp/2018/02/26/a-coal-exec-sued-john-oliver-for-calling-him-a-geriatric-dr-evil-a-judge-tossed-the-case/

Wang, B. B. and Zhou, Q. (in press) Comparing Chinese and American Public Opinion about Climate Change and Our Collective Future. In Jingfang Liu and Phaedra C. Pezzullo (eds.), *Green Communication and China: On Crisis, Care, and Global Futures*. East Lansing, MI: Michigan State University Press

Wang, B. B., Sheng, Y. T., Ding, M., et al. (2017). Climate change in the Chinese mind. China Center for Climate Change Communication.

Watercutter, A. (2018). I learned about climate change by watching Fortnite on Twitch. *Wired Magazine*, October 9. Available at: www.wired.com/story/fortnite-twitch-climate-scientists/

Watts, A. (2018). Michael Mann gets award for climate activism. *Watts Up with That* blog, February 14. Available at: https://wattsupwiththat.com/2018/02/14/michael-mann-gets-award-for-climate-activism/amp/

Weeman, K. (2018). Unprecedented ice loss in Russian ice cap. Cooperative Institute for Research in Environmental Sciences University of Colorado, 18 September 18. Available at: https://cires.colorado.edu/news/unprecedented-ice-loss-russian-ice-cap

Weigel, D. (2017). Rohrabacher, on jokes, recalls one on dinosaur flatulence. *Washington Post*, May 17. Available at: www.washingtonpost.com/politics/2017/live-updates/trump-white-house/trump-comey-and-russia-how-key-washington-players-are-reacting/rohrbacher-on-jokes-recalls-one-on-dinosaur-flatulence/?utm_term=.f9682387f5ff

Weingart, P., Engels, A. and Pansesgrau, P. (2000). Risks of communication: Discourses on climate change in science, politics, and the mass media. *Public Understanding of Science*, 9: 261–83.

Weintraub, K. (2017). Politics-wary scientists wade into the Trump fray at Boston rally. *Scientific American*, February 20. Available at: www.scientificamerican.com/article/politics-wary-scientists-wade-into-the-trump-fray-at-boston-rally/

Weintrobe, S. (2013). The difficult problem of anxiety in thinking about climate change. In *Engaging with Climate Change: Psychoanalytic and Interdisciplinary Perspectives*, pp. 33–47. London: Routledge.

Weisskopf, M. (1988). Two Senate bills take aim at "greenhouse effect". *Washington Post*, July 29, A17.

Weitz, E. (2009). *The Cambridge Introduction to Comedy*. Cambridge: Cambridge University Press.

White, R. (2011). Climate change, uncertain futures and the sociology of youth. *Youth Studies Australia*, 30(3): 13.

Whitmarsh, L. and Corner, A. (2017). Tools for a new climate conversation: A mixed-methods study of language for public engagement across the political spectrum. *Global Environmental Change*, 42: 122–135.

Whitmarsh, L. (2008). What's in a name? Commonalities and differences in public understanding of "climate change" and "global warming." *Public Understanding of Science*, 1: 1–20.

Wiest, S. L., Raymond, L. and Clawson, R. A. (2015). Framing, partisan predispositions, and public opinion on climate change. *Global Environmental Change*, 31: 187–198.

Wigley, T. M. L. (1999). *The Science of Climate Change: Global and US Perspectives*. Washington, DC: Pew Center on Global Climate Change.

Wilkinson, K. K. (2012). *Between God and Green: How Evangelicals Are Cultivating a Middle Ground on Climate Change*. Oxford: Oxford University Press.

Williams, N. (2016). Creative processes: From interventions in art to intervallic experiments through Bergson. *Environment and Planning A*, 48(8): 1549–1564.

Williams, R. (1973). *The City and the Country*. London: Chatto and Windus.

Williamson, P. (2016). Take the time and effort to correct misinformation. *Nature News*, 540(7632): 171.

Willis, M. J., Zheng, W., Durkin, W. J., IV, et al. (2018). Massive destabilization of an Arctic ice cap. *Earth and Planetary Science Letters*, 502: 146–155.

Wilson, Kris M. (2007). Television weathercasters as potentially prominent science communicators. *Public Understanding of Science*, 17: 73–87.

Wine, D. H., Phillips, W. J., Drive, A. B. and Morrison, M. (2017). Audience segmentation and climate change communication. In M. Nisbet (ed.), *Oxford Research Encyclopedia of Climate Science*, Vol. 1, pp. 66–94. Oxford: Oxford University Press.

Winn, B. M. (2009). The design, play, and experience framework. In R. E. Ferdig (ed.), *Handbook of Research on Effective Electronic Gaming in Education*, 1010–1024. London: IGI Global.

Wodak, J. (2018). Shifting baselines: Conveying climate change in popular music. *Environmental Communication*, 12(1): 58–70.

Wolf, J. and Moser, S. C. (2011). Individual understandings, perceptions, and engagement with climate change: Insights from in-depth studies across the world. *Wiley Interdisciplinary Reviews: Climate Change*, 2(4): 547–569.

Wozniak, A., Lück, J. and Wessler, H. (2015). Frames, stories, and images: The advantages of a multimodal approach in comparative media content research on climate change. *Environmental Communication*, 9(4): 469–490.

Wynne, B. (2006). Public engagement as a means of restoring public trust in science: hitting the notes, but missing the music? *Community Genet*, 9: 211–220.

Wynne, B. (2008). Elephants in the rooms where publics encounter "science"? *Public Understanding of Science*, 17: 21–33.

Yachnin, J. (2017). Trump dubs himself "an environmentalist". *Energy and Environment Daily*, January 24. Available at: www.eenews.net/eenewspm/2017/01/24/stories/1060048903

Yong, E. (2017). How coral researchers are coping with the death of reefs. *The Atlantic*, November 21. Available at: www.theatlantic.com/science/archive/2017/11/coral-scientists-coping-reefs-mental-health/546440/

Young, J. (2016). A comedian and an academic walk into a podcast. *Chronicle of Higher Education*, December 7. Available at: www.chronicle.com/article/A-Comedianan-Academic/237715

Young, N. and Coutinho, A. (2013). Government, anti-reflexivity, and the construction of public ignorance about climate change: Australia and Canada compared. *Global Environmental Politics*, 13(2): 89–108.

Yuan, S., Besley, J. C. and Dudo, A. (2018). A comparison between scientists' and communication scholars' views about scientists' public engagement activities. *Public Understanding of Science*, 28(1): 101–118.

Zaller, J. R. (1992). *The Nature and Origins of Mass Opinion*. Cambridge: Cambridge University Press.

Zaval, L. (2016). Culture and climate action. *Nature Climate Change*, 6(12): 1061.

Zaval, L. and Cornwell, J. F. M. (2017). Cognitive biases, non-rational judgments and public perceptions. In M. Nisbet (ed.), *Oxford Research Encyclopedia of Climate Science*, Vol. 1, pp. 643–657. Oxford: Oxford University Press.

Zaval, L., Markowitz, E. M. and Weber, E. U. (2015). How will I be remembered? Conserving the environment for the sake of one's legacy. *Psychological Science*, 26 (2): 231–236.

Zezima, K. (2018). Florence: At least 17 deaths reported as storm slogs across Carolinas. *Washington Post*, September 16. Available at: www.washingtonpost .com/news/post-nation/wp/2018/09/16/florence-several-deaths-reported-as-storm-swamps-carolinas/

Zhang, B., van der Linden, S., Mildenberger, M., et al. (2018). Experimental effects of climate messages vary geographically. *Nature Climate Change*, 8(5): 370.

Zhao, J. (2017). Influencing policymakers. *Nature Climate Change*, 7(3): 173.

Zhu, Y. and Dukes, A. (2015). Selective reporting of factual content by commercial media. *Journal of Marketing Research*, 52(1): 56–76.

Zillmann, D. (2006). Exemplification effects in the promotion of safety and health. *Journal of Communication*, 56: S221–S237.

Zillmann, D., Rockwell, S., Schweitzer, K. and Sundar, S. S. (1993). Does humor facilitate coping with physical discomfort? *Motivation and Emotion*, 17(1): 1–21.

Name Index

General Index

Page numbers in *italics* refer to Figures.